THE CAUSE AND EVOLUTION

OF THE UNIVERSE

THE CAUSE AND EVOLUTION

OF THE UNIVERSE

Fact and Myth in Modern Astrophysics

JOHN AUPING

Universidad Iberoamericana
Mexico City
in collaboration with
AESOP Publications
Oxford, UK

First edition published in 2018 by
Universidad Iberoamericana, Mexico City
In collaboration with AESOP Publications

Copyright © 2018 John Auping

AESOP Publications
Martin Noble Editorial / AESOP
28 Abberbury Road, Oxford OX4 4S, UK
www.aesopbooks.com

Universidad Iberoamericana, Mexico City
Prolongación Paseo de la Reforma 880
Santa Fe, Contadero, 01219
Ciudad de México, CDMX, Mexico
www.http:/ibero.mx/

ISBN: 978-1-910301-52-4

CONTENTS

PREFACE

Seven years have passed since the Spanish publication of John Auping Birch's *Origin and Evolution of the Universe*, and one year since its second edition, and its main ideas remain unaltered. Modern cosmology is characterized by the existence of measurements that allow the testing of different models that in principle are capable of describing the evolution of the Universe. On the other hand, modern cosmology tends to make several assumptions, sometimes very speculative, that need to be clarified to the readers who want to follow recent developments in this field. The work of Dr Auping seeks to assist readers to differentiate observationally verified aspects of cosmology from ideas whose verification is distant, or perhaps impossible. Such a task is performed by using a careful application of the orthodox scientific method.

This English edition is a part of his original work especially devoted to the description of the dynamics of stars, and the analysis of the *Big Bang*, steady state and multiverse models from a critical point of view. The author approaches different aspects of the evolution of the Universe using different branches of astrophysics, Newtonian mechanics, nuclear physics, thermo-dynamics, quantum physics and general relativity, with a clear and concise narrative. Mathematical boxes support the deeper study of mathematical-physical relations, which can be omitted by readers who are not specialized.

The mix of science, science fiction and metaphysics in modern cosmology is analyzed with strict hard core scientific arguments. The history of cosmology reveals ideas, many times antagonistic, both at the level of the interpretation of astrophysical observations, and at the level of the speculations about the origin of the Universe and the fine-tuning of its physical constants, that made it possible for us to be here to discuss it.

The search for the truth about the origin of the Universe necessarily touches on philosophical issues. Firstly, starting from Popper's philosophy of science, the author clarifies where exactly the frontier lies between science, science fiction and metaphysics. It then appears that in the final analysis of the scientific fact of fine-tuning present in the Big- Bang, we are left with only two rational options to explain it: a multiverse, which the author shows to be science fiction, sometimes with a-theological intentions; or an intelligent cause, which is part of the discourse in the frontier of physics and metaphysics, with obvious theological implications. The dialogue between faith and science is expressed clearly and objectively in this work, where the observable and the logically demonstrable, set the pattern of what is true.

This text is recommended for all those who are looking for an overview of all the contemporary currents of cosmology, including an incisive, historical approach to the subject that will be of particular interest to discerning readers who will thus be able to form their own opinion.

The author takes an objective point of view, trying to be independent of positions based on arguments of authority, which makes this work especially original and valuable.

Dominique Brun, PhD
Director
Physics and Mathematics Department
Iberoamericana University

INTRODUCTION

There is a tale by Hans Christian Andersen, *The Emperor's New Clothes*, about a tailor who offers the emperor a new suit of clothes that he says is invisible to those who are unfit for their positions, or incompetent. When the emperor walks before his subjects in his new suit, no one dares to say that they don't see any suit at all for fear that they will be dismissed as unfit for their positions. Finally, a child cries out, *"But he isn't wearing anything at all!"*

In this book, I play the role of this child. During the glorious days of Einstein, Hubble, and Lemaître, astrophysics was on top of its game. It was then part of pure science, where theories, like Einstein's general relativity and Lemaître's idea of an expanding universe were confronted with hard facts, provided by scientists like Hubble. Famous scientists like Einstein were willing to publicly correct their theories when the facts proved them wrong.

Since then, however, astrophysics has been dressing itself with the clothes of the Emperor in Anderson's tale, and, as a consequence, as George Ellis said, *"Physics ain't what it used to be"*, which means, in the words of Sheldon Glashow, that *"the historical connection between experimental physics and theory has been lost..."*[1]. Sometimes experimental physics is absent, as in the case of superstring theory, and sometimes theoretical acumen is absent, as in the case of some quantum physical theories, leading to what Popper called a *"schism in physics"*.[2]

A few astrophysicists not only see this, but speak out forcefully, leading the way out of the morass, for example, David Albert, Fred Cooperstock,

---

[1] Sheldon Glashow, *Interactions. A Journey through the Mind of a Particle Physicist and the Matter of this World* (1988): 335.
[2] Karl Popper, *Quantum Theory and the Schism in Physics* (1982).

George Ellis, Richard Feynman, Daniel Friedan, Sheldon Glashow, David Gross, Alister McGrath, Hans Ohanian, Roger Penrose, John Polkinghorne, Karl Popper, Lee Smolin, Gerard 't Hooft, David Wiltshire, Peter Woit, and others. Following their footsteps, I am greatly in debt to these authors.

Of course, they are not the only authors whose publications I read: they are just 16 among the 290 authors, whom I consulted.

Modern astrophysics is so plagued by myths that they have become mainstream. The central pillars of today's standard ΛCDM model are two enormous myths, dark matter CDM and dark energy Λ.

It is here the place to explain what I mean by 'myth'. In the case of dark matter and dark energy 'myth' refers to a theory that intends to explain certain observed phenomena by supposedly existing causes that have not yet been observed in reality or experiment, where, as a matter of fact, orthodox general relativity is necessary and sufficient to explain the same phenomena. In this case, Ockham's razor is the criterion to prefer general relativity over these speculations. Ockham's razor is also the reason why the explanation of certain phenomena by causes within our Universe is to be preferred over conformal cyclical cosmology, which pretends to explain them by causes existent in a universe of a previous aeon. In the case of some theories that pretend to explain the fine tuning of physical constants, 'myth' means: a scientific theory that has been falsified by the facts. And finally, in the case of multiverse theories, 'myth' refers to a speculative theory about the existence of something where, as a matter of principle, this theory can never be falsified. What these different uses of the word 'myth' have in common is that they refer to speculations that have not been corroborated by the facts, as would be desirable from the point of view of Popper's philosophy of science.

It is unfortunate that of all sciences, precisely astrophysics, which in the days of Kepler, Galileo and Newton paved the way for modern science, has lately – the last 80 years or so – got swamped in myths. How did this come about? In the case of dark matter and dark energy, as I shall explain in this book, the problem is that though many astrophysicists pay lip-service to Einstein, they have not been able or willing to think through all the implications of his theory of general relativity, nor have they been able or willing to use his mathematics in the solution of certain astrophysical problems, applying instead Newtonian mechanics, which leads to wrong conclusions, because, as a matter of fact, general relativity is necessary and sufficient to solve these problems.

In the case of eternal inflation, superstrings and the multiverse, however, something quite different is going on. In part, it is fascination with mathematical miracles, in part, it is outright denial. Confronted with the undeniable fine-tuning of the physical constants of the axiomatic laws that determine the evolution of the Universe, favorable to the evolution of complex life, the science fiction of the multiverse serves the purpose of avoiding the only other rational explanation of fine-tuning, i.e. an intelligent cause.

The first six chapters of the book have the following structure. In Chapters 1, 3 and 5, I present undeniable astrophysical facts about the evolution of the Universe, the fine-tuning of physical constants and the increasing entropy of the Universe, respectively. Then, in Chapters 2, 4 and 6, respectively, I expose the myths that arose in order to explain some of these phenomena, tracking in detail the origin of these myths, and simultaneously give what I prove to be the true explanation of certain astrophysical phenomena.

Chapter 7 explains Popper's philosophy of science, which is the only serious one in town these days, and Chapter 8 treats the problem of the cause of the Universe, which is part astrophysics, part metaphysics.

To serve the reader who is taken aback by the mathematics, I put them in Math boxes, which can be skipped, without losing track of the argument in the written text. The only exception is Chapter 5, where the integrals are in the main text.

This book would not have been possible without the help and criticism of my colleagues at the Universidad Iberoamericana in Mexico City. I would like to especially thank Dominique Brun, Salvador Carrillo, Leopoldo García-Colín Scherer (since deceased), Guillermo Fernandez, José Heras (since moving on to another university), Alejandro Mendoza, Alfredo Sandoval, Roberto Serna and Erich Starke. Suffice to say that any mistakes in this book are on my account, and any merits are shared with them.

I also thank the Universidad Iberoamericana and its Rector, David Fernandez, for their help, both financial and editorial, in making possible this publication. Finally, last but not least, I want to thank Blaise Machin and Martin Noble for their proofreading and critical commentaries.

Any corrections and suggestions the reader might have for a second edition are welcome.

John Auping,
casaexodo2014@hotmail.com
Mexico City, September 2017

PART I

ASTROPHYSICS

CHAPTER 1

FACT: THE EVOLUTION OF THE UNIVERSE

In this chapter, we shall first give an account of the discovery of the expansion of the Universe (Section 1.1); then present a short history of the expanding Universe starting from the Big Bang (1.2); then analyse the production of elements in the stars (1.3) and the life cycles of different types of stars (1.4); and finally, the emergence and evolution of life on Earth (1.5).

Section 1.1 The discovery of the expansion of the Universe

I will first explain why a static Universe, as conceived by Isaac Newton (1642-1727) and, initially, Albert Einstein (1879–1955), would necessarily collapse (Section 1.1.1), then present the Friedman-Lemaître model of an expanding Universe (Section 1.1.2), and finally show how Hubble's Law corroborates the Friedman-Lemaître model (Section 1.1.3).

Section 1.1.1 A static Universe will necessarily collapse

A logical consequence of Newton's theory of gravity was that a static Universe would have to collapse and could not exist. Newton was reluctant to accept this conclusion. According to Newton, the Universe had no centre. He maintained that the Universe is infinite and that an infinite, isotropic and homogeneous Universe would not collapse. Every star was equal to every other star (the Universe is homogeneous) and the stars were distributed in an equidistant and uniform way (the Universe is isotropic). Since the Universe is also spatially infinite, every individual star would be attracted by an equal number of other stars in all directions, keeping it in its place, and therefore the Universe would not collapse, and would therefore be static. As a matter of fact, Newton's conjecture about a static Universe can be refuted by Newton's law of gravity. In Math box 1.1, I prove that different universes, supposing they have the same mass density, will collapse in the same amount of time, *independently of their radius*.

Math box 1.1 A homogeneous and static universe is unstable

Suppose we have two spherical, homogeneous universes, U_1 with radius R_1 and U_2 with radius R_2, such that $R_2 = 2R_1$. Since we suppose they are homogeneous, it follows that both volume and mass of universe U_2 are eight times the volume and mass of universe U_1:

(1) $V = \frac{4}{3}\pi R^3$.

(2) $R_2 = 2R_1$.

From equations (1) and (2), we obtain:

(3) $V_2 = 8V_1$.

Since we suppose homogeneous universes with the same mass density, it follows that

(4) $M_n \propto V_n$.

From (3) and (4), we obtain:

(5) $M_2 = 8M_1$.

At the periphery of either Universe, there is an object Q with mass m. In universe U_1 the force F_1 with which Q is gravitationally attracted to the centre of mass of the universe is:

(6) $F_1 = \frac{-GM_1 m}{R_1{}^2}$.

Now acceleration a is defined as:

(7) $a = \frac{F}{m}$.

It follows from (6) and (7) that:

(8) $a_1 = \frac{-GM_1 m}{mR_1{}^2} = \frac{-GM_1}{R_1{}^2}$.

In the same way, we obtain the acceleration in universe U_2:

(9) $a_2 = \frac{-GM_2 m}{mR_2{}^2} = \frac{-GM_2}{R_2{}^2}$.

From (2), (5) and (9), it follows that:

(10) $a_2 = \frac{-8GM_1}{4R_1{}^2} = \frac{-2GM_1}{R_1{}^2}$.

From (8) and (10) it follows that:

(11) $a_2 = 2a_1$.

This means that the acceleration in universe U_2 is twice that of universe U_1, which implies that at any time the velocity of the object Q in universe U_2 is twice that of the same object in universe U_1. The object Q, travelling in universe U_2 twice the distance it has to travel in universe U_1, with twice its velocity, will arrive at the centre of mass at the same moment.

A spatially infinite universe will also collapse, contrary to what Newton thought, but it will take an infinite time to do so, since acceleration will be zero once the implosion velocity has reached that of light.

Newton's idea of infinite space does not prevent it from collapsing, but the process would take infinite time, since acceleration will be zero once the implosion velocity has approached that of light. Anyhow, a slight disturbance in the gravitational field, or some slight inhomogeneity, would create a situation where things would start falling towards each other.

Einstein, who initially conceived a static universe, introduced the famous cosmological constant Λ into the equations for his general theory of relativity to neutralise the effect of gravitational attraction. This cosmological constant was thought to be a repulsive force that allowed the Universe (at that time identified with our galaxy, the Milky Way) to remain in a static equilibrium. Nobody knew what the cosmological constant really represented, but, in the words of Einstein, it was *"necessary only for the purpose of making a quasi-static distribution of matter"* [3] in the Universe.

As a matter of fact, Einstein's static Universe is as impossible as Newton's. The exact value of the cosmological constant, arbitrarily chosen to explain the initial equilibrium of a static Universe, does not guarantee that this equilibrium will be maintained. Since the repulsive force increases linearly with respect to distance and gravity falls by the inverse square of distance, an equilibrium between the two forces in a static universe is not possible. A tiny variation in the gravitational field would cause the Universe to start collapsing or exploding. By postulating a fragile equilibrium between the gravitational and the repulsive force, Einstein fell into the same kind of trap that had blinded Newton when he postulated a static distribution of matter in an infinite space. [4]

Section 1.1.2 The Friedmann-Lemaître model of an expanding Universe

Einstein published his general theory of relativity, complete with the cosmological constant, in 1917, under the title *Cosmological Considerations in the General Theory of Relativity*. Having read Einstein's essay, the Russian mathematician Alexander Friedmann (1888–1925) came up with an alternative theory, which he published in 1922. Making use of the equations of Einstein's general theory of relativity, his model included the idea of the

---

[3] Cited in Simon Singh, *Big Bang. The Origin of the Universe* (2004): 148.
[4] These are only two of many mistakes Einstein made. See Hans Ohanian, *Einstein's Mistakes. The Human Failings of a Genius* (2008).

escape velocity, defined as the velocity an object needs to escape from the gravitational field of an object with mass M and radius R. Friedmann showed that even when the cosmological constant was zero, the Universe would not necessarily collapse, provided it was in a state of expansion. There are three possible scenarios for the interaction between the gravitational force, which makes the Universe prone to collapse, and the expansive kinetic energy:

a. Gravity gradually overcomes the expansive force and the Universe will eventually collapse: here the velocity of expansion is less than the escape velocity.

b. The expansion has enough kinetic energy to overcome gravity: the velocity of expansion is greater than the escape velocity and thus ends up being positive and constant forever. In this case, the Universe does not collapse.

c. When there is enough kinetic energy to prevent gravitational collapse but not enough to escape once and for all from the gravitational field, gravity acts as a brake on expansion without ever reversing it: the velocity of expansion is equal to the escape velocity. In this case, there is also no collapse.

Math box 1.2 explains these three scenarios in the language of mathematics:

Math box 1.2 The Friedmann-Lemaître model of an expanding Universe

According to Newton, the escape velocity v_{esc} is the velocity an object requires to escape from a gravitational field of another object with mass M and radius R is:

$$(1) \quad v_{esc} = \sqrt{\frac{2GM}{R}} \implies v_{esc}^2 = \frac{2GM}{R},$$

where G is the gravitational constant from Newton's universal law of gravitation with the value: $G = 6.67408 * 10^{-11} m^3 s^{-2} kg^{-1}$.

We now define the expansion velocity v_{ex} as the escape velocity v_{esc} with a constant χ, such that $\chi > 0$, when the expansion velocity v_{ex} is bigger than the escape velocity v_{esc}; $\chi < 0$, when the expansion velocity v_{ex} is smaller than the escape velocity v_{esc}; or $\chi = 0$, when the expansion velocity v_{ex} is equal to the escape velocity v_{esc}:

$$(2) \qquad v_{ex}^2 = v_{esc}^2 + \chi = \frac{2GM}{R} + \chi.$$

The mass M of a spherical object is equal to the product of the volume of the sphere $(V = \frac{4}{3}\pi R^3)$, where R is the radius of the sphere, and its density ρ:

$$(3) \qquad M = \frac{4}{3}\pi R^3 \rho.$$

Combining (2) and (3), we obtain:

$$(4) \qquad v_{ex}^2 = \frac{8}{3}\pi G R^2 \rho + \chi.$$

Einstein published his treatise on general relativity in 1917. In 1922 and 1927, respectively, Alexander Friedmann and Georges Lemaître, one independently of the other, used Einstein's general relativity to transform the Newtonian escape and expansion velocities and apply them to the Universe, and they obtained:

$$(5) \qquad \chi = -kc^2,$$

where k is the Friedmann-Lemaître-Robertson-Walker constant of the curvature of the Universe. Combining (4) and (5) we obtain:

$$(6) \qquad v_{ex}^2 = \frac{8}{3}\pi G R^2 \rho - kc^2,$$

such that if $k = +1$, the Universe is closed, and it collapses before its radius R reaches infinity; if $k = 0$, the Universe is flat and it will expand forever, with the expansion velocity reaching zero when its radius R reaches infinity and its density ρ, zero ($v_{ex} \rightarrow 0$, when $R \rightarrow \infty$ and $\rho \rightarrow 0$); and if $k = -1$, the Universe is open and it will expand forever, with the expansion velocity being constant when R and t reach infinity.

I will now introduce the Hubble constant. The Hubble constant is the ratio of the recession velocity of a galaxy (which is equal to the expansion rate of the Universe v_{ex} at that point in the Universe) and its distance d from Earth:

(7) $H = v_{ex}/d \Longrightarrow v_{ex} = Hd.$

Suppose that we are looking at a galaxy with a high recession velocity, at the limit of the observable Universe. In this case, the distance d is equal to the radius R of the observable Universe. So, equation (7) becomes:

(8) $H = v_{ex}/R \Longrightarrow v_{ex} = HR \Longrightarrow v_{ex}^2 = H^2 R^2.$

From equation (6) and (8), we deduce that the density of the Universe ρ is:

(9) $\rho = 3H^2/8\pi G + 3kc^2/8\pi GR^2.$

The critical density ρ' is obtained when $k = 0$:

(10) $\rho' = 3H^2/8\pi G.$

The constant Ω is the ratio of the Universe's density and its critical density:

(11) $\Omega = \frac{\rho}{\rho'}.$

Combining equations (9), (10) and (11), we get Ω as a function of the constant k, speed of light c, radius of the Universe R and Hubble constant H:

(12) $\Omega = \frac{\rho}{\rho'} = 1 + \frac{kc^2}{H^2 R^2}.$

Equation (12) is equation (366) from my treatise on general relativity in my Spanish language e-book on the Universe.[5]

---

[5] John Auping, "La construcción de la geodésica y el tensor de Einstein", e-book: www.ibero.mx, publicaciones, publicaciones electrónicas, *El Origen y la Evolución del Universo* (2016): 639–696

Einstein was not impressed by Friedmann's ideas about an expanding Universe. He initially objected that *"the results concerning the non-stationary world, contained in [Friedmann's] work, appear to me suspicious, in reality it turns out that the solution given in it does not satisfy the equations [of general relativity]"*, but after Friedmann defended the correctness of his calculations, Einstein was forced to retract his criticism and admitted that Friedmann's dynamic model was indeed mathematically correct: *"Mr Friedmann's results are both correct and clarifying: they show that in addition to the static solutions to the equations [of general relativity] there are time varying solutions with a spatially symmetric structure."*[6]

Nonetheless, Einstein would not accept that Friedmann's mathematical model described *reality*, and the scientific community backed Einstein. Friedmann died young. However, a few years later, the Belgian astrophysicist and Catholic priest Georges Lemaître (1894–1966) gave new life to the dynamic model. Unaware of Friedmann's work, he developed his own model of an expanding Universe based on Einstein's general relativity.

When Lemaître published his essay (in French, in 1927), a recent publication (1926) by Edwin Hubble (1889–1953) had established that the so-called nebulae were not intra-galactic dust clouds, but faraway galaxies outside our galaxy, the Milky Way,[7] and another recent publication (1925) by Gustaf Strömberg (1882–1962) had established, on the basis of the radial velocities of these galaxies, that they were receding.[8] In his 1927 essay, Lemaître referred to their work. At that time, Hubble had not yet published his data relating the recession velocities of these galaxies with

---

[6] Cited in Simon Singh, *Big Bang. The Origin of the Universe* (2004): 153, 155.

[7] Edwin Hubble, "Extragalactic nebulae", in: *Astrophysical Journal*, vol. 64 (1926): 321–369.

[8] Gustaf Strömberg, "Analysis of radial velocities of globular clusters and non-galactic nebulae", in: *Astrophysical Journal*, vol. 61 (1925): 353–362.

their absolute distance and though Strömberg had explored a correlation between radial velocity and absolute distance, in his sample it was so weak ($R_{PEARSON}$ = 0.23), that only 5% of recession velocity was explained by absolute distance, which could easily be attributed to measuring errors, which is why Strömberg himself deemed this correlation to be quite insignificant.

Lemaître's contribution to our understanding of the Universe is admirable for three reasons, of which the first two ones are well known:

1) He attributes the recession velocities of faraway galaxies to the expansion of the Universe, *"L'éloignement des nébuleuses extra-galactiques est un effet cosmique dû a l'expansion de l'espace"*.[9] This was translated a few years later into English by Arthur Eddington (1882–1944): *"The recession velocities of extragalactic nebulae are a cosmic effect of the expansion of the universe."*[10]

2) Lemaître went much further than Friedmann, postulating not only the expansion of the Universe, but its origin in the explosion of an original super quantum, which was later christened the *Big Bang*: *"If we go back in the course of time we must find fewer and fewer quanta, until we find all the energy of the universe packed in a few or even in a unique quantum."*[11] A few years later, he would elaborate this point in a book.[12]

3) He predicted, two years before Hubble proposed and corroborated that hypothesis empirically, that *"the Doppler effect [of the light of*

---

[9] George Lemaître, "Un univers homogène de masse constante et de rayon croissant rendant compte de la vitesse radiale des nébuleuses extragalactiques", in: *Annales de la Société Scientifique de Bruxelles*, vol. 47ª (1927): 58.

[10] George Lemaître, "A Homogeneous Universe of Constant Mass and Growing Radius Accounting for the Radial Velocity of Extragalactic Nebulae", in: *Monthly Notices of the Royal Astronomical Society*, vol. 91 (1931): 489; (2013): 1645.

[11] George Lemaître, "The Beginning of the World from the Point of View of Quantum Theory", *Nature*, vol. 127, no. 3210 (1931): 706.

[12] George Lemaître, *L'Hypothèse de l' atome primitive. Essay de cosmogenie* (1946).

receding galaxies is] due to the variation of the radius of the Universe",[13] establishing a linear relationship between this Doppler effect (also known as redshift) and absolute distance from Earth.

Lemaître is well known because of the first two contributions, but he is not known for the third one, though it is, as a matter of fact, quite remarkable. Lemaître starts his essay commenting that both the De Sitter model and Einstein's static model of the Universe have merits and deficiencies. The De Sitter model does allow for an explanation of the fact *"that extragalactic nebulae seem to recede from us with a huge velocity"*, but it *"ignores the existence of matter and supposes its density equal to zero"*, whereas, on the other hand, Einstein's model adequately allows for *"the obvious fact that the density of matter is not zero and it leads to a relation between this density and the radius of the universe,"* but does not explain the recession velocity of faraway galaxies.[14] After establishing that existing *"theory can provide no mean between these two extremes"*, Lemaître proposes *"an intermediate solution"*,[15] which requires that he elaborates mathematically his own theory that has four important elements: a) an *"Einstein universe of variable radius"*; with b) *"constant total mass"*, where c) *"the density, uniform in space, varies with time"*, and d) a *"Doppler effect due to the variation of the radius of the universe"*.[16]

I will further develop some of his equations and their implications in Math box 1.3, showing that Lemaître discovered Hubble's law, deriving it from his dynamic model of the Universe, *two years before* Hubble discovered it on the basis of empirical observations of some nearby receding galaxies.

---

[13] George Lemaître, *A Homogeneous Universe of Constant Mass and Growing Radius Accounting for the Radial Velocity of Extragalactic Nebulae* (2013): 1642.
[14] *Ibidem*: 1636.
[15] *Ibidem*: 1637.
[16] *Ibidem*: 1638, 1639, 1638, 1641, respectively.

Math box 1.3 Lemaître discovered Hubble's law before Hubble did

I integrated Lemaître's model on the relations between the variable radius of the Universe, variable mass density, expansion velocity and gravitation, into Friedmann's theory of the interaction of escape velocity and expansion velocity, in Math box 1.1. I will now explain another part of Lemaître's theory: the relationship between the varying radius of the expanding Universe and Doppler effect of light coming from receding galaxies. As a matter of fact, this part of Lemaître's model, normally not recognised in the literature, occupies the third part of his eleven-page essay.[17] He neatly establishes Hubble's law two years before Hubble did.

In Lemaître's model, R is the expanding radius of the Universe; R_0 is the distance of a receding object; and R_E is the radius of the curvature in Einstein's static Universe, the value of which Lemaître took from Hubble's 1926 essay, where Hubble gave estimates of the values of the curvature radius, volume, mass, and mass density of *"the finite universe of general relativity"*:[18]

(1) $R_E = 2.7 * 10^{10}$ *parsecs* $= 88.065 * 10^9$ *light years.*[19]

Hubble grossly underestimated these values,[20] but this does not really affect Lemaître's exercise, because he does not pretend to present the empirical values of the variables he uses in his theory, but rather an instantiation of a theoretical scenario that establishes the relationship between absolute distance (R_0) of a receding galaxy and the ensuing

---

[17] George Lemaître, *A Homogeneous Universe of Constant Mass and Growing Radius Accounting for the Radial Velocity of Extragalactic Nebulae* (2013): 1641–1645.
[18] Edwin Hubble, "Extragalactic nebulae", in: *Astrophysical Journal*, vol. 64 (1926): 368–369.
[19] Edwin Hubble, *Ibidem*: 369 and George Lemaître *Ibidem*: 1641.
[20] Edwin Hubble, "Extragalactic nebulae", in: *Astrophysical Journal*, vol. 64 (1926): 368–369, has the following values: radius $R = 2.7 * 10^{10}$ *parsecs*; volume $V = \pi^2 R^3 = 3.5 * 10^{32}$ *cubic pcs*;   mass $M = 1.8 * 10^{57}$ *grams* $= 9 * 10^{22} \odot$; mass density $\rho = 1.5 *$ *grams per cubic centimeter*.

'Doppler effect' or redshift, represented as v/c. He defines "*the apparent Doppler effect due to the variation of the radius of the universe [as] the ratio of the radii of the universe at the instants of observation and emission, diminished by unity*".[21]

In his essay, Lemaître manages R_E as a limit, with constant value, to which R and R_0 tend with the passage of time. As a matter of fact, he conceives a closed Universe, where "*the lines starting from a same point come back to their starting point*" [22] (a positive curvature, with $k > 0$).

Lemaître does not give the actual, variable values of R_0, but rather the values of $n = R/R_0$ and R_E. Using his fundamental equation, through which he relates R, R_0 and R_E, and which he reproduced in his essay three times, in different forms, I obtained the values for R and R_0. The first time he writes this fundamental equation in the following way: [23]

(2) $R^3 = R_E{}^2 R_0 \rightarrow R = \sqrt[3]{R_E{}^2 R_0{}^{1/3}}$ and

(3) $n = \dfrac{R}{R_0}$.

From (2) and (3) we obtain;

(4) $\dfrac{R}{R_0} = \sqrt[3]{R_E{}^2 R_0{}^{-2/3}} = n \;\rightarrow$

(5) $R_0{}^{-2/3} = \dfrac{n}{\sqrt[3]{R_E{}^2}} \rightarrow \sqrt[3]{R_0{}^2} = \dfrac{\sqrt[3]{R_E{}^2}}{n} \;\rightarrow$

(6) $R_0{}^2 = \dfrac{R_E{}^2}{n^3} \;\rightarrow$

(7) $R_0 = \dfrac{R_E}{\sqrt[2]{n^3}}$.

---

[21] Meaning that the radius at the moment of emission is shorter than at the moment of observation. George Lemaître, *A Homogeneous Universe of Constant Mass and Growing Radius Accounting for the Radial Velocity of Extragalactic Nebulae* (2013): 1642.
[22] *Ibidem*: 1636.
[23] *Ibidem*: 1641 (eq. 18).

From (3) and (7) we obtain the value of R:

(8) $R = n * R_0$.

The second time he presents his fundamental equation of R, he does so by way of two equations that we can combine:[24]

(9) $y = \dfrac{R_0}{R}$ and

(10) $R_0{}^2 = R_E{}^2 y^3$.

From (9) and (10), we obtain:

(11) $R_0{}^2 = R_E{}^2 \dfrac{R_0{}^3}{R^3} \rightarrow R = \sqrt[3]{R_E{}^2 R_0}{}^{1/3}$,

which is our equation (2), and from there we get to (7) and (8). The third time he writes his fundamental equation of R, is at the end of his essay:[25]

(12) $R = R_E \sqrt[3]{\dfrac{R_0}{R_E}}$,

which leads to our equation (2) and from there to (7) and (8).

Perhaps Lemaître wrote the same equation of the radius R of the Universe in three different ways, depending on the context. Since Lemaître gives us the values of R_E and n,[26] we can obtain the value of R_0, through equation (7) and the value of R through (8) and this way complement the values he published. In Table 1.1, columns one and four are Lemaître's,[27] and columns two and three contain the values which I derived from the values of n in the first column and from R_E, which is a constant.

Columns 2 (implicit in Lemaître's data) and 4 (explicit in his data) are the basis for the Pearson correlation in Graph 1.1.

---

[24] George Lemaître, *A Homogeneous Universe of Constant Mass and Growing Radius Accounting for the Radial Velocity of Extragalactic Nebulae* (2013): 1643 (eqs. 26, 27).
[25] *Ibidem*: 1645 (equation at bottom of page).
[26] *Ibidem*: 1641 (value of R_E) and 1644 (1st column of table at bottom of page).
[27] *Ibidem*: table at the bottom of page 1644.

$n = \dfrac{R}{R_0}$	Distance of receding galaxy R_0 (10^9 ly)	Radius of expanding universe R (10^9 ly)	Indicator Doppler effect v/c
1	88.065	88.065	19
2	31.1357	62.271	9
3	16.9481	50.844	5 2/3
4	11.0081	44.033	4
5	7.8768	39.384	3
10	2.7849	27.849	1
15	1.5159	22.738	1/3
20	0.9846	19.692	0

Table 1.1 Lemaître discovered Hubble´s law before Hubble did

The correlation between distance and redshift is $R_{PEARSON} = 0.9876$, which means that 97.5% of the variation in redshift is explained by the absolute distance of the galaxy. We have a perfectly linear relationship, as in Hubble's law. Though Lemaître did not graph his values, we may do so, measuring the absolute distance of the galaxy (column 2) in the X axis, and the redshift (column 4) in the Y axis. By doing so, the linear relationship between distance and redshift becomes visible.

One should consider that these are not observed values, but *theoretical* values which he obtained *extrapolating* present day empirical values, taken from Strömberg and Hubble's sample of 42 galaxies, at an average distance of $\pm10^6$ $pc = \pm3.26 * 10^6$ ly,[28] and with a redshift of almost zero. The zero value of the redshift at such a short distance is a consequence of his rounding off these empirical values in the graph's origin, the first point on the very large, theoretical distance scale he extrapolates in his model. At these larger, theoretical distances, the redshift increases correspondingly.

---

[28] George Lemaître, *A Homogeneous Universe of Constant Mass and Growing Radius Accounting for the Radial Velocity of Extragalactic Nebulae* (2013): 1642.

Graph 1.1 is based on the theoretical values of Lemaître's long term extrapolations. Only the value in the left lower corner is empirical.

Graph 1.1 Lemaître discovered Hubble's law before Hubble did

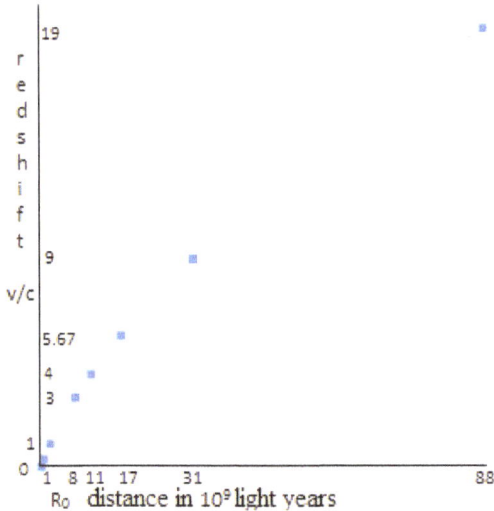

Lemaître estimated the present value of the radius of the Universe to be $0.20 * R_E = 17.6 * 10^9 \ ly$,[29] which is, of course, also the age of the Universe, and comes quite close to what I myself, following Wiltshire, found: 15.8 billion years (see the end of Math box 1.6).

From Lemaître's estimate of the age of the Universe, we can derive his implicit 'Hubble constant', two years before Hubble published his estimate of $H = \frac{1}{t} = 558 \ km \ s^{-1} \ Mpc^{-1}$. Since one Mega parsec (Mpc) is $3.26 * 10^6 \ ly$, one light year is $9.46 * 10^{12} \ km$, and one year has $31.536 * 10^6 \ s$, Lemaître's implicit Hubble constant is:

(13) $H = (3.26 * 10^6) * (9.46 * 10^{12})/[(31.536 * 10^6) * (17.6 * 10^9)] \cong 55.56 \ s^{-1}$.

[29] George Lemaître, *A Homogeneous Universe of Constant Mass and Growing Radius Accounting for the Radial Velocity of Extragalactic Nebulae* (2013): 1645.

> This value comes remarkably close to what I found to be the correct value, if we measure the age of the global-average Universe with our terrestrial clock: $H_0(\tau_w) = 66.5\ s^{-1}$.[30]

The important thing to notice in Math box 1.3 is the fact that Hubble's law and the correct value of Hubble's constant are implicit in Lemaître's theory, *two years before* Hubble derived his law and constant from his empirical observations of nearby receding galaxies. This fact certainly pays additional tribute to Lemaître's genius.

After publishing his model in a French language journal, in 1927, Lemaître was met by the same deafening silence that had greeted Friedmann's model. That same year, he presented his model to Einstein at a conference. Einstein introduced him to Friedmann's work but rejected his model, arguing that *"your calculations are correct, but your physics is abominable"*.[31] Rejection by Einstein was tantamount to rejection by the scientific community as a whole and Lemaître stopped insisting. There is a profound irony to all of this. The rebel who had challenged the academic establishment of his era, was now, in 1927, its *"dictator"*, as he himself noticed: *"to punish me for my contempt for authority, fate made me an authority myself"*.[32] Einstein had developed the theory and equations that would have allowed him to predict the expansion of the Universe, yet he clung to his static model. If Einstein had believed what his equations, in their original form, were telling him, he would have been able to predict the expansion of the Universe before it was observed.[33] When Hubble discovered that galaxies were receding, as predicted by Lemaître, Einstein

---

[30] Time is different, if measured with clocks in walls, in voids or global-average clocks. See Chapter 2.
[31] Cited in Simon Singh, *Big Bang. The Origin of the Universe* (2004): 160
[32] *Ibidem*: 160.
[33] John Hawley & Katherine Holcomb, *Foundations of Modern Cosmology* (1998): 280.

publicly changed his mind, regretting what he called *"the greatest blunder of his entire life"*.[34]

Section 1.1.3 Hubble's data corroborate the Friedman-Lemaître model

Edwin Hubble (1889–1953), after proving that the Universe is bigger than the Milky Way[35], made another, even more important discovery, which settled the dispute between the Friedmann–Lemaître theory of an expanding Universe, and the Newton–Einstein model of a static Universe. To understand the implications of his discovery, we need to recall the Lorentz transformation of the wavelength of light and the redshift of the light spectrum that occurs when objects move away from us at speed.

When Harlow Shapley (1885–1972) and Heber Curtis (1872–1942) debated their respective theories about the size of the Universe, the majority of *nebulae*, which originally were not distinguished from dust clouds in our galaxy, and finally identified as other, much more distant galaxies, were already known to exhibit redshift or blueshift. In 1912, the US astronomer Vesto Slipher (1875–1969) applied the non-relativistic Doppler effect formula to the Andromeda nebula, which exhibited blueshift. Slipher calculated that it was approaching the Earth at a speed of 125 miles per second. We now know that Andromeda and the Milky Way constitute a galaxy cluster, with the two galaxies trapped in a single gravitational field, just like the Earth and the Moon. This means that Andromeda oscillates between periods of moving closer or further away, although it has been moving closer for the last few million years. In 1922, Slipher published the results of research showing that 36 nebulae out of a sample of 41 exhibited redshifts.

---

[34] Cited in John Hawley & Katherine Holcomb, *Foundations of Modern Cosmology* (1998): 280.
[35] Edwin Hubble, "Extragalactic nebulae", in: *Astrophysical Journal*, vol. 64 (1926): 321–369.

However, nobody connected these results to the Friedmann–Lemaître theory, even though they did corroborate it.

When Hubble and Milton Humason (1891–1972) graphed the recession velocity of galaxies (on the vertical axis) and their absolute distance (on the horizontal axis), they discovered a linear relationship between redshift and the absolute distance of a galaxy. Their initial results, based on the observation of 20 nearby galaxies (from 0 to 7 million light years), were published in 1929, and are reproduced in Graph 1.2. A parsec is 3.0857 ∗ $10^{16}m$.

In a subsequent study, in 1931, Hubble and Humason added another eight galaxies at distances of between 7 and 100 million light years.[36] Their observations conclusively confirmed the hypothesis of the linear relationship between recession velocity and absolute distance and gave rise to Hubble's Law, as explained in the Math box 1.4.

Graph 1.2 Hubble and Humason's original 1929 graph[37]

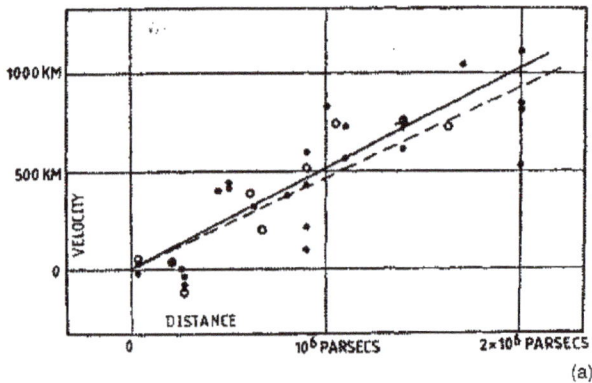

(a)

---

[36] Edwin Hubble & Milton Humason, "The Velocity-Distance Relation among Extra-Galactic Nebulae", in: *Astrophysical Journal*, vol. 74 (1931): 43–80.

[37] Edwin Hubble, "A Relation between Distance and Radial Velocity among Extra-Galactic Nebulae", in: *Proceedings of the National Academy of Sciences of the United States of America*, vol. 15 (1929): 168–173.

Math box 1.4. Hubble's Law

We should recall that in this case the velocity of expansion v_{ex} is equal to the recession velocity v and the distance d is equal to r:

(1) $H = \frac{v_{ex}}{d} \Rightarrow v_{ex} = Hd \underset{d=r}{\Longrightarrow}$

(2) $v = Hr$ (in km/s).

By (2) we obtain:

(3) $H = \frac{v}{r} = \frac{r/t}{r} = 1/t$ (in s^{-1}).

By (3), we can calculate t, the age of the Universe:

(4) $t = \frac{r}{v} = \frac{1}{H}$.

Here t is the age of the Universe since the Big Bang. However, Hubble and Humason's measurements suffered from systematic errors, with varying degrees of severity. The first error was that they had used the non-relativistic formula for redshift to derive the recessional velocity. However, even though this procedure is incorrect, for $z < 0.1$ it does not result in serious errors. Using the two formulae above, you will find that the non-relativistic formula for redshift of $z = 0.1$ is:

(5) $z = \frac{v}{c} = 0.1 \Rightarrow v = 0.1c$,

in other words, 10% of the speed of light. On the other hand, the relativistic formula gives the following result:

(6) $1.1 = \sqrt{\frac{1+v/c}{1-v/c}} \Rightarrow 1.21 = \frac{1+v/c}{1-v/c} \Rightarrow 2.21\frac{v}{c} = 0.21 \Rightarrow v = \frac{0.21}{2.21}c \cong$
    $0.095\ c$,

in other words, 9.5% of the speed of light. From equations (5) and (6), we obtain the difference between the correct, relativistic redshift and the incorrect, non-relativistic redshift, which is only 5.2%.

The maximum speed of a galaxy recorded by Hubble and Humason at a distance of 100 million light years, was 20,000 km/s, in other words, 6.7% of the speed of light. At such slow speeds, the difference between the relativistic and non-relativistic results is not significant, as I just showed in Math box 1.4. This implied that the associated systematic margin of error in the redshift values, though real, was not enough to undermine the conclusions of their research.

However, the error in the measurement of the absolute distance of galaxies was much more serious. The method was based on the relationship between the luminosity and the period of Cepheid variable stars. Cepheids, like Polaris, the North Star, vary in luminosity. Today we know that this variation, which can last from about a week to about a month, is caused by the fact that these stars suffer gravitational collapse and then expand again, since the compression of the collapse reignites the nuclear fusion at their core. In the period of collapse, they are dimmer, and in the period of expansion, brighter. John Herschel (1792–1871) used daguerreotypes to register the alternation in the observed luminosity L_o of the Cepheids, at Harvard College's observatory. One of his female collaborators, Henrietta Leavitt (1868–1921), discovered that "*there is a simple [linear] relation between the brightness of the variables and their periods*".[38] If two Cepheids, number 1 and number 2, with the same period, and therefore with the same real luminosity L_r, have different observed luminosities, this difference is due to the fact that these Cepheids are at different distances. If $L_{o_1} = 25\,L_{o_2}$, but both have the same period, so that $L_{r_1} = L_{r_2}$, this means that Cepheid 2 lies at a distance five times farther away than Cepheid 1. This method makes it also possible to know the exact distance from Earth of Cepheid variable stars in the sky, as shown in Math box 1.5.

---

[38] Henrietta Leavitt & Edward Pickering, "Periods of 25 Variable Stars in the Small Magellanic Cloud", in: *Harvard College Observatory Circular*, vol. 173 (1912): 1–3.

Math box 1.5 A galaxy's distance from Earth calculated with a Cepheid

A sphere has a surface of

(1) $A = 4\pi r^2$.

An observer at a distance r from a galaxy with a Cepheid variable star receives a quantity of light that is equal to its real luminosity L_r divided by the sphere's surface:

(2) $L_o = \dfrac{L_r}{A} = \dfrac{L_r}{4\pi r^2}$.

From equation (2), we obtain:

(3) $r = \sqrt{\dfrac{L_r}{4\pi L_o}}$.

Let us consider an example of how this is done. A telescope on Earth with a radius of two metres has a surface of $2\pi * 2\ m^2$. It receives light from a distant star with an energy of $4 * 10^{-8}\ W$ per square metre. This means the observed luminosity per square metre is:

(4) $L_o/m^2 = \dfrac{4*10^{-8}}{4\pi}\ W/m^2 = \dfrac{10^{-8}}{\pi}\ W/m^2 \to L_o = \dfrac{10^{-8}}{\pi} W$.

Now we know from the period of the Cepheid variable star that the real luminosity is:

(5) $L_r = 4 * 10^{30}\ W$.

Combining (3), (4) and (5), we obtain, in the case of this example:

(6) $r = \sqrt{\dfrac{4*10^{30}}{4\pi * \frac{10^{-8}}{\pi}}}\ m = \sqrt{10^{38}}\ m = 10^{19}m \cong 324.08\ pc \cong 1{,}057\ ly$.

The Cepheid variable stars allowed Leavitt and Hubble to calibrate an exact distance scale of galaxies in the nearby Universe.

However, at that time, it was not known that there are actually two types of Cepheid variable stars. Hubble's observations of several galaxies pertained to type I Cepheids, which are four times brighter than their type II counterparts and had been used by Leavitt and Shapley used to calibrate their scale for absolute distance. As a result, Hubble underestimated the distance of type I Cepheid galaxies by a factor of two. Besides, the Cepheid distance scale could not be used for more distant galaxies (over 10 million light years), although today with the Hubble Space Telescope it can be used at distances of up to 50 million light years. As such, Hubble had to make approximate estimates based on the assumptions that, for mid distances, the absolute luminosity of the brightest stars in different galaxies is constant and that, for longer distances, the absolute luminosity of galaxies of a certain type is also constant. As a consequence of these accumulated measurement errors, observing galaxies with a high recession velocity, appearing to be so close, Hubble overestimated the value of the Hubble constant and underestimated the age of the Universe.

Math box 1.6 The age of the Universe according to Hubble´s constant

Hubble calculated the value of the Hubble constant to be $558\ kms^{-1}Mpc^{-1}$. He then used the formula we saw above to calculate the age of the Universe:

$$(1)\ t = \frac{1}{H} = \left(\frac{1}{558\ kms^{-1}Mpc^{-1}}\right) = \frac{1}{558}\ km^{-1}s\ Mpc.$$

Since one Mega parsec (Mpc) is $3.26*10^6$ light years, one light year is $9.46*10^{12}\ km$ and one year contains $31.536*10^6\ s$, according to Hubble, the age of the Universe was:

$$(2)\ t = (3.26*10^6)*(9.46*10^{12}) / (31.536*10^6*558) = 1.75\ \text{bill. years.}$$

Hubble underestimated the age of the Universe by 85%! With time, more precise estimates were made based on more exact measurements. A sample of these is the following. In the 1970s and 1980s, according to Malcolm Longair, different astrophysicists made estimates of the Hubble constant H_o and the age of the Universe t_0,[39] which varied from:

(3) $50 < H_o < 80\ kms^{-1}Mpc^{-1}$ $\rightarrow t_0 = \pm 15$ billion years.

In 1994, Edward Kolb and Michael Turner[40] made the following very crude estimate:

(4) $40 < H_o < 100\ kms^{-1}Mpc^{-1}$ $\rightarrow t_0 = 9.8 * 10^9 \leq t_0 \leq 24.4 * 10^9$ years.

In 1997, Wendy Freedman[41] estimated:

(5) $H_0 = 73 \pm 10\ kms^{-1}Mpc^{-1}$ $\rightarrow t_0 = \pm\ 13.4$ billion years.

That same year, in the same book, Allan Sandage and Gustav Tammann[42] estimated:

(6) $H_0 = 55 \pm 10\ kms^{-1}Mpc^{-1}$ $\rightarrow t_0 = \pm\ 17.8$ billion years.

In 1999, Tripp and Branch[43] estimated:

(7) $H_0 = 63 \pm 10\ kms^{-1}Mpc^{-1}$ $\rightarrow t_0 = \pm 15.5$ billion years.

Accurate observations from the Hubble Space Telescope supported the following estimate by Freedman and 14 other astrophysicists in 2001:[44]

---

[39] Malcolm Longair, *The Cosmic Century* (2006): 344.
[40] Edward Kolb & Michael Turner, *The Early Universe* (1994): 503.
[41] Wendy Freedman, "Determination of the Hubble Constant", in: Neil Turok ed., *Critical Dialogues in Cosmology* (1997): 92–129.
[42] Allan Sandage and Gustav Tammann, "The evidence for the Long Distance Scale with $H_0 < 65$" in: Neil Turok ed., *Critical Dialogues in Cosmology* (1997): 130–155.
[43] Robert Tripp & David Branch, "Determination of the Hubble Constant Using a Two-Parameter Luminosity Correction for Type Ia Supernovae", in: *The Astrophysical Journal* vol. 525 (1999): 209–214.

> (8) $H_0 = 70 \pm 8 \ kms^{-1}Mpc^{-1}$ $\rightarrow t_0 = \pm 13.97$ billion years.
>
> Fourteen billion years is the age of the Universe accepted by most astrophysicists who subscribe to the ΛCDM model. The problem is that dark energy (Λ) and dark matter (CDM) are myths, not facts, as I explain in Chapter 2. David Wiltshire proved what we already should have known, taking Einstein's general relativity seriously, that there is no such thing as absolute time. The Universe is like a sponge, with holes (the voids) and walls (galaxy clusters). Clocks run at different rates in walls and voids, and in the Universe at large, which is an average of walls and voids. The measurements of time in the Universe at large give different results, when done with wall clocks, like our own galaxy cluster clock, or average clocks. Since wall clocks run slower than global-average clocks, when used to measure the time it took the Universe to expand, the resulting age of the Universe is smaller than would be the case if a faster running, global-average clock were to be used. All this is explained in detail in Chapter 2. As a consequence, there are different Hubble constants according to the type of clock we use. There is no such thing as absolute time, or an absolute Hubble constant, or an absolute age of the Universe! It all depends on the clock we use to measure it. When measured with the global-average clock, we obtain the following Hubble constant and age of the Universe at large:
>
> (9) $H_0(t) = 48.2 \rightarrow t_0(t) = 20.3$ billion years.
>
> But we obtain a different result for the Hubble constant and the age of the Universe at large, when we measure it with a wall clock, like the one we use here on Earth, in our own galaxy cluster:
>
> (10) $H_0(\tau_w) = 66.5 \rightarrow t_0(\tau_w) = 14.7$ billion years.

---

[44] Wendy Freedman *et al.*, "Final results from the Hubble Space Telescope Key Project to measure the Hubble constant", in: *Astrophysical Journal*, vol. 533 (2001): 47–72, they reported $H_0 = 70 \pm 7 \ kms^{-1}Mpc^{-1}$.

After consulting Math box 1.6, the reader might wonder what is the 'real' age of the Universe. As a matter of fact, both ages that I just mentioned in equations (9) and (10) are 'real', it all depends on how we measure it, with wall clocks, void clocks, or global-average clocks (see Section 2.2 of Chapter 2). Our Universe at large (on average) has an age of 14.7 billion years, when measured with clocks we use on our planet Earth, in our galaxy.

I will now come back to the main argument of this section, the corroboration of the Friedmann–Lemaître model of an expanding Universe by Hubble's law. Hubble's considerable measurement errors gave detractors of the Big Bang model powerful ammunition, since geological research into certain rock formations on Earth had dated the planet to 3.4 billion years, which was incompatible with Hubble's calculation dating the Universe to 1.75 billion years. These errors were not corrected until 1948, when Walter Baade (1893–1960) conducted more accurate observations with a new 200-inch (5 m) telescope installed on Mount Wilson. He discovered there are two types of Cepheids, which had been confounded by Hubble, distorting the scale for measuring absolute distances.[45]

For a long time, the lack of accurate observations meant that the empirical value of the Hubble constant was not correctly estimated and its value was the object of considerable debate for many years.[46] As we saw in Math box 1.6, a wide variation of estimates of the Hubble constant and the age of the Universe developed over time, until Wiltshire, taking general relativity seriously, did away with the idea of absolute time.

Hubble was never interested in the theoretical implications of his observations and did not go beyond reporting their results, even though they corroborated the main claims of the Friedmann–Lemaître theory. In his 1927 essay, Lemaître had not only proposed his model of the Big Bang but

---

[45] Walter Baade, "A revision of the extra-galactic distance scale", in: *Transactions of the International Astronomical Union*, vol. 8 (1952): 397–398.
[46] Malcolm Longair, *The Cosmic Century* (2006): 343.

had also predicted the relation between redshift, recession velocity and distance... just like Hubble observed a few years later! When a scientific theory makes predictions that are later empirically corroborated, it is strengthened and stands as true, as long as new facts do not falsify it.

Einstein, who had already read Hubble and Humason's results, travelled to Mount Wilson in February 1931. Hubble and Humason showed him the photographic plates revealing the redshift of distant galaxies. Einstein immediately addressed the journalists who had gathered at the observatory library, renouncing his own static model of the Universe and accepting Friedmann and Lemaître's dynamic model. He had the courage to publicly rectify his errors.

Sir Arthur Eddington (1882–1944), a Quaker pacifist and English astronomer, as well as being one of the few people who could fully grasp the mathematics of Einstein's general theory of relativity, also made amends. Eddington had received and ignored a letter from Lemaître in 1927, in which he had enclosed his recent publication that derived the dynamic model of the Universe from Einstein's equations. Now that Hubble's observations were in the headlines, Lemaître wrote to Eddington again, who sent a letter to the prestigious journal *Nature* in June 1930, drawing the scientific community's attention to Lemaître's work. In 1931, Lemaître himself was given the opportunity twice to discuss his theory in *Nature*.[47] He presented his idea of the original super quantum in the following terms:

"Sir Arthur Eddington states that, philosophically, the notion of a beginning of the present order of Nature is repugnant to him. [But] I think that the present state of quantum theory suggests a beginning of the world very different from the present order of Nature. Thermodynamical principles from the point of view of quantum theory may be stated as

---

[47] George Lemaître, *Nature*, vol. 127, no. 3210 (1931): 706; George Lemaître, "The Evolution of the Universe: Discussion", *Nature,* vol. 128, no. 3234 (1931): 699–701.

follows: (1) Energy of constant total amount is distributed in discrete quanta. (2) The number of distinct quanta is ever increasing. If we go back in the course of time we must find fewer and fewer quanta, until we find all the energy of the universe packed in a few or even in a unique quantum."[48]

Eddington also translated Lemaître's work, which was originally written in French, into English and it was published in the *Monthly Notices of the Royal Astronomical Society*.[49] From 1930, Lemaître went on to give lectures throughout the world and received a number of international prizes. In 1933, he met Einstein at a seminar on Hubble's observations and the Big Bang model in Pasadena, California. This time, instead of berating his physics as abominable, as he had done six years earlier, Einstein heaped praise on Lemaître: *"This is the most beautiful and satisfactory explanation of creation to which I have ever listened."*[50]

Section 1.2 A brief history of the Universe following the *Big Bang*

In this section, I describe the history of the expanding Universe, starting from the *Big Bang*. Unlike human history, whose reconstruction depends on records left behind by eyewitnesses and historic objects that can be dated using techniques such as C-14 radiocarbon dating, we can actually see the Universe's history with our own eyes. This is possible because the further away cosmic phenomena and objects are in space, the further back they are in time. Even if electromagnetic records travel at the speed of light, the vastness of the Universe means that we receive electromagnetic waves from all past epochs. Today, we can see objects that originated in a distant past, for example, we can see lighter elements of the periodic table throughout the

---

[48] George Lemaître, "The Beginning of the World from the Point of View of Quantum Theory", *Nature*, vol. 127, nr. 3210 (1931): 706.
[49] George Lemaître, "A Homogeneous Universe of Constant Mass and Growing Radius Accounting for the Radial Velocity of Extragalactic Nebulae", in: *Monthly Notices of the Royal Astronomical Society*, vol. 91 (1931): 483–490.
[50] Cited in Simon Singh, *Big Bang. The Origin of the Universe* (2004): 276.

entire Universe, such as hydrogen (protons), deuterium, helium and beryllium and lithium isotopes whose nuclei were produced during the first 15 minutes of the Universe. In Graph 1.3, the observable objects and events are in red, future ones in black.

Graph 1.3 The observable Universe

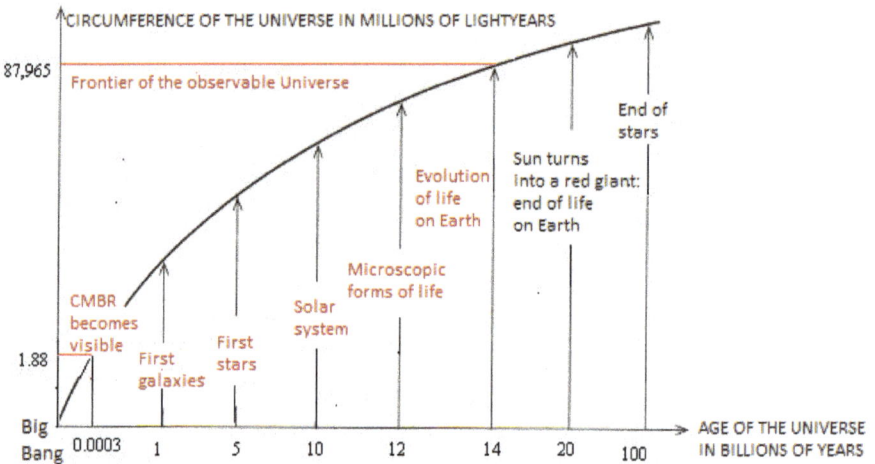

Astrophysicists have shown great interest in the Universe's first few minutes. However, since this epoch cannot be directly observed, a series of hypothetical reconstructions have been undertaken instead, first by Ralph Alpher and Robert Herman, then by Steven Weinberg and Jonathan Allday.[51] I have divided these initial moments into eight stages (I to VIII).

I. Scientists believe the Universe was extremely small, hot and dense to begin with, before expanding rapidly. According to Weinberg, it is

---

[51] Ralph Alpher, James Follin & Robert Herman, "Physical Conditions in the Initial Stages of the Expanding Universe", in: *Physical Review*, vol. 92 (1953): 1347–1361; Steven Weinberg, *The First Three Minutes: A modern view of the origin of the universe* (1977); Jonathan Allday, *Quarks, Leptons and the Big Bang. Second edition* (2002): 244–266.

impossible to say for sure what really happened during the time interval from $t = 0$ to $t = 10^{-43}$:

"We do not know enough about the quantum nature of gravitation even to speculate intelligently about the history of the Universe before this time. We can make a crude estimate that the temperature of $10^{32^o}\,K$ was reached some 10^{-43} seconds after the beginning, but it is not really clear that this estimate has any meaning. Thus, whatever other veils may have been lifted, there is one veil, at a temperature of $10^{32^o}\,K$, that still obscures our view of the earliest times."[52]

We do not have a theory that would allow a hypothetical reconstruction of this phase, as John Barrow points out:

"Suppose we take the whole mass inside the visible Universe and determine its quantum wavelength. We can ask when this quantum wavelength of the visible Universe exceeds its size. The answer is when the Universe is smaller than the Planck length in size (10^{-33} cm), less than the Planck time in age (10^{-43} secs) and hotter than the Planck temperature (10^{32} degrees). Planck's units mark the boundary of applicability of our current theories. To understand what the world is like on a scale smaller than the Planck length we have to understand fully how quantum uncertainty becomes entangled with gravity... The constants of Nature mark out the frontiers of our existing knowledge..."[53]

To understand the point Barrow makes, one must know that in quantum physics there are minimum quantities of time and length that cannot be subdivided further. These are two of the so-called Planck units. The Planck unit of time is $t = 5.4*10^{-44}\,s$ and of length, $1.616 *10^{-35}\,m$. When it comes

---

[52] Steven Weinberg, *The First Three Minutes: A modern view of the origin of the universe* (1977): 125.
[53] John Barrow, *The Constants of Nature* (2002): 43.

to temperature, there is a minimum, which is, of course, zero degrees K, but also a maximum, $1.4*10^{32}$ K. Math box 1.7 contains all the Planck units.

Math box 1.7 Planck units of mass, length, time and temperature[54]

In the following equations, c is the velocity of light; G is the gravitational constant; and k is the Boltzmann constant.

Planck mass: $m_{pl} = \sqrt{\dfrac{\hbar c}{G}} = 2.1*10^{-5}$ $g = 2.1*10^{-8}$ kg.

Planck length: $l_{pl} = \sqrt{\dfrac{\hbar G}{c^3}} = 1.616*10^{-35}$ m.

Planck time: $t_{pl} = \sqrt{\dfrac{\hbar G}{c^5}} = 5.391*10^{-44}$ s.

Planck temperature: $T_{pl} = \dfrac{m_{pl}c^2}{k} = \sqrt{\dfrac{\hbar c^5}{Gk^2}} = 1.416808*10^{32}$ K.

Planck constant: $h = 6.6260755*10^{-34}$ $Js = kgm^2s^{-1}$ and $\hbar = h/2\pi = 1.0546*10^{-34}$ Js.

The Planck length and the Planck time are the smallest possible units. The Planck temperature is the hottest possible temperature. The Planck mass, however, is not the smallest possible mass. Many things weigh less, like, for example, a flea's egg or a proton. The Planck mass is big, because the gravitational force constant G is relatively very weak.

Three of these units (length, time and temperature) are crucial for reconstructing the first fraction of a second of the Universe. George Smoot makes the same point as Weinberg and Barrow: that currently there is no theory of physics that allows us to have a look behind the veil that covers

---

[54]Max Planck, "Über irreversible Strahlungsvorgänge", in: *Sitzungsberichte der Königlich Preußischen Akademie der Wissenschaften zu Berlin* (1899): 478–480.

what happened before the Universe was $t = 5.3*10^{-43}$ old.[55] Smoot calls this stage from $t = 0$ to $t = 10^{-43}$ the quantum gravity epoch, the physical laws of which are unknown.[56] Roger Penrose too, in the context of his criticism of theories that see the Big Bang as a big bounce in an infinite succession of big bangs and big crunches, argued, as recently as 2016, that a theory of quantum gravity, the Holy Grail of quantum physics, still eludes us:

"We cannot expect any reasonable classical equations of state to provide us with a bounce within the context of Friedmann-Lemaître-Robinson-Walker models, and the issue must be raised as to whether the equations of quantum mechanics will enable us to fare better... space-time curvature radii would become indefinitely small near the classical singularity, eventually becoming smaller even than the Planck scale of $\sim 10^{-33}$ cm... At this scale, there would have to be drastic departures from the normal smooth-manifold picture of space-time... the procedures of general relativity will necessarily have to be modified in order to fit in with those of quantum mechanics in the vicinity of such a violently curved space-time geometry. That is to say, some appropriate quantum gravity theory appears to be needed to cope with those situations where the classical procedures of Einstein lead to singularity... A trouble here is that, even now, there is no generally accepted quantum-gravity proposal."[57]

II. From 10^{-43} to 10^{-34} seconds, the temperature fell from 10^{32} to 10^{27} degrees Kelvin. The Universe was pure energy. Three of the four great forces of the Universe were still unified: the strong nuclear force, the weak nuclear force and the electromagnetic force. The gravitational force was a separate force.

---

[55] George Smoot & Keay Davidson, *Wrinkles in Time* (1993): 283.
[56] George Smoot & Keay Davidson, *Wrinkles in Time* (1993): 150–151.
[57] Roger Penrose, *Fashion, Faith and Fantasy: on the New Physics of the Universe* (2016): 228.

III. From 10^{-34} to 10^{-10} seconds, the temperature fell from 10^{27} to 10^{15} degrees Kelvin. This epoch saw the separation of the strong nuclear force from the weak nuclear force and the electromagnetic force, which were still unified. In terms of matter, quantum physicists speculate that quarks and anti-quarks abounded at this stage. Since these elementary particles break the rule that charges must be whole numbers, some physicists doubt about their independent existence.[58] However, as a mathematical construct, the concept of the quark is extremely useful. As well as quarks, the Universe also abounded in highly energised photons, electrons, positrons and other particles and anti-particles that annihilated each other.

IV. From 10^{-10} to one second, the temperature fell from 10^{15} to 10^{10} degrees Kelvin. The weak nuclear force and the electromagnetic force separated. The annihilation of matter and anti-matter ended, leaving just one billionth of the original matter, which is all the matter that exists today. Matter was organised into protons and neutrons. Alpher and Herman envisaged the generation of protons and neutrons in two phases. First, high-energy photons produced protons and anti-protons and neutrons and anti-neutrons, and vice versa, in a period referred to by Alpher and Herman as the 'inter-conversion' of radiation and matter.[59] In their study of the early Universe, Kolb and Turner estimated the ratio of baryons and photons resulting from this inter-conversion to be $4 * 10^{-10} < \eta = \frac{n_B}{n_\gamma} < 7 * 10^{-10}$.[60] This number gives the *relative* abundance of baryons, not the absolute figure. According to Alpher, the matter–radiation inter-conversion phase was followed by a proton–neutron inter-conversion phase in which neutrons and protons were

---

[58] Jonathan Allday, *Quarks, Leptons and the Big Bang* (2002): 167–180.
[59] Ralph Alpher, James Follin & Robert Herman, "Physical Conditions in the Initial Stages of the Expanding Universe", in *Physical Review*, vol. 92 (1953): 1347–1361; see also Jonathan Allday, *Quarks, Leptons and the Big Bang* (2002): 244–266.
[60] Edward Kolb & Michael Turner, *The Early Universe* (1994): 16, 127.

kept in thermodynamic equilibrium by certain interactions, as defined in Math box 1.8.[61]

Math box 1.8 Matter-radiation and neutron–proton inter-conversion

Inter-conversion (as indicated by $\leftrightarrow$) of radiation and matter occurs as follows. High energy photons collided and produced protons and anti-protons, and neutrons and anti-neutrons, and vice-versa, matter and anti-matter annihilated, when colliding, producing gamma-ray photons:

(1) $\gamma + \gamma \leftrightarrow p^+ + \bar{p}^-$.

(2) $\gamma + \gamma \leftrightarrow n + \bar{n}$.

The radiation field also generated pairs of electrons and positrons, which then annihilated each other when colliding:[62]

(3) $\gamma + \gamma \leftrightarrow e^- + e^+$.

The production of protons and neutrons stopped at $10^{-6}\,s$ after temperature had dropped to $10^{13}\,K$, and the production of electrons stopped a few seconds later, when temperature had dropped to $10^9\,K < T < 10^{10}\,K$. The inter-conversion of radiation and matter was followed by the inter-conversion of neutrons (n) and protons (p^+): [63]

---

[61] See Ralph Alpher, James Follin & Robert Herman, "Physical Conditions in the Initial Stages of the Expanding Universe", in *Physical Review*, vol. 92 (1953): 1354–1358.

[62] Fred Hoyle & Roger Tayler, "The Mystery of the Cosmic Helium Abundance", in: *Nature*, vol. 203 (1964): 1108–1110.

[63] See Jeremy Bernstein, *Kinetic Theory in the Expanding Universe* (1988): 109; Edward Kolb & Michael Turner, *The Early Universe* (1994): 89. Alpher writes $n + v \to p + e^+$, although in the context it is clear this should be $n + v \to p + e^-$. See also See Fred Hoyle & Roger Tayler, "The Mystery of the Cosmic Helium Abundance", in: *Nature*, vol. 203 (1964): 1108–1110; Jonathan Allday, *Quarks, Leptons and the Big Bang* (2002): 262; Malcolm Longair, *The Cosmic Century* (2006): 322 and John Barrow & Frank Tipler, *The Anthropic Cosmological Principle* (1986): 398; Michael Barnett *et al.*, *The Charm of Strange Quarks* (2000): 162. Barrow mistakenly writes interaction (4) as $n + e^+ \to p + v_e$, whereas, as a matter of fact, we do not have an electron neutrino, but an electron anti-neutrino $\bar{v}_e$.

(4) $n + e^+ \leftrightarrow p^+ + \bar{v}_e$.

(5) $n + v_e \leftrightarrow p^+ + e^-$.

(6) $n \leftrightarrow p^+ + e^- + \bar{v}_e$.

A small fraction (about one in 1000) of free neutrons decay producing the same particles, but emitting an extra particle in the form of a gamma ray:

(7) $n \leftrightarrow p^+ + e^- + \bar{v}_e + \gamma$.

Note that in all these processes of inter-conversion, nature upholds the law of conservation of charge. A proton has two up quarks and one down quark (uud), with a total charge of $2 * \left(+\frac{2}{3}\right) + \left(-\frac{1}{3}\right) = +1$, a neutron has one up quark and two down quarks (udd), with a total charge of $\left(+\frac{2}{3}\right) + 2 * \left(-\frac{1}{3}\right) = 0$. An electron has a charge of -1, and a positron of +1. Anti-particles have the same mass as their particle counterparts, but a contrary charge.

Table 1.2 The characteristics of quarks

type	up	down	charm	strange	top	bottom
symbol	u	D	c	S	t	b
mass GeV/c^2	0.005	0.01	1.5	0.2	175	4.7
electric charge	+ 2/3	- 1/3	+ 2/3	- 1/3	+2/3	- 1/3
spin	½	½	½	½	½	½

When a neutron decays, to produce a proton, an electron, and an electron anti-neutrino, it so happens that a *down* quark transforms into an *up* quark, in the process emitting an electron:

(8) $udd \rightarrow uud + e^- + \bar{v}_e$

In 1953, Alpher and Herman calculated the ratio of the relative abundance of protons and neutrons in the Big Bang taking into account the instability of the neutron, which they assigned a half-life of 12.8 minutes, yielding a ratio of $4.5 \leq N_{pro}/N_{neu} \leq 6.0$.[64] As a matter of fact, a free neutron has a shorter half-life, of 10.2 minutes, which means the authors slightly overestimated the amount of neutrons produced. In 1977, Steven Weinberg corrected the figure, allowing for slightly more protons, with a ratio of $(N_{pro}/N_{neu}) \cong 6.7$, equivalent to 13% neutrons and 87% protons.[65] In 1994, Kolb and Turner estimated the ratio of neutrons and protons at the end of inter-conversion when the numbers freeze out as $(n/p)_{freeze-out} \cong 1/6$. Since 50% of the neutrons decay after a little more than ten minutes, this ratio falls to $(n/p)_{nucleo-synthesis} \cong 1/7$, when, a few minutes later, the remaining neutrons are 'trapped' and stabilised in helium nuclei by nucleosynthesis.[66]

After the production of protons and neutrons, at $t = 0.01$ s, the temperature was 10^{11} K and the energy density of the Universe $21*10^{44}$ eV per litre, roughly equivalent to 3.8 billion kilograms per litre. At this point, according to Weinberg, the circumference of the Universe was possibly "*about four light years*", which implies a radius of about $\frac{4}{2\pi}ly = 0.64$ light years.[67] Such an estimate would imply that during the first hundredth of a second after the Big Bang, the expansion velocity of the Universe was $2*10^7$ times the velocity of light! With an expansion velocity equal to the velocity of light, the radius of the Universe would have been only 3,000 km.

---

[64] Ralph Alpher, James Follin & Robert Herman, "Physical Conditions in the Initial Stages of the Expanding Universe", in *Physical Review*, vol. 92 (1953): 1357–1358.

[65] Steven Weinberg, *The First Three Minutes: A modern view of the origin of the universe* (1977): 98.

[66] Edward Kolb & Michael Turner, *The Early Universe* (1994): 88–89, 95.

[67] Steven Weinberg, *The First Three Minutes: A modern view of the origin of the universe* (1977): 94.

In the Math box 1.9, I give a more conservative estimate of this rapid expansion, maintaining, however, Weinberg's implicit point that during a fraction of a second, expansion velocity was faster than light. This point is not to be confounded with Alan Guth's inflationary period (see Chapter 4).

Math box 1.9 The volume and radius of the Universe at $t = 0.01\ s$

The estimate of the quantity of protons and neutrons at $t = 0.01\ s$ is the following:

(1) 10^{80} protons $+ 0.22 * 10^{80}$ neutrons $= 1.22 * 10^{80}$ baryons.

Photons do not compete for space, but baryons do. The volume of one proton or neutron is approximately:

(2) $V_{baryon} = (10^{-15})^3\ m^3$.

From (1) and (2), we obtain a total volume of the Universe of:

(3) $V_{universe} = 1.22 * 10^{80} * (10^{-15})^3\ m^3 = 0.122 * 10^{36}\ m^3$.

From the volume of a sphere, we obtain its radius:

(4) $V = \frac{4}{3}\pi r^3 \rightarrow r = \sqrt[3]{\frac{3V}{4\pi}}$.

From (2) and (4), we obtain the radius of the Universe at $t = 0.01\ s$:

(5) $r = \sqrt[3]{\frac{0.366*10^{36}}{4\pi}}\ m = 0.30767372 * 10^{12}\ m$.

Since this distance was covered in one hundredth of a second, the expansion velocity per second was:

(6) $v_{ex(t_0 \rightarrow t=0.01)} = 30.767372 * 10^{12}\ m/s$.

Since light travels at a lesser speed:

(7) $c = 299{,}792{,}458\ m/s$,

it follows from (6) and (7) that during one hundredth of a second after $t_0 = 0$, the expansion velocity was a hundred thousand times the velocity of light:

(8) $v_{ex(t_0 \to t=0.01)}/c = 102{,}629 \cong 10^5.$

This estimate is a bit more conservative than Weinberg's $v_{ex(t_0 \to t=0.01)}/c \cong 2 * 10^7$, which is ten million times the velocity of light. The reader must take into account that these values and figures are estimates, based on 'reverse engineering' of the 'machinery' of the Universe at a later point in time where it can be observed today, looking backwards in time. It is not exact science, rather an approximation.

V. From one second to three minutes, the temperature fell from 10^{10} to 10^9 degrees Kelvin. Neutrinos separated from matter and electrons and positrons were annihilated, leaving an excess of electrons equal to the quantity of protons. The cosmic background radiation separated from matter but remained invisible, since photons were continuously colliding with different particles and the Universe was opaque. The ratio of neutrons to protons fell slightly because, on the one hand, neutrons decayed into protons, and, on the other, the temperature had fallen to a point of no return, preventing the transformation of protons into neutrons. There would have been no neutrons left at all if nuclear reactions had not started at the end of this period, producing stable helium nuclei, in which neutrons were trapped.[68]

VI. From around three minutes to between 15 and 20 minutes, the temperature fell from 10^9 to 10^8 degrees Kelvin, and the Universe transformed into a giant hydrogen bomb with a radius of almost one light year, producing large quantities of helium nuclei and leaving a large

---

[68] Edward Kolb & Michael Turner, *The Early Universe* (1994): 16.

quantity of hydrogen that did not undergo fusion.[69] The production of helium in the Big Bang was discovered by Gamow, Alpher and Herman, as we shall now see.

George Gamow (1904–1968) was a Russian physicist who first discovered Friedmann's work in 1923 during his time in Leningrad. In 1933, Gamow and his wife escaped from the Soviet Union to continue his work in cosmology at George Washington University in the United States. He remained sceptical of Lemaître's model of the repeated nuclear fission of a primitive super quantum, since both nuclear fusion and fission ultimately give rise to nuclei in the middle of the Mendeleev periodic table, converging upwards (fusion) or downwards (fission) on iron, which has 26 protons and 30 neutrons and is extremely stable. It is simply not feasible that abundant quantities of light elements such as helium and hydrogen were formed by the repeated fission of a super-heavy atom, as argued by Lemaître.

Gamow understood that while fission was impossible, nuclear fusion at the heart of stars also failed to account for the volumes of helium observed in the Universe. Current estimates are similar to the values that Gamow considered: for every 10,000 hydrogen atoms, there are 1,000 helium atoms, 6 oxygen atoms, 1 carbon atom and less than 1 atom of the remaining elements. Based on the speed at which helium is produced in the heart of stars, it would take 27 billion years to produce the quantity of helium observed in the Universe,[70] so most of the helium must have been produced during the Big Bang, before the formation of the stars.

Gamow was not a strong mathematician and was helped by Ralph Alpher and Robert Herman, a student born in 1921. The trio realised there was a window of just 15 minutes in the evolution of the Universe, shortly after the Big Bang, in which it would have been possible to produce helium. At temperatures in excess of millions of millions of degrees, particles travel too

---

[69] See Jeremy Bernstein, *Kinetic Theory in the Expanding Universe* (1988): chapter 9.

[70] Simon Singh, *Big Bang. The Origin of the Universe* (2004): 310.

fast to undergo fusion and when they fall below a few million degrees they are already too slow to bond. Protons must come into frequent contact at a high enough speed to allow the strong nuclear force (which joins the baryons in the nucleus) to overcome the electromagnetic repulsion of the positive charges.

Another limit on this critical time is the average half-life of neutrons of just 10 minutes. After years of calculations and verifying the data, in April 1948 they published a mathematical proof in the journal *Physical Review* showing that the Big Bang model produced hydrogen and helium in the same proportions as currently observed. In 1953, the relative abundance of helium atoms was estimated to be between 9% and 12.5% by Alpher and Herman, making the corresponding figure for hydrogen between 91% and 87.5%.[71] In 1964, Hoyle and Tayler arrived at the figures of 12.3% and 87.7% ($He/H \cong 0.14$).[72]

In a classic study from 1974, the US astrophysicist Wagoner calculated the abundance of primordial helium mass, estimating it to be between 26% and 32%.

Wagoner also calculated the relative abundances of other light elements and isotopes produced by nucleon-synthesis shortly after the Big Bang and compared them with the levels currently observed in the Universe.

After examining Table 1.3, it is hard not to admire the ability of modern astrophysics to reconstruct the initial conditions of the early Universe, though there are also serious problems due to the difficulty in calculating the relative abundances of the light elements and the ratio of baryons and photons during the Big Bang.

---

[71] Ralph Alpher, James Follin & Robert Herman, "Physical Conditions in the Initial Stages of the Expanding Universe", in *Physical Review*, vol. 92 (1953): 1358.
[72] Fred Hoyle & Roger Tayler, "The Mystery of the Cosmic Helium Abundance", in: *Nature*, vol. 203 (1964): 1109.

Table 1.3 Relative abundance of light elements according to Wagoner[73]

Element	Fraction of total mass	Location	Production P and destruction D	Produced in Big Bang
Deuterium ^2H	$\leq 4.1*10^{-4}$	Planets, meteorites, interstellar medium	Some P in solar system, 10–75% D in stars	$(0.3 \text{ to } 5) *10^{-4}$
Tritium ^3He	$\leq 2.46 *10^{-4}$	Meteorites, solar wind	P possible in stars, 10–75% of D in stars	$\leq 1*10^{-4}$
Helium ^4He	0.26 to 0.32	Galaxies, stars, interstellar medium	1–4% P in stars, $\approx$0% D in stars	0.22 to 0.32
Lithium-6 ^6Li	$0.4*10^{-9}$	Earth, meteorites	Sufficient P from cosmic radiation	$\leq 1*10^{-9}$
Lithium-7 ^7Li	$< 2.35 *10^{-8}$	Meteorites, stars, interstellar medium,	P possible in stars, 10–75% D in stars	$\leq 2*10^{-8}$
Beryllium ^8Be	$< 5.9*10^{-10}$	Meteorites, interstellar medium	Sufficient P from cosmic radiation	$\leq 3*10^{-10}$
Boron-10 ^{10}B	$0.3*10^{-9}$	Meteorites	Sufficient P from cosmic radiation	$\leq 1*10^{-9}$
Boron-11 ^{11}B	$\leq 3*10^{-9}$	Meteorites, Sun	Sufficient P from cosmic radiation	$\leq 3*10^{-9}$
$A \geq 12$	$1.5 *10^{-2}$	Stellar photosphere	Sufficient P in stars	$\leq 10^{-5}$

---

[73] Robert Wagoner, "Big Bang Nucleosynthesis Revisited", in: *Astrophysical Journal*, vol. 179 (1973): 353.

The method used by astrophysicists to establish Big Bang nuclear synthesis has five steps: i) measure the relative abundance of deuterium D / H at present; ii) use this data to estimate the relative abundance of deuterium in the Big Bang; iii) use this data to estimate the ratio of baryons to photons in the Big Bang; iv) use this data to calculate the relative abundances of helium and lithium–7 in the Big Bang; and v) compare these values with those currently observed. The margin of error increases with each step.

In 1994, Prantzos and others reached a more accurate estimate of the abundance of helium mass of between 22.0% and 23.7%.[74] An outstanding problem of considerable difficulty, however, was the fact that the relative abundances of deuterium and lithium appeared not to coincide with the same baryon/photon ratio, as Steigman explained in 1996.[75] In 1998, more precise measurement of deuterium in five quasars, where deuterium is not burnt like in stars, yielded a primordial quantity of deuterium of $D/H = 3.4$ *10^{-5}.[76] This did not solve the problem, however, but only revealed it much more clearly than before, since the abundances of deuterium and lithium-7 were not concordant.

Another problem also arose at the same time. A new method for deriving the baryon/photon ratio, by measuring cosmic background radiation, gave a baryon density of $\Omega_B \cong 0.0653$, which was incompatible with the density of $\Omega_B \cong 0.041$ derived from the observation of the relative abundances of light elements (both figures having a confidence interval of 95%). Burles, Nollett and Turner warned that the relative abundances of helium, deuterium and lithium-7 derived from this new estimate of Ω_B and η based on the cosmic

---

[74] Nikos Prantzos, Elisabeth Vangioni-Flam & Michel Cassé, eds, *Origin and Evolution of the Elements* (1994): 92, Table I.

[75] Gary Steigman, "Testing Big Bang Nucleosynthesis", in: *Cosmic Abundances*, Stephen Holt & George Sonneborn, eds, ASP Conference Series, vol 99 (1996).

[76] David Tytler, John O'Meara, Nao Suzuki & Dan Lubin, "Deuterium in quasar spectra", in: "Review of Big Bang Nucleosynthesis and Primordial Abundances", in: *Physica Scripta* (2000): T 85–103.

background radiation *"would conflict significantly with observed abundances"*.[77] Thus, instead of solving the problem, advances in modern observation technology in fact exacerbated it, according to Juan Lara.[78]

Graph 1.4, from Steigman's 2006 paper, shows that the permitted range of η (where $\eta_{10} \equiv 10^{10} * \eta$), derived from the cosmic background radiation (the Wilkinson Microwave Anisotropy Probe, or *WMAP*) was compatible with the ranges of the relative primordial abundances of tritium (^{3}He) and of deuterium (D), but not of lithium-7 (^{7}Li) and helium (^{4}He). It also shows that the ranges for lithium-7 and deuterium were mutually incompatible, as well as the ranges for helium and deuterium.

Graph 1.4 The baryon/photon ratio and light elements abundances[79]

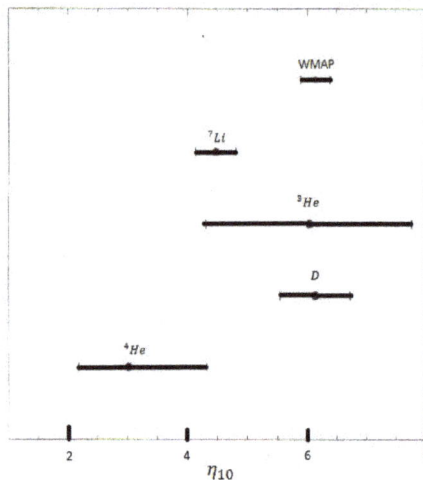

---

[77] Scott Burles, Kenneth Nollett & Michael Turner, "What is the Big Bang Nucleosynthesis Prediction for the Baryon Density and How Reliable Is It?", *Physical Review D* (2001): 63–69 (quote on p. 68).

[78] Juan Lara, "Deuterium and ^{7}Li Concordance in Inhomogeneous Big Bang Nucleosynthesis models", *Frontier in Astroparticle Physics and Cosmology: Proceedings of the 6th International* Symposium, K. Sato and S. Nagataki eds, (2004): 87 ss.

[79] Gary Steigman, "Primordial Nucleosynthesis: Successes and Challenges", in: *International Journal of Modern Physics E* vol. 15 (2006):1–36.

New models of atomic physics have helped close the gap between cosmic background radiation (*WMAP*) observations and the relative abundances of deuterium and helium, though the conservative range for the baryon/photon ratio remains as large as Kolb and Turner described it 20 years ago ($3*10^{-10} < \eta < 10*10^{-10}$).[80]

The problem of helium lying outside the range of the baryon/photon ratio derived from the *WMAP*, as can be seen in Steinberg's graph, was solved in 2007 by the Peimberts (father and son), astrophysicists at the Universidad Nacional Autónoma de México, as can be seen in the results of Table 1.4. They calculated the abundance of primordial helium mass in the Universe to be 24.77%, making the proportion of hydrogen 75.23%. Since helium makes up 25% of the total mass of baryonic material, this implies that approximately 8% of the atoms are helium atoms.

Table 1.4 Baryon/photon ratio η and the baryon density Ω_B[81]

Method	$Y_P(\equiv \dfrac{^4He}{mass_{bar}})$	$D_P(\equiv \dfrac{10^5 D}{mass_{bar}})$	$\eta_{10}(\equiv 10^{10}*\eta)$	Ω_B [c]
Y_P	0.2477 ± 0.0029[a]	$2.78^{+2.28}_{-0.98}$[b]	5.813 ± 1.81[b]	0.0433 ± 0.0135[b]
D_P	0.2476 ± 0.0006[b]	2.82 ± 0.28[a]	5.764 ± 0.360[b]	0.0429 ± 0.0027[b]
WMAP	0.2482 ± 0.0004[b]	2.57 ± 0.15[b]	6.116 ± 0.223[b]	0.0456 ± 0.0017[a]

Notes: [a] observed value; [b] theoretically expected value; [c] it is assumed $h \cong 0.7$

---

[80] Edward Kolb & Michael Turner, *The Early Universe* (1990): 106.
[81] Manuel Peimbert, Valentina Luridiana & Antonio Peimbert, "Revised Primordial Helium Abundance", *Astrophysical Journal*, vol. 666 (2007): 636–646.

Only one incompatibility remains, e.g., the range for lithium-7, which was therefore left out Peimbert's table. Only by abandoning the Standard Big Bang Nucleosynthesis (*SBBN*) and adopting a new model that does not assume the Universe to be homogeneous, as proposed in Juan Lara's Inhomogeneous Big Bang Nucleosynthesis (*IBBN*), can we allow for "*a larger range of acceptable ^7Li depletion factor, to bring deuterium and ^7Li in concordance with each other*".[82] This model, however, is still in its early stages.

Let us return to our brief history of the Universe after the Big Bang. After little more than half an hour, the temperature was $3*10^8 K$ and the mass to energy density of the Universe was just 10% greater than the normal density of water. The nuclear reactions had come to an end, and would only restart in the hearts of the first stars, one billion years later.

VII. From half an hour to 300,000 years after the Big Bang, the temperature fell from 100 million to 3,000 degrees Kelvin. Hydrogen and helium nuclei existed as plasma because electrons were still free and continued to prevent the free passage of photons, meaning that the Universe was still opaque. At the end of this period, an event referred to as 'recombination' occurred, when the electrons were trapped by the nuclei to form hydrogen and helium atoms and the cosmic background radiation became visible, lighting up the entire Universe. This radiation is now observed as the cosmic microwave background radiation (*CMBR*), as the wavelength of light was stretched out because of the expansion of the Universe.

The history of the discovery of *CMBR* has been covered by various authors[83] and I shall now summarise it briefly. Ralph Alpher was set on an

---

[82] Juan Lara, "Neutron Diffusion and Nucleosynthesis in an Inhomogeneous Big Bang Model", arxiv. org/abs/astro-ph/0506364v2

[83] John Hawley & Katherine Holcomb, *Foundations of Modern Cosmology* (1998): 319–352; Simon Singh, *Big Bang. The Origin of the Universe* (2004): 401–463; George

academic career and the muted reception of his theory on the relative abundance of helium based on the Big Bang convinced him to pursue another line of enquiry. Working with a fellow scientist, Robert Herman, he explored the events that took place after the first 20 minutes of the Universe's evolution, concluding that although the critical temperature and pressure required for the nuclear fusion of new elements other than helium were not present, the temperature (a few million degrees Kelvin) was nonetheless high enough to allow hydrogen and helium to behave as a plasma. Plasma is a hot ionised gas state in which the high speed of nuclei and electrons stops them bonding together and prevents the free passage of photons, resulting in an optical 'fog'.

They calculated that around 300,000 years after the Big Bang, the temperature had fallen to 3,000 degrees Kelvin, allowing for *recombination*, in which *plasma* transitioned to a *normal gas*, to take place. During this transition phase, electrons had slowed down enough to be trapped by the positive charge of the nuclei in the gas and started to rotate around them at different fixed distances, just like in normal atoms. Some 300,000 years after the Big Bang, the electrons were trapped by hydrogen and helium nuclei and from this point on photons could travel freely in all directions, without colliding with free electrons. The 'cloud' cleared and the Universe was illuminated by a bright light that was none other than the relic of the incandescent radiation from the explosion of the Big Bang itself.

Alpher and Herman argued that if we accept that the Universe is in continuous expansion, the wavelength of this radiation has been increasing as space itself is stretched by the Universe's expansion, while the frequency of this radiation has been falling at the same time. They calculated that weak radiation with the frequency and wavelength of electromagnetic microwaves and a temperature of 5 degrees Kelvin should *still* be detectable as an

---

Smoot, *Wrinkles in Time* (1993); Steven Weinberg, *The First Three Minutes: A Modern View of the Origin of the Universe* (1977): 47–72.

ancient relic of the light originally emitted by the Big Bang, stretched by the expansion of space for some 13 to 14 billion years. Alpher and Herman published the results of their research on the *CMBR* in 1948,[84] but unfortunately the scientific community continued to ignore them and attempts were not made to detect this *CMBR* in the cosmos. Disappointed, Gamow, Alpher and Herman published a final paper in 1953, summarising and improving in their calculations, results and predictions, before turning their backs on cosmology to pursue research in other areas.[85]

Corroboration of Gamow, Alpher and Herman's theory, however, would come forward in the most unexpected manner. Arno Penzias, a German-US physicist born in 1933, and Robert Wilson, a US astronomer born in 1936, were both working at Bell Laboratories on a project to determine the properties of radio waves originating from the outermost layers of our galaxy. In 1964, they discovered an inexplicable 'noise': while controlling all possible sources of error (including pigeon excrement inside the antenna), they realised there was an independent source of radiation with a wavelength of 1 mm coming from everywhere in equal quantities.

At an astronomy conference in Montreal in 1964, Penzias casually mentioned this mysterious radiation to Robert Burke, an astronomer at MIT who had read a draft paper on the work of Robert Dicke and James Peebles, two astronomers at Princeton University who – unaware of Gamow, Alpher and Herman's theory – had independently predicted the existence of *CMBR* with a wavelength of 1 mm. Penzias rang Dicke, dragging him out of a meeting on the possibility of constructing a *CMBR* detector, and informed him that he had discovered the *CMBR*… By sheer coincidence, Bell Labs had discovered the *CMBR*!

---

[84] Ralph Alpher & Robert Herman, "Evolution of the Universe", in: *Nature*, vol. 162 (1948): 774–775.
[85] Ralph Alpher, James Follin & Robert Herman, "Physical Conditions in the Initial Stages of the Expanding Universe", in: *Physical Review*, vol. 92 (1953): 1347–1361.

Penzias and Wilson published the results of their experiments in *Astrophysical Journal* in 1965,[86] with Dicke and his team publishing the theoretical explanation of the phenomena in the same issue.[87] Dicke and Peebles made no reference to the previous work by Alpher and Herman. In 1978, Penzias and Wilson received the Nobel Prize for their discovery of *CMBR*.

More recent measurements have arrived at a more accurate temperature of the *CMBR* of 2.728 ±0.004 degrees Kelvin, with a confidence interval of 95%.[88] It is in the micro-wave range of frequencies at 160.23 GHz, with a peak wavelength of 1.063 mm. The very low photon energy is about 6.626534×10^{-4} eV.

Image 1.1 The *CMBR* in the microwave range of frequencies, and its anisotropy[89]

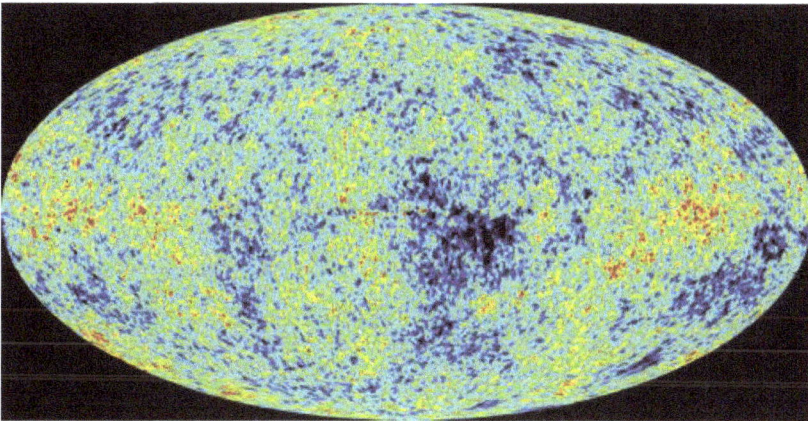

---

[86] Arno Penzias and Robert Wilson, "A measurement of excess antenna temperature at 4080 MHz", in: *Astrophysical Journal*, vol. 142 (1965): 419–421.

[87] Robert Dicke, James Peebles *et al.*, "Cosmic black-body radiation", in: *Astrophysical Journal*, vol. 142 (1965): 414–419.

[88] Dale Fixsen *et al.*, "The cosmic microwave background spectrum from the full COBE FIRAS data set", in: *Astrophysical Journal*, vol. 473 (1996): 576–587.

[89] Copyright NASA, http://lambda.gsfc.nasa.gov/product/cobe.

As the expansion of space stretched the wavelength of the *CMBR* beyond the lengths of visible light, the Universe became dark. These so-called "Dark Ages" ended with the formation of the first stars, four hundred million years after the Big Bang, when the processes of nuclear fusion inside the stars illuminated the Universe once again. At about that time, the first galaxies emerged, which can be observed in Image 1.2, that allows us to look back in time 13.2 thousand million years!

Image 1.2 Photo of eXtreme Deep Field, a sequel to the original Hubble Ultra Deep Field[90]

The eXtreme Deep Field or XDF, seen in the photo, was assembled by combining 10 years of NASA Hubble Space Telescope photographs taken of a patch of sky at the centre of the original Hubble Ultra Deep Field, a picture the Hubble Space Telescope took in 2003 and 2004, which collected

[90] Credit: NASA, ESA, G. Illingworth, D. Magee, and P. Oesch (University of California, Santa Cruz), R. Bouwens (Leiden University), and the HUDF09 Team. Image released September 25, 2012.

light over many hours to reveal thousands of distant galaxies in what was the deepest view of the universe so far. The XDF is a small fraction of the angular diameter of the full Moon.

Why did these galaxies form where they did? The *anisotropy* of the *CMBR*, referring to variations in its temperature and wavelength, is indicative of the irregularities of the original gravitational field, which lie at the basis of the 'lumpiness' of the Universe, i.e., the concentration of matter in certain areas (the 'walls' of galaxy clusters) and not in others (the voids) giving the Universe the structure of a sponge. In a lecture in Mexico City, in June 2007, George Smoot conjectured that both the anisotropy of the *CMBR* and the perturbations in the gravitational field, which concentrates matter in a hundred thousand million galaxies, originated in a hundred thousand million quantum fluctuations during the first fraction of a second of the Big Bang. Half a billion years after the Big Bang, dust clouds started to condense, embedded in this structure, producing galaxy clusters and galaxies.

The discovery of the *anisotropy* of the *CMBR* owes much to the tenacity of George Smoot who dedicated many years of his life to finding and mapping it.[91] He first tried to send his instruments to the stratosphere with a balloon, but to no avail. With the help of NASA, he then constructed a satellite, name *COBE* (*Cosmic Background Explorer*) that had three instruments, two for observing the background radiation in the infrared wavelength, and one *Differential Microwave Radiometer*, to observe this radiation, in the micro-wave frequency discovered by Penzias and Wilson, simultaneously in two different parts of the sky, with an angle of $60^0 = \frac{1}{3}\pi \, radians$, in order to detect the variation of the *CMBR* in these two parts of the heavens. By taking these photos millions of times, it would be possible to map the anisotropy of the *CMBR*.

---

[91] George Smoot & Keay Davidson, *Wrinkles in Time. The Imprint of Creation* (1993).

The *COBE* was programmed to be launched in 1988, but when the Space Shuttle *Challenger* was consumed in flames, in a tragic accident in January 1986, everything was postponed. However, Smoot found an old rocket, and his 1000-man strong team, coordinated by another astrophysicist, John Mather, was finally able to launch their three *COBE* instruments into space, in 1989. The *COBE* made the journey around the Earth 14 times mapping the *CMBR* anisotropy. After eliminating 'noise' and error in the data, Smoot was finally able to publish the results at a Conference of the *American Physical Society*, in 1992, and then in the *Astrophysical Journal*.[92] Smoot and Mather won the 2006 Nobel Prize in Physics for their work with *COBE* that led to the *"discovery of the black body form and anisotropy of the cosmic microwave background radiation"*.[93]

Later measurements further improved on Smoot's original map. In 2001, Charles Bennett, Gary Hinshaw and their team launched a NASA Explorer Satellite (the *Wilkinson Microwave Anisotropy Probe*), which mapped the *CMBR* anisotropies with much more precision. After publishing their results from 2002 to 2009,[94] they received *"the 2012 Gruber Cosmology Prize for their exquisite measurements of anisotropies in the relic radiation from the Big Bang: the Cosmic Microwave Background [Radiation]"*.[95]

---

[92] George Smoot, John Mather, Charles Bennett *et al.*, "Structure in the COBE differential microwave radiometer first-year maps", in: *Astrophysical Journal*, vol. 396 (1992): L1–L5.

[93] Nobel Prize Announcement, October 3, 2006.

[94] Charles Bennett, *et al.* "The Microwave Anisotropy Probe (MAP) Mission", in: *Astrophysical Journal,* vol. 583 (2003a): 1–23; Charles Bennett *et al.*, "First-Year Wilkinson Microwave Anisotropy Probe (WMAP). Observations: Foreground Emission", in: *Astrophysical Journal Supplement.* vol. 148 (2003b): 97–117; Gary Hinshaw *et al.*, "Three-Year Wilkinson Microwave Anisotropy Probe Observations: Temperature Analysis", in: *Astrophysical Journal Supplement*, vol. 170 (2007): 288–334; Gary Hinshaw *et al.*, "Five-Year Wilkinson Microwave Anisotropy Probe Observations: Data Processing, Sky Maps and Basic Results", in: *Astrophysical Journal Supplement*, vol. 180 (2009): 225–245.

[95] 2012 Gruber Cosmology Prize Press Release.

Section 1.3 Production and evolution of the elements in the stars

The discovery of the structure of the atom (see Math box 7.1 in Chapter 7) gave rise not only to quantum physics, but also to nuclear physics. The latter area is important to astrophysics because of the fusion of elements in the Big Bang and inside stars. In 1900, the German physicist Friedrich Dorn (1848–1916) showed that radium, first discovered by Marie and Pierre Curie, was not only radioactive, but transformed into another hitherto unknown element called radon, an inert gas. The Scottish chemist Sir William Ramsay (1852–1916) had already discovered other inert gases (those whose valence is zero, meaning they do not combine with other elements to form molecules), specifically helium, neon, argon, krypton and xenon, in the very column predicted by Mendeleev, who determined the atomic mass of radon. Furthermore, in 1903, Ramsay and the English chemist Frederick Soddy (1877–1956) showed that the gas helium was formed during the radioactive process of the nuclear fission of uranium and radium. When radium, which has 88 protons and 138 neutrons, decays, it transforms into radon, with 86 protons and 136 neutrons, and helium, which has a nucleus with 2 protons and 2 neutrons. A small loss of mass occurs during this slow transformation (the element has a half-life of 1,600 years), which is released as a vast amount of energy,[96] in line with Einstein's famous equation $E = mc^2$.

Following up on previous work on quantum tunnelling by George Gamow (1904–1968), Ronald Gurney (1898–1953) and Edward Condon (1902–1974)[97], Fritz Houtermans (1903–1966) and Robert d'Escourt Atkinson (1898–1982) were the first ones to explain the nuclear fusion that

---

[96] A radium nucleus has 226.025402 amu (atomic mass units), a radon nucleus radon, 222.01757, and helium 4.002602, so that 0.00523 amu are released as energy. One amu is equivalent to $1,66 * 10^{-19}$ *joules*.

[97] Ronald Gurney & Edward Condon, "Quantum Mechanics and Radioactive Disintegration", in: *Nature*, vol. 122, number 3073 (1928): 439 ss. and in: *Physical Review*, vol 33 (1929): 127–140.

occurs in main sequence stars like the Sun.[98] I will come back to the tunnel effect shortly. Houtermans was captured twice, first by the KGB, then by the Gestapo, but survived on both occasions. During his imprisonment, the German physicist Hans Bethe (1906–2005) completed the work Houtermans and Atkinson had begun.[99] Bethe, who received the 1967 Nobel Prize, was born to a Jewish mother in 1906 and escaped from the Nazis in 1933. He showed how hydrogen is transformed into helium in stars.

Nuclear physics plays a major role in cosmology's understanding of the production of helium during the Big Bang, which we have already analysed in Section 1.2 of this chapter, as well as in the production of helium from hydrogen inside stars, and the formation of carbon and oxygen from helium and of other heavier elements, as we shall now see.

In what follows, I will first analyse how the gravitational force overcomes the electromagnetic force in the centre of stars allowing the strong nuclear force to start the process of nuclear fusion (Section 1.3.1); then we will see the production of helium from hydrogen in the centre of the stars (Section 1.3.2); and finally, the production of carbon and oxygen from helium (Section 1.3.3).

Section 1.3.1 How gravity overcomes the electromagnetic force

Let us consider proto-stars and see how the gravitational force, which on its own is extremely weak, can overcome the electromagnetic force, responsible for the repulsion of two protons, giving rise to the strong nuclear force, which holds protons together in the nuclei of atoms.

The electromagnetic force, responsible for the repulsion of two protons, is much stronger than the gravitational force, which causes them to collide. However, the strong nuclear force, which binds protons together in the

---

[98] Fritz Houtermans & Robert Atkinson, "Zur Frage der Aufbaumöglichkeit der Elemente in Sternen", in: *Zeitschrift für Physik*, vol. 54 (1929):656–665.

[99] Simon Sing, *Big Bang. The Origin of the Universe* (2004): 301–303.

nuclei of atoms, only overcomes the electromagnetic force at very short distances. To do so, it needs gravity's help. Graphs 1.5 and 1.6 give us an idea of the ratio of the electromagnetic and gravitational forces (Graph 1.5), and the ratio of the strong nuclear and electromagnetic forces (Graph 1.6):

Graph 1.5 The ratio of the electromagnetic and the gravitational force

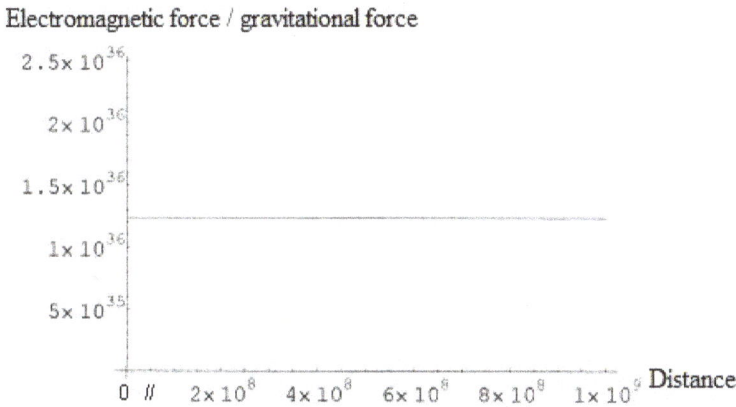

Electromagnetic force / gravitational force

2.5×10^{36}

2×10^{36}

1.5×10^{36}

1×10^{36}

5×10^{35}

0 // 2×10^{8} 4×10^{8} 6×10^{8} 8×10^{9} 1×10^{9} Distance

Graph 1.6 The ratio of the strong nuclear and the electromagnetic force

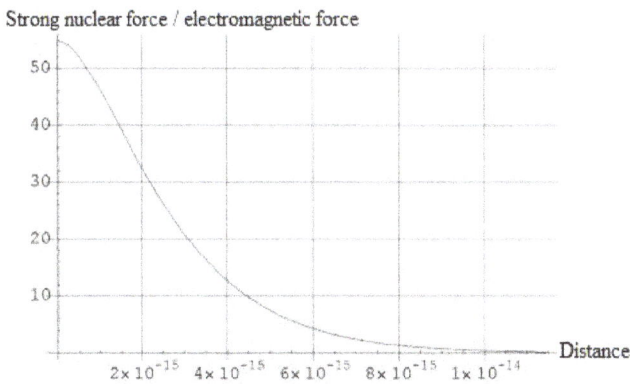

Strong nuclear force / electromagnetic force

50

40

30

20

10

2×10^{-15} 4×10^{-15} 6×10^{-15} 8×10^{-15} 1×10^{-14} Distance

The increasing pressure of the gravitational force is needed to bring protons close enough together to allow the nuclear force to overcome the electromagnetic force and bind protons together within nuclei. Under normal circumstances, when a cloud of gas is in hydrostatic equilibrium, the kinetic energy of the particles exerts an expansive pressure equivalent to the gravitational energy that causes them to collapse. In dense clouds, the gravitational energy produces so much kinetic energy that collisions between protons become so strong that the strong nuclear force overcomes the electromagnetic repulsion between protons. However, for this to happen, it is first necessary to reach a critical mass (see Math box 1.10).

Math box 1.10 Critical mass of a star needed for start of nuclear fusion

The critical mass at which the gravitational energy produces sufficient kinetic energy so that the strong nuclear force overcomes the electromagnetic repulsion is the Jeans mass M_J, which is a function of the pressure P, the gravitational constant G and the mass density ρ,[100] or of the Jeans length l_J and mass density ρ.[101] Both functions are identical.

$$(1)\ M_J = C\, P^{3/2}/(G^{3/2}\rho^2) = C\rho l_J{}^3.$$

Since $l_j = \dfrac{v_{sound}}{\sqrt{G\rho}}$ and $v_{sound} \approx \sqrt{P/\rho} \Rightarrow$

$$(2)\ l_J = P^{\frac{1}{2}}\rho^{-\frac{1}{2}}G^{-\frac{1}{2}}\rho^{-\frac{1}{2}} \Rightarrow l_J = P^{\frac{1}{2}}\rho^{-1}G^{-\frac{1}{2}} \Rightarrow$$

$$(3)\ \rho l_J{}^3 = P^{3/2}/(G^{3/2}\rho^2).$$

If the mass of the gas cloud does not reach this critical threshold, the gravitational force will not be strong enough to overcome the thermal pressure and initiate the production of helium through the nuclear fusion of hydrogen, resulting in the death of the proto-star and its transformation into a brown dwarf, similar like our planet Jupiter. However, if the star's mass exceeds the critical threshold, the situation changes completely.

---

[100] Hannu Karttunen, Pekka Kröger, Heikki Oj, Markku Poutanen, Karl Donner, *Fundamental Astronomy* (2003): 123–124.
[101] Sergio Mendoza, *Astrofísica relativista*, www.mendozza.org/sergio (2007): 113–116.

Gravitational force is proportional to mass and inversely proportional to distance squared. The work that can be done by the gravitational force is a function of the total mass of a gas cloud divided by the radius of its sphere. As gravity exerts itself within the gas cloud, the density of ionised atoms in the sphere increases. Furthermore, as gravity causes the density of the gas cloud to increase, gravitational energy decreases. The conservation of energy (see Chapter 5) means that as gravitational energy decreases, the kinetic energy of protons and neutrons increases, which is the same as saying that their velocity—or, in other words, their temperature—increases.

When the temperature and density pass a critical threshold, a phenomenon known as the 'tunnel effect' occurs. The tunnel effect refers to the small probability that two colliding protons combine with two neutrons to form a helium nucleus. The probability is low because the repulsive electromagnetic force between two protons normally prevents this from happening, but above a certain critical mass, gravity, which increases the pressure, density and temperature of the sphere, can overcome the electromagnetic repulsion between protons, which is $1.15 * 10^{36}$ times stronger.

Let us imagine a series of gas clouds, or spheres, filled with hydrogen atoms, each with 10 times more atoms than the previous one. The first sphere has just 10 atoms, the next 100, the next 1000, and so on. The electromagnetic force that prevents two protons meeting has a big advantage in the first sphere but doesn't change as the number of atoms in the sphere increases. On the other hand, the work done by the gravitational force is squared every time the sphere's mass is cubed. This means that when the number of hydrogen atoms in the sphere reaches approximately 10^{57}, the work done by the gravitational force results in collisions between protons that are so strong that the distance between them is sufficiently reduced to allow the strong nuclear force to overcome the electromagnetic force (which makes the positive charged protons repel each other) as I shall now explain in Math box 1.11.

Math box 1.11 The critical number of particles required for the gravitational force to overcome the electromagnetic force

The gravitational force must overcome the electromagnetic repulsion between protons, which have a positive charge. However, the electromagnetic force is $1.15 * 10^{36}$ times stronger than the gravitational force. How can the gravitational force, being so weak, overcome the electromagnetic force? Based on empirical observations, we know that the critical threshold at which nuclear fusion begins is where the mass of the sphere is greater than or equal to 80% of the mass of the Sun:

(1) $M_E \geq 0.8 M_S$.

How can we explain this theoretically? The gravitational force is proportional to mass and inversely proportional to distance squared:

(2) $F_G \propto (m/r^2)$.

The work of the gravitational force is equivalent to the gravitational force multiplied by distance. In this case, the distance is the radius of the sphere (the gas cloud). Hence, the work that can be done by the gravitational force is a function of the total mass of a given gas cloud divided by the radius of the sphere, specifically:

(3) $W_G \propto (m/r^2)r = m/r$.

In a sphere with constant density, mass is proportional to volume (equation 4) and, since volume is proportional to radius cubed (equation 5), the radius is therefore proportional to the cube root of the mass (equation 6):

(4) $m \propto V$ &

(5) $V \propto r^3 \Rightarrow r \propto V^{1/3}$.

From equations (4) and (5), we obtain:

(6) $r \propto m^{1/3}$

By (3) and (6), the work done by gravity must *increase* in proportion to the cube root of the mass squared:

(7) $W_G \propto \sqrt[3]{m^2} = m^{2/3}$.

Hence, for a sphere with 10^n gravitationally colliding atoms of mass m, the work done by the gravitational force is:

(8) $W_G \propto 10^n * 10^{2/3} = 10^{(2/3)n}$.

The work done by the gravitational force is a function of the total mass of the gas cloud. We reach a point at which the gravitational force exceeds the electromagnetic force. How is this possible? Imagine a series of gas clouds, or spheres filled with hydrogen atoms. Each sphere has 10 times more atoms than the previous one. The first sphere has just 10 atoms, the next one 100, the next one 1,000 and so on. The electromagnetic force, which prevents two protons from colliding, begins in the first sphere with an advantage of:

(9) $E_1/W_1 \propto 1.1506 * 10^{36}/10^{2/3} \cong 2.479 * 10^{35} \approx 10^{36} \Rightarrow$

(10) $E_1 \approx 10^{36} * W_1$.

However, this electromagnetic force does not change as the sphere grows $(E_n = E_1)$. In contrast, the work done by the gravitational force is squared every time the mass of the sphere is cubed:

(11) $W_n = W_1 * 10^{(2/3)n} \Rightarrow$
(12) $if \ [(E_n = W_n) \wedge (E_n = E_1)] \Rightarrow E_n = W_1 * 10^{(2/3)n} \approx 10^{36} * W_1$.

And so, it follows from equation (12) that:

(13) $(2/3)n \approx 36 \Rightarrow n \approx 54$.

So, when the sphere has approximately 10^{54} hydrogen atoms, the work done by the gravitational force exceeds the force with which the protons repel each other, and the strong nuclear force can take over.

Once gravity 'defeats' the electromagnetic force, there will be enough protons that collide with sufficient velocity to set the processes of nuclear fusion in motion. When there are many collisions of protons at an extremely high speed, two will occasionally combine to form a helium nucleus. If the probability that two protons combine is $p = 10^{-n}$ and there are 10^{n} high-speed proton collisions per second per cubic centimetre, a helium atom will be formed every second in every cubic centimetre.

Section 1.3.2 The fusion of helium from hydrogen at the centre of stars

To understand the language of nuclear physics we have to become familiar with a few units, as shown in Math box 1.12.

Math box 1.12. Some units used in nuclear physics		
Concept	**Unit**	**metre, kilogram, second (MKS)**
Force	N (Newton)	$N = kgms^2$
Energy	$J = Nm$ (Joule)	$J = kgm^2s^{-2}$
Mega-electronvolt	MeV	$10^{-6}J$
Electronvolt	$eV = 1.60217733 * 10^{-13}MeV$	$1.60217733 * 10^{-19}J$
Neutron mass	m_n	$1.674929 * 10^{-27}kg$
Proton mass	m_p	$1.672623 * 10^{-27}kg$
Electron mass	m_e	$.000910938356 * 10^{-27}kg$
Atomic mass unit	$amu = \dfrac{1}{12}m_{^{12}C^6}$	$1.6605402 * 10^{-27}\ kg$ $= 931.4943333\ MeV$
Mass carbon nucleus $^{12}C^6$	$m_{^{12}C^6} = 12\ amu$	$11{,}177.932\ MeV$
Velocity of light	c	$299{,}792{,}458\ ms^{-1}$

Nuclear fusion is the main source of energy produced in stars, particularly the fusion of $^1H^1$ hydrogen nuclei (protons) into $^4He^2$ helium (alpha particles). The atomic number at the right represents the number of protons in a nucleus, for example, two protons in the helium nucleus. The atomic number at the left represents the number of neutrons plus protons, for example two neutrons and two protons in the case of helium. The mass of the nucleus is less than the sum of the masses of the nucleons (protons and neutrons). The difference is referred to as the binding energy.

Math box 1.13 The production of energy in the Sun

In the fusion of atomic nuclei, from hydrogen to helium and from helium to carbon and oxygen, and from there all the way up to iron, part of the original mass of the nuclei is released as energy. The transformation of mass into energy occurs according to Einstein's famous formula $E = mc^2$. For example, the mass of four $^1H^1$ atoms is 4.0313 *amu*; that of one $^4He^2$ is 4.00268 *amu*. So, 0.02862 *amu*, which is about 0.7% of the original mass of the four hydrogen nuclei, is released as energy, about $4.27 * 10^{-12}$ *J*. Hence, in the transformation of one kilogram of hydrogen, 0.993 kg becomes helium and 0.007 kg is released as energy:

(1) $E = mc^2 = (0.007)(299{,}792{,}458)^2 kgm^2s^{-2} = 6.2913 * 10^{14}$ *J*.

In the Sun, 584 million tonnes of hydrogen are transformed into 580 million tonnes of helium every second, with 4.27 million tonnes of mass released as energy, an amount that is more or less equivalent to:

(2) $E = mc^2 \cong (4.27 * 10^9)(300 * 10^3)^2 kgm^2s^{-2} \cong 3.838 * 10^{26}$ *J*.

The mass of the Sun at present is about $2 * 10^{30}$ *kg*, of which about 70% is hydrogen, 28% helium, and 2% consists of other elements. Only about a seventh part of that hydrogen mass is available at any time for hydrogen fusion in the core of the Sun. So, it still has enough hydrogen fuel to burn for about ten thousand million years.

We must now turn our attention to the several ways how hydrogen can be fused into helium. It has three pathways:

1) From hydrogen, through deuterium and an isotope of helium, to helium (ppI).
2) From hydrogen, through beryllium and lithium to helium (ppII).
3) From hydrogen, through various isotopes of beryllium and borium, to helium (ppIII).

In Math box 1.14, we find the exact description of these different paths.

Math box 1.14 Three ways how helium is produced in the stars

The first proton–proton chain reaction has three steps, forming helium ($^4He^2$) from hydrogen ($^1H^1$), via deuterium ($2H^1$ = one proton + one neutron):

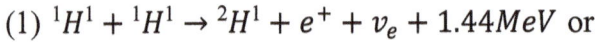

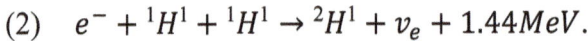

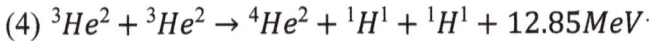

(1) $^1H^1 + {}^1H^1 \rightarrow {}^2H^1 + e^+ + v_e + 1.44 MeV$ or

(2) $e^- + {}^1H^1 + {}^1H^1 \rightarrow {}^2H^1 + v_e + 1.44 MeV.$

(3) $^2H^1 + {}^1H^1 \rightarrow {}^3He^2 + \gamma + 5.49 MeV.$

(4) $^3He^2 + {}^3He^2 \rightarrow {}^4He^2 + {}^1H^1 + {}^1H^1 + 12.85 MeV.$

The first reaction (equations 1 and 2), in which two protons undergo fusion to form a deuterium nucleus, has an extremely low probability. A proton in the centre of the Sun takes an average of 10^{10} years to collide with another proton and form deuterium. If the process was quicker, the Sun would have spent all its fuel a long time ago.

The neutrino v_e from the first step escapes from the Sun and the positron e^+ and an electron e^- annihilate each other, causing the emission of two gamma rays. In contrast, the second reaction (equation 3), in which deuterium and hydrogen fuse to produce a helium isotope occurs frequently. This is why there is so much hydrogen (protons) and so little deuterium in the centre of stars.

The first steps of the ppII and ppIII chain reactions are the same as for ppI, but then they follow different routes. In the ppII chain reaction, helium is formed from hydrogen through the intermediate steps of the production of beryllium and lithium. In the ppIII, the same is achieved through the intermediate production of beryllium and boron.

In the ppII chain reaction helium is produced through $^7Be^4$ and $^7Li^3$:

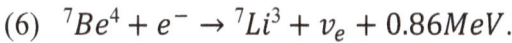

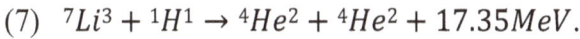

(5) $^4He^2 + {}^4He^2 \rightarrow {}^7Be^4 + \gamma + 1.59 MeV.$

(6) $^7Be^4 + e^- \rightarrow {}^7Li^3 + v_e + 0.86 MeV.$

(7) $^7Li^3 + {}^1H^1 \rightarrow {}^4He^2 + {}^4He^2 + 17.35 MeV.$

In the ppIII chain reaction, a beryllium isotope is produced, then a boron isotope, then another beryllium isotope, which then immediately disintegrates into two helium nuclei, since this isotope has a half-life of just $7*10^{-17}$ seconds:

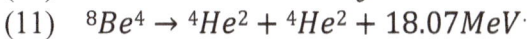

(8) $^4He^2 + {}^4He^2 \rightarrow {}^7Be^4 + \gamma + 1.59 MeV.$

(9) $^7Be^4 + {}^1H^1 \rightarrow {}^8B^5 + \gamma + 0.14 MeV.$

(10) $^8B^5 \rightarrow {}^8Be^4 + e^+ + v_e.$

(11) $^8Be^4 \rightarrow {}^4He^2 + {}^4He^2 + 18.07 MeV.$

Section 1.3.3 The production of carbon and oxygen from helium in stars

While Bethe solved the problem of the formation of helium from hydrogen, it was Fred Hoyle (1915–2001) who solved the enigma of the formation of carbon from helium.

In the *triple alpha process*, three $^4He^2$ helium nuclei combine in such a way as to produce a $^{12}C^6$ carbon nucleus. Historically, seemingly insurmountable obstacles had prevented scientists from explaining the formation of carbon in stars. To understand this, we first need to familiarise ourselves with the concept of *nuclear resonance*. Resonance is said to occur when the sum of the intrinsic and kinetic energy of particles A and B undergoing fusion is equal to or slightly less than the energy of the new $^{12}C^6$ nucleus that is formed. The fusion of new nuclei from other lighter nuclei is much easier when there is resonance because it significantly reduces the time required for fusion. If, on the other hand, there is a slight shortfall in the mass of the new nucleus, the surplus mass from the nuclei undergoing fusion is transformed into energy in line with Einstein's famous formula, in a transformation that releases vast amounts of energy and takes relatively more time. Because it slows the fusion process down, the half-life of one of the nuclei undergoing fusion may be less than the duration of the fusion process itself, stopping the process from taking place. This is precisely what happens when beryllium $^8Be^4$ and helium $^4He^2$ are about to fuse to produce carbon $^{12}C^6$: the process is stopped in its tracks.

In the triple alpha process, two $^4He^2$ alpha particles must first combine to form a beryllium $^8Be^4$ isotope, which must then combine with another alpha particle to form carbon $^{12}C^6$ with six protons and six neutrons. While normal beryllium $^9Be^4$ has four protons and five neutrons, the beryllium $^8Be^4$ isotope has only four protons and four neutrons. The seemingly insurmountable obstacle in the triple alpha process is that the beryllium isotope's half-life is $7*10^{-17}$ seconds, much less than the time required for the fusion of the beryllium $^8Be^4$ nucleus and the helium $^4He^2$. In other words, before the carbon $^{12}C^6$ carbon nucleus has formed, the $^8Be^4$ nucleus will already have dissolved due to the absence of resonance in the fusion of an $^8Be^4$ nucleus and a $^4He^2$ nucleus. This makes it impossible to form carbon, which is the missing link in the creation of all elements heavier than carbon and of carbon-based life on Earth: "*There would be no carbon, nor carbon-based life in the Universe.*"[102] Before carbon has the chance to form, the beryllium isotope disintegrates into two alpha particles.

---

[102] John Barrow & Frank Tipler, *The Anthropic Cosmological Principle* (1986): 253.

The English astrophysicist Fred Hoyle argued that the existence of carbon implies there must be a way for it to produce it in the centre of the stars. He concluded that there must be an excited state of carbon, with 7.6549 MeV more energy than normal carbon in an intermediate stage, providing the time required for normal, non-excited carbon to form. If this higher-energy state of carbon existed, when $^8Be^4$ and $^4He^2$ collide, instead of a surplus of $+7.365$ *MeV*, there would be a shortfall of -0.29 *MeV* (subtracted from the kinetic energy of the neighbouring particles with extremely high temperatures) and *resonance would occur*: the excited $^{12}C^6$ carbon would form before the $^8Be^4$ isotope disintegrated into two particles.

In 1953, shortly after having postulated the existence of the excited carbon nucleus, Hoyle spent a year's sabbatical at the California Institute of Technology, close to the Kellogg Radiation Laboratory, where Willy Fowler had risen to fame as a prestigious experimental nuclear physicist. Hoyle asked Fowler to look for this excited state of carbon, which Fowler managed to find in just 10 days. The excited state of carbon had 7.65 *MeV* more energy than carbon in its normal state.[103] In Math box 1.15, I first analyse the bottleneck that seems to prevent carbon from being produced and then the actual process of nuclear resonance and an excited state of carbon, that as a matter of fact permits carbon to be produced in the stars.

Math box 1.15 How carbon is produced in the Sun and other stars

I first analyse the bottleneck which would occur if no excited state of carbon were to exist. Let us have a look at the numbers. Beryllium $^8Be^4$ has an atomic mass of 8.005305 amu and helium $^4He^2$, 4.002602 amu.[104]

---

[103] Simon Singh, *Big Bang. The Origin of the Universe* (2004): 395–396.
[104] David Lide ed., *CRC Handbook of Chemistry and Physics* (1994-95): 11–36.

(1) $^8Be^4 + {}^4He^4 = 12.007907amu = 11,185.297MeV$.

(2) $^{12}C^6 = 12amu = 11,177.932MeV$.

From (1) and (2), we obtain a *surplus* mass of :

(3) $11,185.3MeV - 11,177.9MeV = 0.0079amu = +7.4MeV$.

The surplus mass-energy of $0.007907amu = +7.365MeV$ is released as gamma rays. Since the surplus atomic mass is transformed into energy *before* the $^{12}C^6$ carbon nucleus is produced, the window of opportunity for the $^8Be^4$ beryllium isotope, with its short half-life of $7*10^{-17}s$, to fuse with helium $^4He^2$, to produce carbon $^{12}C^6$, disappears before it can be used to actually produce carbon. So, what were to happen if no excited state of carbon would exist, is the following:

(4) $^8B^5 \rightarrow {}^8Be^4 + e^+ + v_e$ or

(5) $^4He^2 + {}^4He^2 \rightarrow {}^8Be^4$.

(6) $^8Be^4 \rightarrow {}^4He^2 + {}^4He^2 + 18.07MeV$

No carbon would be produced! Let us now see what actually happens thanks to nuclear resonance. Whether carbon is produced depends on the existence of its excited state:

(7) $^8Be^4 + {}^4He^2 = 12.007907amu = 11,185.297MeV$.

(8) excited state of carbon: $^{12}C^6_{excited} = 11,185.587MeV$.

The missing energy is subtracted from the surrounding kinetic energy:

(9) $11,185.297MeV - 11,185.587MeV = -0.290MeV$.

The absence of the transformation of surplus mass into energy, gives beryllium time to fuse with helium, before its very short life comes to an end, and produce an excited state of carbon $^{12}C^6_{excited}$:

(10) $^8Be^4 + {}^4He^2 \rightarrow {}^{12}C^6_{excited}$.

Once the carbon has 'calmed down,' its survival faces another threat. Carbon $^{12}C^6$ often fuses with helium $^4He^2$ to produce oxygen $^{16}O^8$. If resonance were to occur during nuclear fusion (if the carbon and helium nuclei had a combined atomic mass less than that of the oxygen nucleus), the carbon would be very short-lived and would immediately produce oxygen, once again leaving the Universe without carbon. However, the presence of surplus atomic mass that is transformed into energy in this nuclear fusion process means the process slows down significantly, allowing a large part of the carbon to survive. Once again, the carbon is 'lucky,' this time because of the *absence* of nuclear resonance. To conclude, the carbon is 'lucky' twice: first because there is *resonance* in the fusion of the beryllium isotope and helium to form the *excited* carbon and then because of the *lack of resonance* in the fusion of carbon and helium to form *oxygen*. This double dose of 'luck' means there is enough carbon in the Universe to allow carbon-based life on Earth.

Math box 1.16 The production of oxygen in the stars

The absence of resonance in the nuclear fusion process means all the available carbon does not disappear when oxygen is produced:

(1) $^{12}C^6 + {^4}He^2 \rightarrow {^{16}}O^8 + 7.160 MeV$, *since*

(2) $^{12}C^6 + {^4}He^2 \rightarrow 16.002602 amu = 14{,}906.333 MeV.$

(3) $^{16}O^8 = 15.994915\ uma = 14{,}899.173 MeV.$

(4) *surplus:* $0.007687 amu = +7.160 MeV.$

I will come back to the production of carbon and oxygen in the stars in Chapter 3, where I will analyse the fine-tuning of physical constants in the axiomatic laws of physics, necessary to create the conditions that make the evolution of complex life in some solar system possible.

Table 1.5 summarises the process of the production of elements in a star with a mass of $M_E = 25 \, M_S$, with a notably short lifecycle.

Table 1.5 Creation of elements in stars with 25 times mass of the Sun[105]

Nuclear fusion	temper ature	density g/cm³	lifetime	protons nucleus	neutrons nucleus
hydrogen → helium	$4*10^7$ °C	5	10^7 years	2	2
helium → carbon	$2*10^8$	$7*10^2$	10^6 years	6	6
carbon → neon + magnesium	$6*10^8$	$2*10^5$	600 years	10 12	10 12
neon → oxygen + magnesium	$1.2*10^9$	$5*10^5$	1 year	8 12	8 12
oxygen → sulphur + silicon	$1.5*10^9$	$1*10^7$	½ year	16 14	16 14
silicon → iron	$2.7*10^9$	$3*10^7$	1 day	26	30
collapse of centre	$5.4*10^9$	$3*10^{11}$	¼ s	N/A	N/A
rebound of centre	$23*10^9$	$4*10^{14}$	0.001 s	N/A	N/A
supernova → heavy elements	$\pm1*10^9$	varies	10 s	>26	>30

Section 1.4 The lifecycle of stars with different masses

In 1930, the astronomer Subrahmanyan Chandrasekhar, who was born in India in 1910 and who worked in the United States and received the Nobel Prize in 1983, discovered that once all the nuclear fuel is spent, the final destiny of a star is highly dependent on its total mass. A brief summary of

---

[105] Simon Singh, *Big Bang. The Origin of the Universe* (2004): 388.

the lifecycle of stars with different masses is given below.[106] The symbol M_S represents a unit of mass equivalent to one solar mass. The different lifecycles of stars depend on their mass and determine the different heavy elements that are produced.

The phases of a life cycle are represented by integer numbers as explained below. The number sequences of Table 1.6 are explained in the text.

Table 1.6 The lifecycle of stars depends on their initial mass

Mass of star	Lifecycle
$0.08\ M_S \leq M \leq 0.26\ M_S$	$1 \to 2 \to 3 \to 4$
$0.26\ M_S \leq M < 1.5\ M_S$	$1 \to 2 \to 5 \to 6 \to 7 \to 8 \to 9 \to 10 \to 11 \to 12$
$M > 1.5\ M_S$	$1 \to 2 \to 5 \to 13 \to 14$
$M \cong 5\ M_S$	$1 \to 2 \to 5 \to 13 \to 14 \to 15 \to 16 \to 17 \to 18$
$M \cong 30\ M_S$	$1 \to 2 \to 5 \to 13 \to 14 \to 19 \to 20 \to 21 \to 22$

The birth of a star occurs when a cold interstellar cloud of gas and hydrogen dust with sufficient volume and a temperature of $10^0 K = -263°C$ begins to collapse in freefall due to the effects of gravity. The critical mass required for this collapse to occur is the Jeans mass, named after James Jeans, the astronomer who discovered the law in 1902 (see Math box 1.1).

---

[106] Hannu Karttunen, Pekka Kröger, Heikki Oj, Markku Poutanen, Karl Donner, *Fundamental Astronomy* (2003): Chapters 8 and 11.

Image 1.3 Interstellar gas clouds and the birth of new stars in M16[107]

Multiple fragments form, each transforming into a globule that turns slowly around its axis. The globule has two different parts: the proto-planetary disc, from which planets are born; and a nucleus, which will give rise to a sun. When the growing density of the nucleus starts to prevent the transport of energy, the gas heats up and ionises, resulting in the emission of radiation waves. A temperature of 10,000 degrees Kelvin is required for the ionisation of hydrogen. At a temperature of tens of thousands of degrees Kelvin, all the gas ionises, depriving the hydrogen atoms of electrons.

1)   A proto-star is formed when the two opposing forces of gravity and the pressure created by the high temperature form a hydrostatic equilibrium. The star's subsequent evolution depends on the mass of this proto-star.

---

[107] NASA/ESA/STScl, latest update Feb. 14, 2017, photo by J. Hester and P. Scowen (Arizona State Univ.).

Math box 1.17 The equation of hydrostatic equilibrium

The following equation describes the point of hydrostatic equilibrium:

(1) $\dfrac{dP}{dr} = -\dfrac{GM_r\rho_r}{r^2}$.

---

2)   If a proto-star's mass (M_E) is greater than or equal to 8% of the Sun's mass ($M_E \geq 0.08\ M_S$) and the temperature of the nucleus is $4*10^6\ K$, protons will begin to collide at fast enough speeds to trigger the processes of nuclear fusion, transforming hydrogen into helium ($^1H^1 \rightarrow\ ^4He^2$, see Math box 1.14). When heat is transported by convection in the layers of the spheres, the hot hydrogen rises to the surface and the cold hydrogen sinks.

3)   Small stars whose mass (M_E) is between 8% and 26% of the mass of the Sun ($0.08\ M_S < M_E < 0.26\ M_S$) will take between 30 and 100 billion years to use up all their fuel before collapsing. Since the Pauli exclusion principle prevents two electrons being in the same quantum state, the electrons exert pressure against the gravitational force. This results in the birth of a white dwarf, which gradually cools and burns out but is nonetheless stable.

4) Stars with a mass of $M_E > 0.26\ M_S$ evolve in a different way. The nucleus where the nuclear fusion processes occur grows bigger as the opaquer outer layer of hydrogen that is not subject to nuclear fusion shrinks.

5) Stars in the range $0.26M_S < M_E < 1.5M_S$ take between 2 and 30 billion years to transform hydrogen into helium via the p–p I, p–p II and p–p III chains that we saw in Section 1.3.

6) Helium is first produced in the centre then, as time passes, in an outer layer, such that there are three spheres: a) an inert helium nucleus; b) a

middle layer where hydrogen is transformed into helium; and c) an inert outermost layer of hydrogen. Since the spectra of the stars show the elements present on their surface, they show few helium spectral lines and mostly hydrogen.

7)   The red giant phase marks the end of the lifecycle of a $0.26 \, M_S < M_E < 1.5 M_S$ star, such as our Sun: the star inflates as the centre collapses and heats up. At a temperature of 10^8 K, carbon production via the triple alpha process that we analysed in Section 1.3 sets in, by mean of a sudden explosion.

8) However, the explosion does not destroy the star and in its nucleus helium continues transforming into carbon.

9)   Eventually, an inert carbon nucleus is produced. The helium continues to undergo fusion to produce carbon in a middle layer, while in another outer layer hydrogen continues to transform into helium, with a final, more opaque outermost layer of inert hydrogen.

Figure 1.1 Layers of a star with two nuclear fusion processes[108]

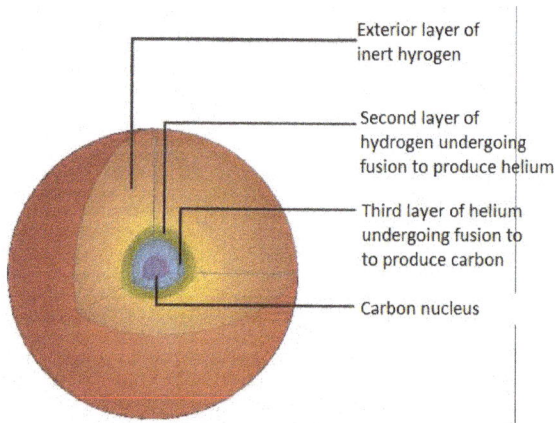

Exterior layer of inert hyrogen

Second layer of hydrogen undergoing fusion to produce helium

Third layer of helium undergoing fusion to to produce carbon

Carbon nucleus

---

[108] Adapted by the author from Habbal Astro 110-01, page 10, 3/18/09.

10) Stars with a mass less than three times that of the Sun ($M_E < 3M_S$) never get hot enough to produce carbon from helium, before the red giant phase. As the red giant inflates, it loses mass from its outer layers, which are thrust into space to form a planetary nebula. As the star contracts, it heats up and the outer layers become visible as a result of the ionisation of the gas once the temperature exceeds 10,000 K, leaving a white dwarf in the centre.

Image 1.4 End of a red giant with a mass $M_E < 3M_S$ [109]

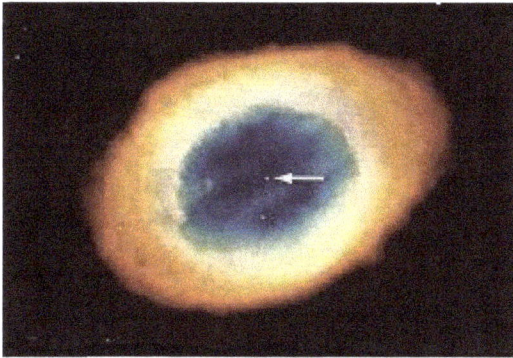

11) The star ends as a white dwarf, with carbon in the nucleus, helium in the middle layer and a thin layer of hydrogen on the surface. In 1980, Hazard and others were unable to provide explanations for the greater-than-expected heavy elements observed in stars at the centre of planetary nebulae (referred to as 'hydrogen deficient stars'). In 1983, Iben proposed the idea of the late helium flash, which occurs according to the following sequence: inert nucleus continues to contract → density of the nucleus increases → high density with triple alpha process → temperature rises to 10^8 °C → helium burns again → layers get mixed up. However, Iben's theory could not be proven because it would require witnessing the explosion of a star and scientists had calculated that such an event can only be seen once every 50 years or so, and even then, only for a few days. There were doubts as to

[109] M57 ring nebula in Lyra NASA/ESA/Hubble & Habbal Astro 110-01, p. 12, 3/18/09.

whether it would be possible to detect an event that occurs in such a short period of time, but it was detected by the Japanese amateur astronomer Sakurai in 1996, and has since been observed and analysed by many scientists. The phenomenon of the late helium flash provides an exceptional opportunity to observe the chemical composition of the star's nucleus.

12) When the mass of the star is $M_E \geq 1.5M_S$, the CNO cycle (carbon $\rightarrow$ nitrogen $\rightarrow$ oxygen) begins. The energy of the star's nucleus is transported to the surface of the nucleus by a convection current in which hot gas moves towards the surface and cold gas moves towards the centre, making the nucleus convective and the outer layer of the star radiative.

13) When all the hydrogen in the centre has been burnt, the star begins to burn the outer layer of hydrogen around a nucleus of helium.

14) For stars with a mass of $M_E \geq 1.5M_S$, the star's helium centre remains convective and the triple alpha process by which carbon is produced begins. There are now four spheres: the centre, where helium is transformed into carbon; a layer of helium; a layer where hydrogen is transformed into helium; and an opaque layer of hydrogen on the surface.

15) After the production of carbon from helium, the helium moves to an outer layer around the existing carbon nucleus.

16) In stars with a mass of $3M_S \leq M_E \leq 15M_S$, the carbon in the centre degenerates due to gravitational pressure and a carbon flash occurs.

17) The carbon flash results in a supernova and the partial destruction of the star.

18) In stars with a mass of $M_E \geq 15M_S$, things occur as in (15) occur, but all the hydrogen is fusing to produce helium, so there is no outer layer of inert and opaque hydrogen.

19) Stars with a mass of $M_E \geq 15M_S$ follow the same process as lighter stars (see 15), although the carbon nuclei in the centre of the star undergo fusion to produce oxygen, magnesium, silicon and iron. Each time a given fuel is exhausted, the star tends to collapse, resulting in the fusion of a new, heavier element, whose temperature stops the gravitational pressure, forming a new hydrostatic equilibrium. Each nuclear fusion process takes place in a different layer: iron, for example, is produced in the centre and silicon, magnesium, oxygen, carbon and helium are produced in successive outer layers, with hydrogen in the outermost layer. In some cases, sulphur and calcium are also produced – the alchemist's dream comes true at last, although not at temperatures that can be reached in earthly laboratories!

20) In the final phase, big stars have approximately six spheres: a) an iron ($^{56}Fe^{26}$) nucleus; followed by b) silicon ($^{28}Si^{14}$); c) oxygen ($^{16}O^8$); d) neon ($^{20}Ne^{10}$); e) carbon ($^{12}C^6$); f) helium ($^4He^2$); and g) hydrogen ($^1H^1$).

Figure 1.2 A massive star is factory of elements, layered like an onion[110]

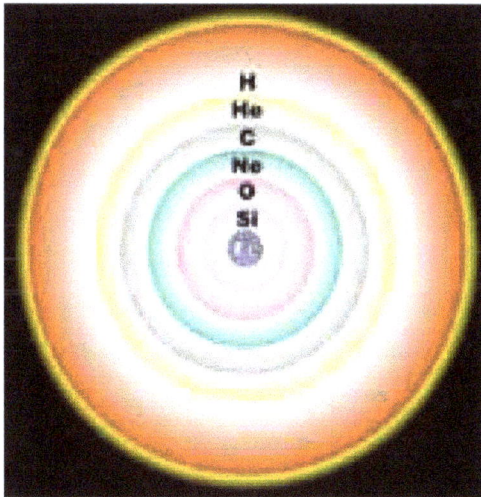

---

[110] Wikipedia, Stellar evolution, author/user: Rursus. Layers not to scale.

21) The final result of the nuclear fusion processes is iron ($^{56}Fe^{26}$). With all the fuel spent and the high temperatures of nuclear fusion no longer present, there is no pressure to counteract gravity. If the nucleus of nickel and iron in the centre is greater than the Chandrasekhar limit, the star can no longer support the outer layers, collapsing at 23% of the speed of light. This collapse results in a supernova, which can shine brighter than the galaxy to which it belongs for a number of weeks. Fred Hoyle was the first to mention the probability of supernova nucleosynthesis.[111]

These supernovas expel the elements produced in the different nucleosynthesis layers when the star was in hydrostatic equilibrium, including oxygen ($^{16}O^8$), neon ($^{20}Ne^{10}$), magnesium ($^{24}Mg^{12}$) and silicon ($^{28}Si^{14}$), and produce other elements, through explosive oxygen ($^{16}O^8$) and silicon ($^{28}Si^{14}$) burning,[112] such as sodium ($^{23}Na^{11}$), sulphur ($^{32}S^{16}$), ($^{35}Cl^{17}$), argon ($^{40}Ar^{18}$), potassium ($^{39}K^{19}$), calcium ($^{40}Ca^{20}$), scandium ($^{45}Sc^{21}$), titanium ($^{48}Ti^{22}$), vanadium ($^{51}V^{23}$), chromium ($^{52}Cr^{24}$), manganese ($^{55}Mn^{25}$), iron ($^{56}Fe^{26}$), cobalt ($^{59}Co^{27}$) and nickel ($^{58}Ni^{28}$). These elements are called 'primary', since they can be fused starting from hydrogen ($^1H^1$) and helium ($^4He^2$).[113]

Elements heavier than iron (except cobalt an nickel), which are much less abundant than the primary ones, are also produced, by fusion and the capture of free neutrons, in a process called the r-process (r for rapid), or even of protons, in a process called rp-process. These processes absorb, rather than release, energy. So much energy is released, however, during these huge supernovas explosions, that there is plenty of it available for

[111] Fred Hoyle, "Synthesis of the Elements from Carbon to Nickel", in: *Astrophysical Journal Supplement*, vol. 121 (1954): 121-146

[112] Stanford Woosley, William Arnett & Donald Clayton, "The Explosive burning of oxygen and silicon", in: *The Astrophysical Journal Supplement*, vol. 26 (1973): 231–312.

[113] Friedrich Thielemann, Ken-ichi Nomoto & Michio Hashimoto, "Explosive Nucleosynthesis in Supernovae", in: Nicos Prantzos, Elisabeth Vangioni-Flam & Michel Cassé, eds, *Origin and Evolution of the Elements* (1994): 297–309

these fusion processes that need energy to fuel them. Besides, there are plenty of free neutron and protons flying around to be able to get stuck to atomic nuclei. The exact nature, however, of the physics of the production of these elements heavier than iron is still not well understood.[114]

22) The pressure of gravity in extremely massive stars is so great that it exceeds the counter pressure of the electrons (see 4). As electrons collide fiercely with protons, an electron and a proton combine to form a neutron, leaving an extremely heavy neutron star with a diameter of around 30 km. Stars that are even more massive give rise to a black hole.

Table 1.7 summarises the stages in the lifecycles of stars of different masses.

Table 1.7 Stages in the lifecycle of a star in millions of years [115]

Mass ($n*M_{SOL}$)	Contraction to main sequence	Main sequence	Main sequence to red giant	Red giant
30	0.02	4.9	0.55	0.3
15	0.06	10	1.7	2
9	0.2	22	0.2	5
5	0.6	68	2	20
3	3	240	9	80
1.5	20	2,000	280	N/A
1 = Sun	50	10,000	680	N/A
0.5	200	30,000	N/A	N/A
0.1	500	10,000,000	N/A	N/A

---

[114] M. Arnold & Kazataka Takahashi, "The synthesis of the nuclides heavier than iron: Where do we stand?", in: Nicos Prantzos, Elisabeth Vangioni–Flam & Michel Cassé, eds, *Origin and Evolution of the Elements* (1994): 395–411.
[115] Hannu Karttunen, Pekka Kröger, Heikki Oj, Markku Poutanen, Karl Johann Donner, *Fundamental Astronomy*, Sixth edition (2003): 264 (Table 12.1).

Contrary to their appearance, the stars we observe in the sky do not have a fixed luminosity and spectrum, but will slowly vary as their lifecycle evolves, in line with their birth, death, re-birth and final death.

Section 1.5 The emergence and evolution of life

The final phase of the evolution of the Universe is the emergence of life and human intelligence on planet Earth. How did this come about? Almost five billion years ago, a star with a mass a dozen times the Sun's mass, exploded in a supernova, creating a cloud of gas and dust with all the elements of the periodic table. This cloud then started to collapse, triggering the formation of our solar system, with the Sun and the planets, and the subsequent emergence and evolution of life on Earth.

The conditions needed to create a planet like the Earth, and trigger the emergence of life, are very special indeed. In Section 8.2 of Chapter 8, I shall argue that at least 20 special circumstances were needed, each of them relatively improbable, and all of them simultaneously. This has been explained in detail by Peter Ward and Donald Brownlee in *Rare Earth. Why Complex Life is Uncommon in the Universe*, and by Guillermo Gonzalez and Jay Richards in *The Privileged Planet*.

In the same chapter, I shall also argue, however, that the immensity of our Universe, through the law of large numbers, produced a probability of almost one for intelligent life to emerge somewhere, in at least one solar system of the Universe, and possibly also in other galaxies.

In *Major Transitions in Evolution*, John Maynard Smith and Eörs Szathmáry have specified the steps needed for life to emerge and evolve on planet Earth. The first step is the transition from individual molecules that replicate themselves to populations of molecules in compartments that replicate themselves. The second step is from nucleic acid molecules that reproduce independently to chromosomes where molecules are integrally replicated. The third step is the transition from RNA as a gen and enzyme to

DNA (the genome or genetic code) and proteins. The fourth step is the transition from ancestors of mitochondria and chloroplast, that live independently as prokaryotes, to eukaryotes, where these organelles live within in the host cell. The fifth step is the transition from replication of the eukaryotes by means of asexual cloning to sexual reproduction that enhances genetic diversity. The sixth step is the transition from unicellular protists to multicellular living organisms (animals, plants, fungi). The seventh step is the transition from individual organisms to non-reproductive colonies of animals. The eighth step is the transition from social primates to human societies, with human intelligence and language, which are necessary to succeed in the complexities of human interaction.

In *The Mind and Its Brain*, Karl Popper and John Eccles explained that the eighth transition is not only a socio-cultural one, but implied a neurological evolution, which produced the emergence of a self-conscious mind that interacts with the brain, which, so they argue in their theory of interactionist dualism, certainly had evolutionary advantages.

CHAPTER 2

MYTH: DARK MATTER AND DARK ENERGY

In the following sections, I shall analyse two well established myths in modern astrophysics: dark matter and dark energy. So many astrophysicists adhere to these myths, that ΛCDM has become the standard model in modern astrophysics, where Λ stands for dark energy, and CDM, for cold dark matter. I will show that these myths arise from using models based on Newtonian gravitational dynamics, and the implicit concept of absolute, homogeneous time, which yield erroneous results. If we use the theory and mathematics of orthodox general relativity, however, it can explain the rotation velocity of galaxies and galaxy clusters, and the apparent acceleration of the expansion of the Universe, with remarkable precision, without any need to speculate about dark matter and dark energy.

Section 2.1 How general relativity refutes the theory about dark matter

I shall first explore how the myth of dark matter came into being, and then explain how the observations of the rotation velocity of galaxies and galaxy clusters can be exactly predicted by general relativity, without resorting to dark matter.

Section 2.1.1 The origin of the dark matter myth

The myth of non–baryonic dark matter came into being in order to explain the apparent discrepancy between the observed visible mass – which is baryonic – and the total mass calculated from certain effects generated by gravitational fields modelled on Newtonian dynamics. The first person to draw attention to the supposedly missing mass in galaxies and galaxy clusters was Fritz Zwicky (1898–1974), in 1933. Zwicky, a Swiss astronomer working in Pasadena, California, compared the redshift of individual galaxies belonging to a cluster with the redshift of the entire cluster, and so was able to establish the proper velocity of a galaxy. Thus he could prove that the orbital velocity of galaxies in a cluster is higher than expected if one would only look into the mass of visible matter (stars and ionized gas) from the point of view of Newtonian gravitational dynamics.[116]

Along this same line of reasoning, Vera Rubin (1928–2016), an American female astronomer, and George Coyne (born 1933), a Jesuit priest and astronomer, speculated that the method of establishing the peculiar velocities of galaxies within a cluster through their redshift, serves to reveal *"the relative distribution of dark and luminous matter"* and *"indicates the existence of large amounts of (dark) matter"*.[117]

Rubin made observations of the orbital velocities of stars in spiral galaxies that seemed to reveal the existence of a halo of non–baryonic dark matter which extends further than the visible disk of the galaxy. Newtonian gravitational dynamics predict that acceleration due to gravity diminishes in proportion with the inverse square of the distance from the central mass and that orbital velocity diminishes in proportion with the inverse square root of the distance. In a series of ten publications, in *The Astrophysical Journal* from 1977 to 1985, she and her team observed about 60 spiral galaxies (20 of type Sa, 20 of type Sb and 20 of type Sc[118]) and reported that the orbital

---

[116] Fritz Zwicky, "Die Rotverschiebung von extragalaktischen Nebeln", *Helvetica Physica Acta,* vol. 6 (1933): 110–127; "On the Masses of Nebulae and of Clusters of Nebulae", in: *The Astrophysical Journal,* vol. 86 (1937): 217–246.

[117] Vera Rubin & George Coyne, eds, *Large Scale Motions in the Universe* (1988): 262, 101–102.

[118] Spiral galaxies type Sa have a big centre, with the arms close to each other; galaxies type Sb, a smaller centre with distinguishable arms; and type Sc, an even smaller centre with arms quite separate from each other.

velocity was almost constant, independent of the distance from the centre of the galaxy.[119]

It is important to distinguish between the astrophysical *observations* of Vera Rubin and her team and the *interpretations* they made of these observations. Two cosmographic observations are beyond any doubt:

1) *First observation*. In the solar system, the orbital velocity of the planets diminishes in proportion with the inverse square root of the distance, so the velocity diminishes as the distance increases.

2) *Second observation*. The rotational velocity of stars about the centre of a spiral galaxy first increases rapidly at a short distance from the galaxy centre, then stops diminishing with distance remaining more or less constant (the curve flattens out: see Image 2.1 with Rubin's original drawings on a photo of spiral galaxy M31)). However, the visible, baryonic mass diminishes rapidly as one moves away from the galaxy centre, as can be inferred from its luminosity.

---

[119] Vera Rubin & Kent Ford *et al.* "Extended rotation curves of high–luminosity spiral galaxies. I. The angle between the rotation axis of the nucleus and the outer disk of NGC 3672", *The Astrophysical Journal*, vol. 217 (1977): L1–L4;
"II. The anomic Sa galaxy NGC 4378", *The Astrophysical Journal*, vol. 224 (1978): 782–795;
"III. The spiral galaxy NGC 7217", *The Astrophysical Journal*, vol. 226 (1978): 770–776;
"IV. Systematic dynamical properties", *The Astrophysical Journal*, vol. 225 (1978): L107–L111;
"V. NGC 1961, The most massive spiral known", *The Astrophysical Journal*, vol. 225 (1979): 35–39;
"VI. Rotational properties of 21 Sc galaxies with a large range of luminosities and radii, from NGC 4605 (R=4 kpc) to UGC 2885 (R=122 kpc)", *The Astrophysical Journal*, vol. 238 (1980): 471–487;
"VII. Rotation and mass of the inner 5 kilo parsecs of the SO galaxy NGC 3115", *The Astrophysical Journal*, vol. 239 (1980): 50–53;
"VIII. Rotational properties of 23 Sb galaxies", *The Astrophysical Journal*, vol. 261 (1982): 439–456;
"IX. Rotation velocities of 16 Sa galaxies and a comparison of Sa, Sb, and Sc rotation properties", *The Astrophysical Journal*, vol. 289 (1985): 81–104.

Image 2.1 The spiral galaxy M31 with Rubin's flat rotation curve[120]

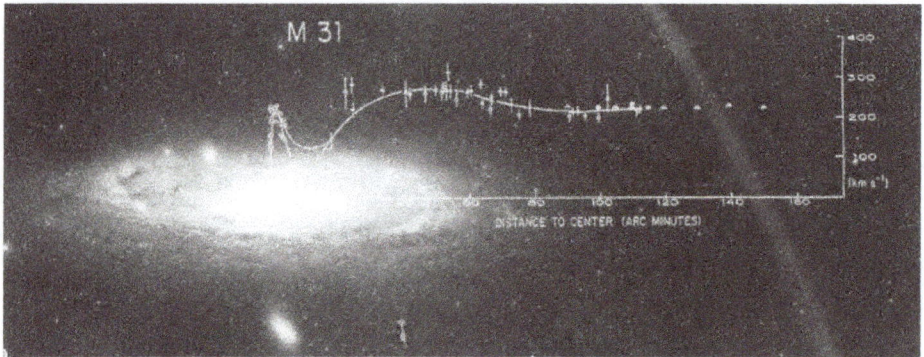

These factual *observations* of facts in spiral galaxies were *interpreted* by Vera Rubin from the point of view of a cosmological model with Newtonian gravitational dynamics:

1) *First part of the interpretation.* Rubin and her team started from the assumption that in spiral galaxies the gravitational dynamics operating are Newtonian. According to Newton, the gravitational acceleration diminishes with the inverse square of the distance, and orbital velocity diminishes with the inverse square root of the distance, as is explained in Math box 2.1.

Math box 2.1 Orbital velocity with Newtonian gravitational dynamics

In Newtonian mechanics, a body orbiting at a distance r from the centre of a spherically symmetrical collection of bodies may be treated as if the total mass M out to the radius r were concentrated at the central point (the Shell Theorem). According to Newton's second law of movement, the acceleration a is given by:

(1) $F = ma = GMm/r^2 \rightarrow a = GM/r^2.$

---

[120] Malcolm Longair, *Galaxy Formation* (2008): 67.

The acceleration of a body in orbit around a big central mass is:

(2) $a = v^2/r.$

From (1) and (2) we obtain:

(3) $\dfrac{GM}{r^2} = \dfrac{v^2}{r}.$

From (3) we deduce the orbital or rotational velocity:

(4) $v = \dfrac{\sqrt{GM}}{\sqrt{r}},$

which means that the orbital velocity is proportional to the square root of the mass and inversely proportional to the root of the distance:

(5) $v \propto \dfrac{\sqrt{M}}{\sqrt{r}}.$

Since it is reasonable to assume that the mass of the galaxy is concentrated at its centre and diminishes if one moves away from the centre, one would expect, according to equation (5) that the rotational velocity v would rapidly decrease as one moves away from the centre, since according to that assumption, the *total* mass M (within the range r) increases more and more slowly and r increases linearly so that M/r diminishes rapidly. The surprise is that one observes the contrary: the orbital velocity, at a certain distance from the centre and beyond, remains constant even as we move away from the centre. The only way to explain this strange phenomenon is to assume that the mass, instead of gradually stopping to increase, actually increases linearly with radius, up to a certain, far away distance. For example, at twice the distance from the centre, we would have twice the mass. That would explain why the rotational velocity remains constant with distance:

(6) $if\ M \propto r \rightarrow \Delta v = 0.$

So, in the context of Newtonian gravitational dynamics there is no other option, but to assume the existence of a halo of non–baryonic dark matter.

2) *Second part of the interpretation.* Within the context of Newtonian gravitational dynamics, an apparent incompatibility arises between the visible galaxy mass (stars and gas), observed as the radiation in some frequency of the electromagnetic spectrum, and the observed constancy of the orbital velocity. This is interpreted as requiring enormous quantities of additional and invisible mass, that increase linearly with distance from the centre. Given the observation of the orbital velocity in these dynamics and given the fact the gravitational constant G and the orbital velocity v appear to be constant, i.e. more or less independent in regard to the distance from the galaxy centre r, it follows that the only way, within the context of these Newtonian dynamics, to resolve this problem is the speculation that the total mass M contained in the sphere with radius r increases linearly with the radius. Graph 2.1 shows the essence of the problem.

Graph 2.1 Observed and expected Newtonian rotation velocity of galaxy

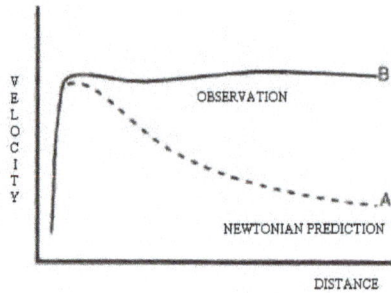

3) *Third part of the interpretation.* Given this speculation of the total galaxy mass increasing linearly with the radius and given the fact that the visible mass decreases rapidly with the distance from the galaxy centre, it follows logically that 'dark matter', which is supposed not to interact with light, and increases with distance from the galaxy centre, is not associated to the visible matter. As a

consequence, the mass luminosity ratio of the galaxy (*M/L*) increases dramatically when one moves away from the galaxy centre.

It is important to point out that in this interpretation, NO observations corroborate the speculations about dark matter as such. Dark matter is not observed, precisely because it is supposed not to interact with light, as Rubin pointed out in 1983:

> "*All attempts to detect a halo by its visual, infrared, radio or X–ray radiation have failed... In sum, the only requirement for the halo is the presence of matter in any cold, dark form that meets the M/L constraint.*"[121]

The conjecture about dark matter depends on the assumption that orbital velocities in spiral galaxies can be explained by Newtonian gravitational dynamics. This assumption then leads to the speculation that the total galaxy mass increases linearly with distance from the centre (see Graph 2.1).

Once the idea of a halo of exotic dark matter was published in the *Scientific American* in 1983, many cosmologists started making references to the speculation about a halo of exotic dark matter in spiral galaxies, dissociated from visible matter, as though it was a scientific fact. For example, in 1994, Kolb & Turner reproduced some flat rotation curves and affirmed that: "*Rotation curve measurements indicate that virtually all spiral galaxies have a dark, diffuse 'halo' associated with them which contributes at least 3 to 10 times the mass of the visible matter*".[122] Stephen Hawking too attributed to dark matter the fact that the stars at the edge of spiral galaxies are kept in their orbits and not thrown into outer space: "*stars on the outskirts of spiral galaxies like our own Milky Way orbit far too fast to be held in their orbits only by the gravitational attraction of all the stars*

---

[121] Vera Rubin, "Dark Matter in Spiral Galaxies", in: *Scientific American* vol. 248 (1983): 98.
[122] Edward Kolb & Turner, *The Early Universe* (1994): 17–18.

that we observe".[123] What Hawking asserts would be true if *Newtonian mechanics* were valid to explain the rotation velocity of galaxies. The problem is that they are *not*: only Einstein's *general relativity* is valid.

Not all cosmologists reflected on the Newtonian assumptions of this theory, but some did. For example, in 2006, Malcolm Longair reproduced Rubin's original *M31* spiral galaxy image with the flat rotation curve and commented on the Newtonian dynamics underlying these speculations:

> "*Vera Rubin and her colleagues pioneered systematic studies of the rotation curves of galaxies... [I]n the outer regions of galaxies, the velocity curves are generally remarkably flat, ($v_{rot} \cong$ constant). The significance of this result can be appreciated from a simple **Newtonian calculation**. If the galaxy is taken to be spherical and the mass within the radius r is M, the circular rotational velocity at distance r is found by equating the inward gravitational acceleration (GM/r^2), to the centripetal acceleration (v_{rot}^2/r), and so $v_{rot} = (GM/r)^{1/2}$. Thus, if v_{rot} is constant, it follows that $M \propto r$, so that the total mass within radius r increases linearly with the distance from the centre. This result contrasts strongly with the variation of the surface brightness of spiral galaxies, which decrease much more rapidly with distance from the centre than as r^{-2}.*"[124]

Jim Peebles went a step further than Longair, stating that Newtonian dynamics may not be applicable in the cases of galaxy clusters and spiral galaxies: "*discovering the nature of the dark matter, or **explaining why the Newtonian mechanics used to infer its existence has been misapplied**, has to be counted as one of the most exciting and immediate opportunities in cosmology today.*"[125] He did not follow up, however, on his own suggestion.

---

[123] Stephen Hawking, *The Universe in a Nutshell* (2001): 186.
[124] Malcolm Longair, *The Cosmic Century* (2006): 248–249, bold characters are mine.
[125] James Peebles, "Dark Matter", in: *Principles of Physical Cosmology* (1993): 417–456 (my bold characters).

The speculation about the existence of non–baryonic dark matter extends to galaxy clusters. Zwicky established that one can determine the total mass of a galaxy through the observation of the curvature of light coming from a star or galaxy that is located behind the Sun or a galaxy cluster. This method, derived from the theory of general relativity, served originally to corroborate this theory. Since it has been corroborated, one now proceeds in the opposite order: "*the angle representing the change of the direction of the light ray is of the order of magnitude [of]* $\varphi = \frac{1}{2}\frac{g*l}{c^2}$ *radians*", which allows us to calculate the total mass of a galaxy that is located between a luminous object and the Earth.[126] According to Zwicky, the observation of these effects of gravitational lensing provides us with the most simple and most exact determination of the masses of galaxies.[127] In Math box 2.2, I explain how total galaxy cluster mass is estimated in the standard ΛCDM model, following a procedure based on Newtonian gravitational dynamics.

Math box 2.2 Total galaxy cluster mass with Newtonian dynamics

The equation for kinetic energy is derived directly from Newton's second law of movement:

(1) $K = \frac{1}{2}mv^2$.

If we assume that the distribution of the velocity is isotropic, in the three directions of the system of coordinates and we assume also spherical symmetry in the galaxy cluster, we obtain:

---

[126] The curvature is $\varphi = \frac{1}{2}\frac{g*l}{c^2}$ radians, where l is the distance travelled by light through a gravitational field, c is the velocity of light and g is the gravitational acceleration. The term g depends directly on the mass of the object that causes the bending of the light. See George Gamov, *Mr. Tompkins in Wonderland* (1958): Lecture 4 "On Curved Space, Gravity and the Universe", equation (7).

[127] Fritz Zwicky, "On the Masses of Nebulae and of Clusters of Nebulae", in: *The Astrophysical Journal,* vol. 86 (1937): 238. Zwicky sees the galaxies as "nebulae".

(2) $K = \frac{3}{2}M\langle v_r{}^2\rangle$,

where $\langle v_r\rangle$ is the average radial velocity. Let us assume also the validity of the virial theorem, that supposes Newtonian gravitational dynamics:

(3) $K = \frac{1}{2}|U_g|$.

and we obtain the equation for potential gravitational energy, derived from Newtonian physics:

(4) $|U_g| = \frac{GM_1M_2}{\langle R\rangle}$,

where $\langle R\rangle$ is the weighted average of the distance between objects with mass M. From equations (3) and (4), we obtain:

(5) $K = \frac{1}{2}GM^2/R_{cl}$.

From equations (2) and (5), we obtain (with Longair[128]):

(6) $M = \frac{3\langle v^2\rangle R_{cl}}{G}$,

where M is the galaxy cluster mass; $\langle v\rangle$ the average rotational velocity of a galaxy; and R_{cl} the average distance between galaxies. From (6) we obtain:

(7) $\langle v_{rot}\rangle = \sqrt{\frac{1}{3}\frac{GM}{\langle R_{cl}\rangle}}$.

The same dependence on Newtonian gravitational dynamics is manifest in the speculation about the *location* of dark matter in galaxy clusters. The Newton–modelled reasoning goes as follows. When galaxies or galaxy clusters collide and cross each other, stars do not collide, but the gas does, so that the heated gas is separated from the stars. Douglas Clowe and his team analysed the case of the galaxy cluster $1E0657 - 558$, also known as

---

[128] Malcolm Longair, *Galaxy Formation* (2008): 66.

the *Bullet Cluster*, which we see side on.[129] The clouds of hot plasma of each cluster collide and reduce their relative velocity, but the stars of the galaxies do not collide physically, so that the visible plasma and the galaxies are spatially separated. The separation of galaxies and plasma permits us to estimate the proportions of visible baryonic matter of both on the basis of their respective luminosities. By observing the effect of weak gravitational lensing – a slight distortion of the elliptic form of the galaxies –, which is more accentuated where galaxies are found (with relatively little visible matter), than in regions with plasma (with relatively more visible matter), the observed gravitational lensing appears to indicate that the location of the dark matter is in and around the galaxies. The variations of gravitational lensing "*are in agreement with the galaxy positions and offset from the gas*",[130] so, according to Clowe, dark matter is associated with the visible matter of the galaxies. The team does not speculate about the character of this dark matter, but it believes it has corroborated its existence.[131]

The analysis of the *Bullet Cluster* $1E0657 - 558$, first realised by Clowe and his team, in 2006, then replicated by Marusa Bradac[132] and her team in the case of another merger of clusters, catalogued as $MACS\,J0025.4 - 1222$, and indirectly corroborated by Richard Massey and his team,[133] who used the observed distortion of the form of half a million galaxies, leads to

---

[129] Douglas Clowe *et al.*, "A Direct Empirical Proof of the Existence of Dark Matter", in: *Astrophysical Journal Letters*, vol. 648 (2006) and *idem*, "Colliding Clusters Shed Light on Dark Matter", in: *Scientific American* (2006).

[130] Douglas Clowe *et al.*, "Catching a Bullet: Direct Evidence for the Existence of Dark Matter", arXiv:astroph/ 0611496, p. 4.

[131] Dennis Zaritsky, a member of Clowe's team, admits that he does not know what this dark matter is. His remarks are referred to in "Colliding Clusters Shed Light on Dark Matter", in: *Scientific American* (August 2006).

[132] Marusa Bradac *et al.*, "Revealing the Properties of Dark Matter in the Merging Cluster $MACS\,J0025.4 - 1222$ ", in: arXiv:0806.2320.

[133] Richard Massey *et al.*, "Dark Matter Maps Reveal Cosmic Scaffolding", in: *Nature* (January 2008).

the conclusion that *"baryons follow the distribution of dark matter even on large scales"*.[134]

Image 2.2 Dark matter (blue) in galaxies, dissociated from plasma (pink)

In a recent survey of galaxy clusters, Hans Böhringer conjectured that the proportions of non–baryonic dark matter and baryonic visible matter are 85% and 15%, respectively and that the 15% corresponding to baryonic visible matter is distributed between stars, 2%, and gas, 13%, in big clusters; and 5% and 10%, respectively in small clusters.[135]

In all these cases, the model used to measure the amount of mass through weak gravitational lensing, is a mixture of general relativity – as far as gravitational lensing is concerned – and *"**Newtonian gravity**"* (Clowe)[136], *"**Newtonian**"* dynamics (Massey)[137], so that this proof of the existence of dark matter rests on the validity of the assumption that Newtonian

---

[134] Richard Massey *et al.*, "Dark Matter Maps Reveal Cosmic Scaffolding", in: *Nature* (January 2008): 5. The colours of Graph 2.2 are not real, but added to the image by Massey.

[135] Hans Böhringer, "Galaxy Clusters as Cosmological Probes", lecture given at the Universidad Iberoamericana, April 16, 2008.

[136] Douglas Clowe *et al.*, "Catching a Bullet: Direct Evidence for the Existence of Dark Matter", arXiv:astroph/0611496, p. 3 (my italics).

[137] Richard Massey *et al.*, "Probing Dark Matter and Dark Energy with Space–Based Weak Lensing", arXiv:astroph/0403229, p. 4, see also Richard Massey *et al.*, "Dark Matter Maps Reveal Cosmic Scaffolding", in: *Nature* (January 2008) (my italics).

gravitational dynamics explain the observations in these cases. In using Newtonian models, however, the total galaxy cluster mass is overestimated by several orders of magnitude, just as is the case with spiral galaxies. In the next section, I shall show that no dark matter is needed, if one uses Einstein's general relativity to explain the rotation velocity of galaxies and galaxy clusters and gravitational lensing.

Section 2.1.2 How the rotation velocity is explained by general relativity

Fred Cooperstock[138] and Steven Tieu, following previous suggestions by Eddington, offer an orthodox solution, along the lines of Einstein's general relativity, by which the discrepancy between observation and prediction disappears. Eddington had mentioned this non–linearity for a system that is variable in time, and the authors extend it to non–linear, but stationary (non–time dependent) problems, as in galactic gravitational dynamics:

"In dismissing general relativity in favour of Newtonian gravitational theory for the study of galactic dynamics, insufficient attention has been paid to the fact that the stars that compose the galaxies are essentially in motion under gravity alone ('gravitationally bound'). It has been known since the time of Eddington that the gravitationally bound problem in general relativity is an intrinsically non linear problem even when the conditions are such that the field is weak and the motions are non–relativistic, at least in the time–dependent case. Most significantly, we found that under these conditions, the general relativistic analysis of the problem is also non–linear for the stationary (non–time–dependent) case at hand. Thus, the intrinsically linear Newtonian–based approach used to this point has been inadequate for the description of galactic dynamics... We demonstrate that via general relativity, the generating potentials

---

[138] Fred Cooperstock, *General Relativistic Dynamics. Extending Einstein's Legacy Throughout the Universe* (2009).

producing the observed flattened galactic rotation curves are necessarily linked to the mass density distributions of the flattened disks [of ordinary baryonic matter], obviating any necessity for dark matter halos in the total galactic composition."[139]

Cooperstock, collaborating first with Tieu and then with Carrick, analysed a total of seven spiral galaxies from the relativistic point of view. In his first publication of 2005, he analysed four spiral galaxies (the Milky Way, NGC 3031, NGC 3198, and NGC 7331).[140] In December 2010 he added another three spiral galaxies proving the same point (NGC 2841, NGC 2903 and NGC 5033).[141] The authors conceive the spiral galaxies as systems that are analogous to *"fluids rotating uniformly without pressure and symmetric around the axis of rotation"*,[142] and explained the rotational dynamics by the gravitational attraction exercised by baryonic matter, within the known form of the visible disk, in relativistic gravitational dynamics (see Math box 2.3).

Math box 2.3 Spiral galaxy mass in relativistic gravitational dynamics

We start from the line element of an object in free fall in general relativity, adapted to the polar, cylindrical coordinates r and z:

(1) $ds^2 = -e^{v-w}(u\,dz^2 + dr^2) - r^2 e^{-w} d\phi + e^w (c\,dt - N d\phi)^2,$

where u, v, w and N are coefficients whose value is a function of the coordinates r and z. For various reasons,[143] one may simplify this equation equating $u=1$ and $w=0$:

(2) $ds^2 = -e^v(dz^2 + dr^2) - r^2 d\phi + (cdt - N d\phi)^2.$

---

[139] Fred Cooperstock & Steven Tieu, "General Relativity Resolves Galactic Rotation Without Exotic Dark Matter" arXiv:astro-ph/0507619 (2005): 2–3.

[140] *Ibidem.*

[141] John Carrick and Fred Cooperstock, "General Relativistic Dynamics Applied to the Rotation Curves of Galaxies", arXiv:1101.3224 (December 2010).

[142] Fred Cooperstock & Steven Tieu, "General Relativity Resolves Galactic Rotation Without Exotic Dark Matter" arXiv:astro-ph/0507619 (2005): 4.

[143] *Ibidem*: 4–5.

We obtain the relation between angular velocity ω, and tangential velocity V and the coefficient N (using $\phi = \phi + \omega(r,z)t$):

(3) $\omega = \dfrac{Nc}{r^2}$ and

(4) $V = \omega r$,

so that by (3) and (4), we obtain:

(5) $V = \dfrac{Nc}{r}$.

We use Einstein's field equations for N and ρ in a weak field with a cloud of particles in rotational motion, subject neither to pressure neither nor to friction:

(6) $N_{rr} + N_{zz} - \dfrac{N_r}{r} = 0$ and

(7) $\dfrac{N_r{}^2 + N_z{}^2}{r^2} = \dfrac{8\pi G\rho}{c^2}$.

Equation (6) can be represented as a function of the gravitational potential ϕ for rotating galaxies:

(8) $\nabla^2 \phi = 0$.

where the zero value is due to the absence of pressure and friction in a system of particles in rotational motion. If there were no rotational motion, the system would need pressure (a non–zero value) to be stable, as in the Poisson equation of Newtonian gravity for weak fields:

(9) $\nabla^2 \phi = 4\pi G\rho$.

In a way analogous to the derivation of the Newtonian gravitational field and potential, we obtain the gravitational potential of a system of particles in rotational motion subject neither to pressure, nor to friction:

(10) $\phi = \int \dfrac{N}{r} dr \;\rightarrow\; \dfrac{\partial \phi}{\partial r} = \dfrac{\partial}{\partial r} \int \dfrac{N}{r} dr = \dfrac{N}{r}$.

From equations (5) and (10), we obtain the rotational velocity:

(11) $V = c\frac{N}{r} = c\frac{\partial\phi}{\partial r}$.

In polar cylindrical coordinates, the solution to equation (8) is:

(12) $\phi = Ce^{-k_n|z|}J_0(kr)$,

where J_0 is the Bessel function $J_m(kr)$ of zero order ($m=0$) and C is an arbitrary constant. We can rewrite equation (12) as a linear summary:

(13) $\phi = \sum_n C_n e^{-k_n|z|}J_0(k_n r)$.

From equations (11) and (12), we obtain:

(14) $V = c\frac{\partial\phi}{\partial r} = -c\sum_n C_n k_n e^{-k_n|z|}J_1(k_n r)$,

and from (11) and (14), we obtain:

(15) $N = \frac{Vr}{c} = -\sum_n C_n k_n r e^{-k_n|z|}J_1(k_n r)$.

By solving $\sum_n C_n k_n$ for $n=1$ to $n=10$, we obtain the theoretical rotational velocity curves, with the Bessel function of order one:

(16) $V(r,z) = -c\sum_{n=1}^{10} C_n k_n e^{-k_n|z|}J_1(k_n r)$ and

(17) $N(r,z) = -\sum_{n=1}^{10} C_n k_n r e^{-k_n|z|}J_1(k_n r)$.

From (16) and (17), we confirm (11):

(18) $V(r,z) = \frac{c}{r}N(r,z)$.

Each galaxy is different, and has its proper coefficients C_n and k_n. Cooperstock and Tieu attached the respective values of these coefficients to their article for four galaxies, among them the Milky Way, for $n=1$ to $n=10$.

What many astrophysicists and cosmologists attribute to a halo of cold dark matter is explained by Cooperstock and Tieu with ordinary baryonic matter in the context of relativistic gravitational dynamics, with the result that the flat velocity curve is produced by a disk mass much smaller than the envisaged halo mass of exotic dark matter: *"The non–linearity for the computation of density inherent in the Einstein field equations for a stationary axially–symmetric pressure–free mass distribution, even in the case of weak fields, leads to correct galactic velocity curves as opposed to the incorrect curves that had been derived on the basis of Newtonian gravitational theory."*[144] The predictions of the relativistic model are corroborated (see Graph 2.2).

Graph 2.2 The corroboration of the general relativity in Milky Way[145]

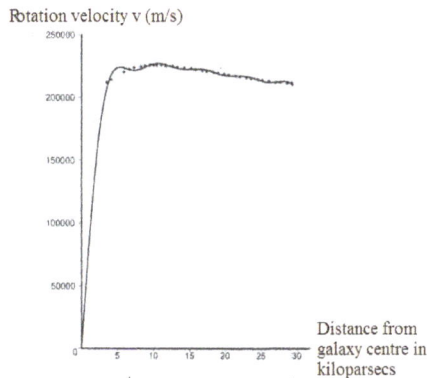

Rotation velocity v (m/s)

Distance from galaxy centre in kiloparsecs

Explanation: the curve is prediction by theory; the points, the observations.

With Wolfram's program *Mathematica*, using equations (16) and (17) of Math box 2.3, Alfredo Sandoval and myself were able to reproduce exactly the same flat rotational velocity curves as obtained by Cooperstock. We discovered, however, that variations in the fourth or fifth or sixth decimal of

---

[144] Fred Cooperstock & Steven Tieu, "Galactic Dynamics via General Relativity", in: *International Journal of Modern Physics A* vol. 22 (2007): 29.
[145] Fred Cooperstock & Steven Tieu, "General Relativity Resolves Galactic Rotation Without Exotic Dark Matter", arXiv:astro-ph/0507619 (2005): 8.

the value of the coefficients C_n and k_n may affect the results in a non–trivial way. This means we cannot use figures that are rounded up to the third or fourth decimal. Graph 2.3 replicates our results of the rotational velocity of the Milky Way:

Graph 2.3 Predictions of general relativity replicated by the author

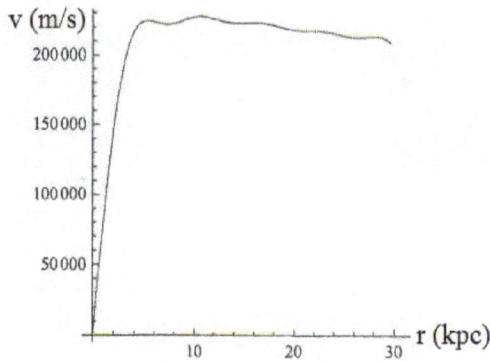

Cooperstock comments that if we have two options, either orthodox physics (general relativity), or inventing something esoteric (dark matter), applying Ockham's razor means that we opt for orthodox general relativity:

> *"The scientific method has been most successful when directed by 'Ockham's razor', that new elements should not be introduced into a theory unless absolutely necessary. If it should turn out to be the case that the observations of astronomy can ultimately be explained without the addition of new exotic dark matter, this would be of considerable significance."*[146]

The work of Cooperstock and Tieu has generated much interest and also some criticism, from Korzynski,[147] Vogt and Letelier,[148] and Garfinkle.[149] Cooperstock and Tieu responded adequately to their critics.[150]

---

[146] Fred Cooperstock & Steven Tieu, "Galactic Dynamics via General Relativity", in: *International Journal of Modern Physics A*, vol. 22 (2007): 30.
[147] Nikolaj Korzynski, "Singular Disk of Matter in the Cooperstock–Tieu Galaxy Model", arXiv:astro-ph/0508377.

Other astrophysicists feel that even with relativistic gravitational dynamics, the baryonic mass in the spiral galaxies is not enough to explain their rotational velocity. One of them (David Wiltshire) wrote to me that *"the masses of the galaxies Cooperstock and Tieu find are greatly in excess of any reasonable estimate of the baryonic mass"*.[151] He referred to Stephen Kent's estimates of spiral galaxy mass.

Actually, Cooperstock and Tieu themselves make that comparison. Kent has three articles on this topic, published between 1986 and 1988.[152] I do not think that Kent's estimates validate the missing mass hypothesis, as I will now show. First, there is no indication in Kent's figures of a 1/6 baryon/total mass ratio. Kent does not use the term 'baryonic mass', but refers to 'stellar mass', being the sum of the stellar 'bulge' mass M_B and stellar 'disk' mass M_D that he obtains by means of estimates of the mass/luminosity ratio M/L (the luminosity being the surface brightness of stars) for bulges and disks. He derives the total mass estimate M_{tot} by means of Newtonian equations that establish a causal relationship between mass and rotational velocity at any chosen radius, where velocity is the observed part (derived from blue and redshifts), radius the chosen part, and mass the inferred part. The

---

[148] Daniel Vogt & Patricio Letelier, "Presence of Exotic Matter in the Cooperstock and Tieu Galaxy Model", arXiv:astro-ph/0510750.

[149] David Garfinkle, "The Need for Dark Matter in Galaxies", arXiv:gr-qc/051182 .

[150] Fred Cooperstock & Steven Tieu, "Perspectives on Galactic Dynamics via General Relativity", arXiv:astro-ph/0512048 and "Galactic Dynamics via General Relativity", in: *International Journal of Modern Physics A*, vol. 22 (2007): 17–28.

[151] E–mail of David Wiltshire, December 2010.

[152] Stephen Kent, "Dark Matter in Spiral Galaxies. I. Galaxies with Optical Rotation Curves", *The Astrophysical Journal,* vol. 9 (June 1986): 1301–1327; "Dark Matter in Spiral Galaxies. II. Galaxies with H1 Rotation Curves", *The Astrophysical Journal,* vol. 9 (April 1987): 816–832; "Dark Matter in Spiral Galaxies. III. The Sa Galaxies", in: *The Astrophysical Journal,* vol. 9 (1988): 514–527.

stellar/total mass ratio, for 7 of the 16 galaxies with complete data sets, is $\frac{M_{B+D}}{M_{tot}} = 0.32$, that is a ratio of 1/3, not 1/6, as Kent asserts.[153]

Of the three galaxies chosen by Cooperstock & Tieu for analysis in their first article (*NGC 3031*; *NGC 3198*; and *NGC 7331*), Kent has complete data sets only for two of them, *NGC 3031* and *NGC 7331*, since he has no bulge mass estimate for NGC 3198. In the case of *NGC 3031* and *NGC 7331*, his stellar/total mass ratios are 1/2 and 1/4, respectively. These ratios, however, have to be corrected, because Kent neither includes estimates of interstellar gas, nor of black holes, but "*the contribution from the stellar component alone*".[154] In galaxies the gas mass is 15% of the baryonic mass, and the stellar mass, 85%. The interstellar gas mass, for example in the Milky Way, is 15% of the total baryonic mass.[155] The additional gas mass changes the Kent estimate of baryon mass for *NGC 3031* and *NGC 7331* to $8.47*10^{10}$ and $14.24*10^{10}$ solar masses, respectively.

Now we have to add the black hole mass M_{BH}, which is originally baryonic matter. The black hole mass is *not* included in Kent's M/L ratios because in both galaxies, he has identical M/L ratios for bulges and disks – 3.76 and 4.04, respectively –, as he explains: "*for NGC 3031 and NGC 7331, the bulge M/L ratio in the full solution was very poorly constrained and so it was kept fixed equal to the disk M/L ratio*".[156] The M/L ratios would have been different for bulges and disks if the black hole had been included in the bulge M/L ratio. In *NG 3031*, the black hole at the centre of the galaxy has a mass of $M_{BH} = 10^8$ solar masses, and Rohlfs and

---

[153] Stephen Kent, "Dark Matter in Spiral Galaxies. II. Galaxies with H1 Rotation Curves", *The Astrophysical Journal,* vol. 9 (April 1987): 827.

[154] Stephen Kent, "Dark Matter in Spiral Galaxies. I. Galaxies with Optical Rotation Curves", *The Astrophysical Journal,* vol. 9 (June 1986): 1301.

[155] Katia Ferrière, "The Interstellar Environment of our Galaxy", arXiv:astro-ph/0106359 (June 2001): 1–56.

[156] Stephen Kent, "Dark Matter in Spiral Galaxies. II. Galaxies with H1 Rotation Curves", *The Astrophysical Journal,* vol. 9 (April 1987): 826.

Kreitschmann correct Kent's estimate of total stellar mass to be $M_{M+B} =$ 7.7 $*$ 10^{10} solar masses,[157] which increases the total baryon mass estimate to 8.97 $*$ 10^{10} solar masses. The *NGC 7331* black hole may be as much as 1.8 $*$ 10^{10} solar masses which increases the baryon mass from 14.24 to 16.04$*10^{10}$ solar masses. Table 2.1 gives the corrected Kent baryon mass estimates, compared to those found by Cooperstock.

Table 2.1 Baryonic mass estimates for *NGC 3031* and *NGC 7331*

Galaxy	Cooperstock	Kent corrected	Cooperstock/Kent
NGC 3031	10.9 $*$ 10^{10}	8.97 $*$ 10^{10}	1.2
NGC 7331	26.0 $*$ 10^{10}	16.04 $*$ 10^{10}	1.6

The table reveals that the baryon masses in both cases are in fact almost identical. In the case of *NGC 7331* there appears to be some missing mass, but it certainly is not the case that the Cooperstock and Tieu mass estimates are *"greatly in excess of any reasonable estimate of the baryonic mass"*.[158] The missing mass may be due to many uncertainties, among other things, as Kent himself points out, the circumstance that: *"the relationship between luminous and dark matter shows significant variation among galaxies"*,[159] and the *"optical rotation curves usually do not place strong constraints on the amount of dark matter in these galaxies...[because] in agreement with Kalnajs, 1983, some rotation curves fit well without the need to assume the existence of any dark halo"*.[160]

---

[157] K. Rohlfs & J. Kreitschmann, "A Two Component Mass Model for M81/NGC3031", *Astronomy & Astrophysics* (1980): 175–182.
[158] David Wiltshire's email to the author, December 2010.
[159] Stephen Kent, "Dark Matter in Spiral Galaxies. II. Galaxies with H1 Rotation Curves", *The Astronomical Journal,* vol. 9 (April 1987): 816.
[160] Kent, Stephen, "Dark Matter in Spiral Galaxies. I. Galaxies with Optical Rotation Curves", *The Astronomical Journal,* vol. 91 (1986): 1301, 1326. Kent is referring to A. J. Kalnajs, *Internal Kinematics of Galaxies*, IAU Symposium (1983).

What is implied in these comments by Kent, though he himself does not make that point, is that there are two ways to estimate the mass of a galaxy, one being the mass/luminosity ratio M/L and the other, the mass inference from rotation velocity interpreted with Newtonian gravitational dynamics. The first method is very insecure, because of the significant variation of the M/L ratio among galaxies, and the second one, based on Newtonian gravitational dynamics, is erroneous. Even so, for some galaxies, both methods – one insecure, and one wrong – sometimes coincide in yielding the same galaxy mass estimate, without resorting to dark matter, as Kent points out. Given the fact that the Newtonian model grossly overestimates the galaxy mass needed to explain the rotation velocity, it follows that, in these cases of coincidence between the two measures, the mass/luminosity model also grossly overestimates the real galaxy mass. The point here is that, given the significant variation of the mass/luminosity model in different galaxies, the M/L ratio may sometimes grossly overestimate the galaxy mass, and sometimes grossly underestimate it. The *NGC 7331* galaxy might be one of these cases, we simply do not know. Here, astrophysics is not so exact as the uncritical adherence to the results of the mass/luminosity M/L ratio appears to assume. The gravitational dynamics of spiral galaxies (see next image) are well explained by general relativity, without any need of speculations about non–baryonic dark matter. Even in the case of weak gravitational fields, we need Einstein.

Can we extend this analysis of galaxy rotational velocity to galaxy clusters? Cooperstock and Tieu think so:[161]

"For the dynamics of clusters of galaxies, the virial theorem is used. This is based on Newtonian gravity theory. It would be of interest to introduce a general relativistic virial theorem for comparison. It is only after possible

---

[161] Fred Cooperstock & Steven Tieu, "Perspectives on Galactic Dynamics via General Relativity", arXiv:astro-ph/0512048 (2005) and Fred Cooperstock, "Clusters of Galaxies", in: *General Relativistic Dynamics* (2009): 135–159.

effects of general relativity are explored that we can be confident about the viability or non–viability of exotic dark matter in nature."[162]

Image 2.3 The dynamics of spiral galaxies are explained by general relativity[163]

Cooperstock and Tieu applied general relativity to the gravitational dynamics of galaxy clusters, and corroborated their theory about total cluster mass and the rotational velocity of galaxies in the cluster[164]. Math box 2.4 summarises their main argument, based on a relativistic model of a weak gravitational field constituted by many bodies under the effect of mutual gravitational attraction with no friction or pressure.

---

[162] Fred Cooperstock & Steven Tieu, "Perspectives on Galactic Dynamics via General Relativity", arXiv:astro-ph/0512048, p. 3. For the virial theorem, see Appendix VIII, Section C 1, in: John Auping, *El Origen y la Evolución del Universo* e–book (2009): 736
[163] The spiral galaxy NGC 6946. Photo by John Duncan, *Astronomía* (2007): 223.
[164] Fred Cooperstock & Steven Tieu, "General Relativistic Velocity", in: *Modern Physics Letters A* vol. 23 (2008): 1745–1755 and Fred Cooperstock, "Clusters of Galaxies", in: *General Relativistic Dynamics* (2009): chapter 10.

Math box 2.4 Galaxy cluster mass in a relativistic gravitational model

Cooperstock starts with Schwarzschild's solution to Einstein's equations, using a metric of spherical coordinates for a spherical mass M,[165] the same that one uses to derive the perihelion rotation of Mercury in a plane:[166]

$$(1) \quad ds^2 = \left(\frac{1}{1-\frac{2GM}{c^2r}}\right) dr^2 + r^2(d\theta^2 + \sin^2\theta d\varphi^2) - \left(1 - \frac{2GM}{c^2r}\right)c^2 dt^2.$$

The terms between parentheses constitute the metric coefficients, that together determine Schwarzschild's metric tensor in four–dimensional space–time, where the mass M is not small. Following Landau and Lifshitz, two Soviet era astrophysicists, but normalizing the equations, so as to make $c=G=1$ and inverting the signs of the metric, Cooperstock obtains:[167]

$$(2) \; ds^2 = -\left(1 - \frac{2m}{r}\right)^{-1} dr^2 - r^2 d\theta^2 - (r^2\sin^2\theta)d\varphi^2 + \left(1 - \frac{2m}{r}\right) dt^2.$$

A big difference between this procedure and the analysis of Mercury's perihelion[168] is that we do not make the simplifying assumption, justifiable in the solar system, that the proper time $d\tau$ of the observed mass and the time dt of the observer are one and the same. Normally, this difference is only considered to be important in the case of strong gravitational fields, but Cooperstock and Tieu show that in the case of a weak gravitational field too, the difference between the proper time $d\tau$ of the observed mass and the time dt of the observer is crucial. The transformation of the coordinates of

---

[165] Fred Cooperstock & Steven Tieu, "General Relativistic Velocity: the Alternative to Dark Matter", in: *Modern Physics Letters A* vol. 23 (2008): 1746, equations (1) and (2).
[166] See equation (382) of Appendix VI B and equation (4) of Appendix VI C of John Auping, *Origen y Evolución del Universo* e–book (2009): 692, 698.
[167] Fred Cooperstock & Steven Tieu, "General Relativistic Velocity: the Alternative to Dark Matter", in: *Modern Physics Letters A* vol. 23 (2008): 1746, equations (1) and (2) and note 6 ($c=G=1$). This is equation (100.2) of L. Landau & E. Lifshitz, *The Classical Theory of Fields* (2002): 321, if one takes into account that r_g (the 'gravitational radius') in Landau and Lifshitz is the mass m in Cooperstock and Tieu.
[168] See Appendix VI C of John Auping, *Origen y Evolución del Universo*, e–book (2009).

the observer (r and t) into the co–moving coordinates of the observed object with its proper time (R and τ) is the following:

$$(3) \quad \tau = t + \int \frac{\sqrt{\frac{2m}{r}}}{1-\frac{2m}{r}}\, dr,$$

$$(4) \quad R = t + \int \frac{1}{\sqrt{\frac{2m}{r}}\left(1-\frac{2m}{r}\right)} \quad \text{and}$$

$$(5) \quad r = \left(\frac{3}{2}(R-r)\right)^{2/3}(2m)^{1/3},$$

which gives us the following transformed Schwarzschild metric, that Cooperstock took from Landau and Lifshitz's *Classical Theory of Fields*, and that depends on the proper time of the massive object:[169]

$$(6) \quad ds^2 = d\tau^2 - \frac{dR^2}{\left(\frac{3}{2(2m)}(R-\tau)\right)^{2/3}} - \left(\frac{3}{2}(R-\tau)\right)^{4/3}(2m)^{2/3} + d\theta^2 +$$

$$+\sin^2\theta\, d\varphi^2.$$

In the case that the value of τ comes close to the value of R, we are in a strong gravitational field and the singularity of a black hole arises where $R = \tau$. But Cooperstock and Tieu are interested in the case of a *weak* gravitational field, where $R \gg \tau$ for all R, implying that $r \gg 2m$ r for all r and the coordinates (r,t). The radial velocity, measured by the external observer is:

$$(7) \quad v_{rad} = \frac{dr}{dt} = -\left(1-\frac{2m}{r}\right)\left(\sqrt{\frac{2m}{r}}\right).$$

---

[169] This is equation (102.3) of L. Landau & E. Lifshitz, *The Classical Theory of Fields*, 4ª revised ed. (2002): 332, if one takes into account that r_g (the gravitational radius) in Landau and Lifshitz is the mass m in Cooperstock and Tieu, and that Landau and Lifshitz normalise only half way (G=1, but $c\neq 1$).

The radial velocity in the proper time of the observed moving object is:[170]

$$(8) \quad v_{rad} = \frac{dR}{d\tau} = -\sqrt{\frac{-g_{11}}{g_{00}}}\frac{dr}{dt} = -\sqrt{\frac{1}{(1-2m/r)^2}}\left(1-\frac{2m}{r}\right)\sqrt{\frac{2m}{r}} \cong -\sqrt{\frac{2m}{r}}.$$

In a weak gravitational field, the radial velocity measured in the proper time of the co–moving object is equal to the radial velocity measured in the time of the terrestrial observer, because the mass m of the field is so reduced that the factor $(1 - 2m/r) \cong 1 - 0 = 1$, as Cooperstock explains: "*the local measures, both proper and external, of the radial velocity are approximately equal in the value of* $-2m/r$".[171]

However, this is only true in the case that almost all the mass of the system is concentrated in the centre of mass, as for example in the solar system, where the weak gravitational field is originated by one massive object.

But things get complicated, when we focus on the collapse of a cloud of particles, where each particle contributes to the total mass and field. It is in this case that the radial velocity as measured in the proper time of the co–moving object, even in the case of non–relativistic velocities, starts differing considerably from the time of the external, terrestrial observer. Parting from the geodesic equation in general relativity, for a cloud of dust particles, taken from the classic work of Landau and Lifshitz,[172]

---

[170] The Schwarzschild metric in Cooperstock and Tieu is $g_{00} = (1 - 2m)/r$ and $g_{11} = -\left(1/(1 - \frac{2m}{r})\right)$. The difference with the Schwarzschild metric in John Auping, *Origen y Evolución del Universo* (2009), Appendix VI B, equation 382, p. 692, is that g_{00} in Cooperstock is my $-g_{44}$, with $g_{44} = -(1 - 2GM/c^2r)$, and g_{11} in Cooperstock is my $-g_{11}$, with $g_{11} = 1/(1 - 2GM/c^2r)$, and Cooperstock normalizing with $G=1$ and $c=1$.

[171] Fred Cooperstock and Steven Tieu "General Relativistic Velocity: the Alternative to Dark Matter", in: *Modern Physics Letters A* vol. 23 (2008): 1748.

[172] The equation (103.1) in L. Landau & E. Lifshitz, *The Classical Theory of Fields*, (2002): 339 is equation (9) in Fred Cooperstock and Steven Tieu, "General Relativistic Velocity: the Alternative to Dark Matter", in: *Modern Physics Letters A* vol. 23 (2008): 1748.

Cooperstock obtains the following geodesic equation for dust particles or objects, as measured by an external (terrestrial) observer:

$$(9) \quad ds^2 = d\tau^2 - e^{\lambda(\tau,R)}dR^2 - r^2(\tau,R)(d\theta^2 + \sin^2\theta \, d\varphi^2).$$

In this case, "*a freely falling dust particle maintains constant space coordinates for all time*", and the exact solution of the four non–trivial Einstein field equations that apply in this case, assumes the form of the following two equations:[173]

$$(10) \quad e^\lambda = \frac{(r')^2}{1+E(R)} \quad \text{and}$$

$$(11) \quad r'^2 = E(R) + \frac{F(R)}{r},$$

where $E(R)$ and $F(R)$ are functions of integration. This leads to the following average radial velocity equation:

$$(12) \quad \frac{dr}{dt} = -\frac{(\alpha+\beta)(1-\beta^2)}{8\pi r^2 \rho^2}\left[\frac{\alpha}{F} + \beta\left(\frac{F''}{(F')^2} - \frac{1}{2F}\right)\right]^{-1}\frac{\partial\rho}{\partial t}, \quad \text{where}$$

$$(13) \quad \beta = \sqrt{\frac{F}{r}},$$

$$(14) \quad \alpha = \frac{rF'}{3F} \quad \text{and}$$

$$(15) \quad \rho = \frac{F'}{8\pi r' r^2}.$$

The factor F is the accumulated mass function conceived as a function of the radius R of the galaxy cluster (wherein the average radial velocity is supposed to be known):

$$(16) \quad F(R) = k_1 R^{k2} \quad \text{and}$$

$$(17) \quad M(R) = F(R)/2.\text{[174]}$$

---

[173] Fred Cooperstock, *General Relativistic Dynamics* (2009): 142.
[174] *Ibidem*: 148–149.

These equations permit us to reconstruct the relation between radial velocity, galaxy cluster mass and galaxy mass density, in a relativistic model, without necessity of non–baryonic dark matter. For example, in the case of the Coma cluster, Cooperstock has the following values of the mass function: $F = 6.641 * 10^{-16} R^{1.453} \rightarrow F' = 9.649 * 10^{-16} R^{0.453} \rightarrow F'' = 4.371 * 10^{-16} R^{-0.547}$ and the radial velocity expressed as the ratio $2m(R_0)/r_0$ is of the order of 10^{-4} *"if we assume, as would a Newtonian, that there exists dark matter present to account for the observed velocities"*, but it is of the order of 10^{-5} *"if we accept only the existence of the matter that we see"*.[175]

Now the problem we face is whether we can reconcile the observed velocities and the baryonic matter that we see, without resorting to dark matter. Cooperstock argues that we can, with the help of the relativistic radial velocity equation and the accumulated mass function (see Math box 2.4). Instead of boosting the mass of the galaxy by inventing dark matter, Cooperstock's model boosts the radial velocity based on visible baryonic matter, using relativistic gravitational dynamics, which is then corroborated by the observed radial velocity. Assuming the baryonic, visible mass is 20% or 30% or 40% of the supposed total mass within a sphere of 3 *Mpc* of $1.3*10^{15}$ solar masses, we obtain a boost factor n of the wrongly inferred Newtonian radial velocity, associated with only observed baryonic mass $(\frac{dr}{dt} = -n\beta)$ of $n = 2.23$, $n = 1.82$ and $n = 1.58$, respectively, to obtain the relativistic radial velocity. Since the observed average radial velocity and all the terms at the right–hand side of the relativistic radial velocity equation (13) are known, we can obtain the value of the change of mass density over time $(\partial\rho/dt)$, that is $2.13*10^{-41}$ *kg* /m^3/*s*, $2.62*10^{-41}$*kg*/m^3/*s* and $3.02*10^{-41}$*kg*/m^3/*s*, respectively. *"Rates of density changes of the order of magnitude*

---

[175] Fred Cooperstock, *General Relativistic Dynamics* (2009): 148.

$10^{-41}kg/m^3/s$ *are quite reasonable as over a period of one billion years*", which is the time of the evolution of the Coma galaxy cluster.

Cooperstock concludes that, while this is only one example, he has been able to account for the observed velocities of galaxies within this cluster, solely within the framework of general relativity and without any extraneous dark matter.[176] Cooperstock draws the following conclusion to his analysis which is worth quoting:

"When the gravity was deduced to be weak within these clusters, astronomers naturally turned to Newtonian gravity to correlate the seemingly anomalously large galactic velocities that they measured with the masses that they believed to be present. They initially deduced that there must be unseen 'dark matter' in the order of 100 times as much as the visible matter to make the mass totals accord with the velocities. However, with the later discovery of very large quantities of gaseous matter, this figure was reduced dramatically but there still remained a large quantity of matter yet to be accounted for. This apparent need is still promoted vigorously by researchers throughout the world [in] a plethora of papers advocating new particles that would conceivably play the role of this exotic missing material. However, insofar as high rotational velocities of stars in galaxies as the basis for the need for dark matter is concerned, the replacement of Newtonian gravity by general relativity removes this requirement. The nonlinearities of general relativity play an important role in systems of freely falling gravitating masses, leading to expressly non–Newtonian behaviour, even when the gravitational field is weak... Had Zwicky done this calculation 70 years ago with general relativity in mind, he might have come to very different conclusions regarding the requirement for vast stores of exotic dark matter."[177]

---

[176] Fred Cooperstock, *General Relativistic Dynamics* (2009): 152.
[177] *Ibidem*: 148, 152–153.

Section 2.2 How general relativity eliminates the myth of dark energy

I shall first explore how the myth of dark energy came into being, and then show how general relativity makes it possible to explain the apparent acceleration of the expansion velocity of the Universe, without resorting to dark energy.

Section 2.2.1 The acceleration of the expansion velocity of the Universe

The cosmological constant was once proposed by Einstein in order to explain why a steady state universe did not collapse. When Einstein conceived the Universe as a stable distribution of matter that did not collapse gravitationally, which is impossible (see Section 1.1.1), he introduced the cosmological constant λ, representing a negative gravitational force to solve the paradox. When Hubble published data on the redshift of galaxies, that corroborated the Friedman–Lemaître model of an expanding Universe, Einstein publicly accepted that the facts refuted his model of a static universe and subscribed to the model of a dynamic one, with no need for gravitational repulsion by a cosmological constant.[178]

Apart from Hubble's observations, an additional problem with Einstein's cosmological constant was pointed out by Steven Weinberg in a 1989 article,[179] ten years before the observations about the apparent acceleration of the Universe's expansion velocity were made public. Weinberg demonstrated that astronomical observations indicate that the cosmological constant is many orders of magnitude smaller than it should be according to modern theories of elementary particles: *"theoretical expectations for the cosmological constant exceed observational limits by some 120 orders of*

---

[178] See Chapter 1.
[179] Steven Weinberg, "The Cosmological Constant Problem", in: *Review of Modern Physics*, vol. 61 (1989): 6–23.

magnitude"[180] and he added, that if no solution is found to overcome this enormous discrepancy between theory and observation, "*we will have to fall back on the anthropic principle to explain why Λ_{eff} is not enormously larger than allowed by observation*".[181] Weinberg's analysis, originally aimed at Einstein's cosmological constant, is also relevant for the modern version of the cosmological constant, which is invoked to explain the apparent acceleration of the expansion velocity of the Universe.

The cosmological constant has recently reappeared in the cosmology of an expanding universe, after the recent discovery of the apparent acceleration of its expansion velocity.[182] I am referring to the observations of luminosity and redshift of supernovae type 1a, discovered in the Supernova Cosmology Project of Saul Perlmutter and his team[183] and the High–z Supernova Search Team of Robert Kirshner and Adam Riess.[184] The observation of the redshift and distance of these supernovae yields a Hubble constant of $\pm 70 \, kms^{-1}/Mpc$. According to these data, type 1a supernovae that are relatively close by have a redshift that is larger than would be expected in the case of a decelerating expansion of the Universe, indicating that in the last thousands of millions of years the expansion is accelerating.

A couple of years later, both teams again presented observations of the same phenomenon, but this time more precise ones, made with the Hubble Space Telescope and reduced the error margin of the observed luminosity considerably. In 2003, Knop and Perlmutter and their team presented data of

---

[180] Steven Weinberg, "The Cosmological Constant Problem", in: *Review of Modern Physics*, vol. 61 (1989): 1.

[181] *Ibidem*: 20.

[182] Perlmutter, Riess and Schmidt received the 2011 Physics Nobel Prize for this finding, leaving out Robert Kirshner who should have been included in this Noble Prize.

[183] Saul Perlmutter, "Measurements of Omega and Lambda from 42 High–Redshift Supernovae", in: *Astrophysical Journal*, vol. 517 (1999): 565–586.

[184] Adam Riess, "Observational Evidence from Supernovae for an Accelerating Universe and a Cosmological Constant", in: *Astronomical Journal*, vol. 116 (1998): 1009–1038; and Robert Kirshner, *The Extravagant Universe* (2002).

11 supernovae of high redshift observed by the Hubble Space Telescope[185] and in 2004,[186] Riess and Kirshner and their team used the same telescope for more precise observations of 16 recent supernovae and revaluated the past evidence of 170 type 1a supernovae and corroborated once again the hypothesis of the recent acceleration of the expansion of the Universe.[187] They also affirmed that the historical transition from deceleration to acceleration occurs at a distance that corresponds to a redshift of $z = 0.46 \pm 0.13$.

Kirshner, Knop, Perlmutter, Riess and Schmidt provide the data of the apparent acceleration of the expansion velocity of the Universe. Many cosmologists interpret these data with the help of a modern version of the ancient cosmological constant Λ, first proposed and then abandoned by Einstein. They speculate about a mysterious dark energy, a negative gravitational force Λ. The sum of the mass density of $\Omega_M \approx 0.3$ and the dark energy density of $\Omega_\Lambda \approx 0.7$ results in a total density of $\Omega_T \approx 1$.

Together with the speculation about dark matter, which I analysed in the previous section, dark matter and dark energy constitute the cornerstones of the standard model in modern astrophysics, named ΛCDM. There are, however, *two different explanations* of this observation, either the speculation on *dark energy*, or orthodox *general relativity*. If we take general relativity seriously, there is no need for dark energy, as we shall now see.

---

[185] Rob Knop, Saul Perlmutter *et al.*, "New Constraints on Ω_M, Ω_Λ and w from an Independent Set of Eleven High–Redshift Supernovae Observed with the HST", 2003, arXiv:astro-ph/0309368.

[186] *Ibidem.*

[187] Adam Riess *et al.*, "Type Ia Supernova Discoveries at z<1 From the Hubble Space Telescope: Evidence for Past Deceleration and Constraints on Dark Energy Evolution", arXiv:astro-ph/0402512 and in: *Astrophysical Journal*, vol. 607 (2004): 665–738.

Section 2.2.2 How to explain the acceleration of the expansion velocity

Some astrophysicists, outstanding among them David Wiltshire, show that the observation of the apparent acceleration of the expansion velocity of the Universe results from the fact that the watch of an observer, embedded in the strong gravitational field of a galaxy cluster, runs more slowly than the watch in a void, or the average watch of the Universe, that has many voids surrounded by walls of galaxies clusters. This means that a receding supernova that travels a certain distance, at the other side of a large void, makes that journey in less time, according to the slower running observer's watch, as compared to the faster running watch mounted on the supernova, or according to the Universe's average watch, creating the perception of a recent acceleration of the expansion velocity.

In the *International Conference on Two Cosmological Models*, held at the Universidad Iberoamericana in 2010, and using the Buchert–Wiltshire paradigm, I presented a physical–mathematical proof of the fact that dark energy is an unnecessary speculation, once we take general relativity seriously,[188] and so did Wiltshire himself.[189] If the reader wants to have a closer look at this physical–mathematical evidence, he can consult the *Proceedings*, available in English, on the internet.[190]

I will now synthesise these arguments, as presented at that International Conference.

General relativity establishes that gravitational fields do not only curve space, but also slow down time. Both space and time are relative, differing

---

[188] See John Auping, "Putting the standard ΛCDM model and the relativistic model in historical context", in: John Auping & Alfredo Sandoval, eds, *Proceedings of the International Conference on Two Cosmological Models,* e–book (2012): 35–115.

[189] David Wiltshire, "Gravitational energy as dark energy: cosmic structure and apparent acceleration", in: John Auping and Alfredo Sandoval, eds, *Proceedings of the International Conference on Two Cosmological Models* (2012): 361–384.

[190] Taking the following steps: 1) *www.ibero.mx*; 2) *publicaciones*; 3) *publicaciones electrónicas*; 4) [choose among various title pages:] *Proceedings of the International Conference on Two Cosmological Models.*

in voids and galaxy clusters. Our galaxy cluster is located in an enormous void of 200 to 300 Mega parsecs (*Mpc*) that expands between 20% and 30% more rapidly than could be expected according to the global Hubble constant; there is a superstructure of 400 *Mpc* known as Sloan's Great Wall, surrounding part of this void; more locally there are two other minor voids of 35 to 70 *Mpc* each, and Shapely's super cluster with a diameter of 40 *Mpc*, at a distance of some 200 *Mpc* from our galaxy. One *Mpc* is 3,261,600 light years, or $3.0857 * 10^{19} km$.

In general, "*some 40–50% of the volume of the universe at the present epoch is in voids of 30 $h^{-1}Mpc$ [≈between 40 a 50 Mpc, JA] in diameter... and there is much evidence for voids 3 to 5 times this size, as well as local voids on smaller scales*".[191]

Wiltshire estimates that today 75.9% of the observable Universe is in voids ($f_v = 0.759$) and 24.1% in walls ($f_w = 0.241$). Obviously, with the passing of time, due to the expansion of the Universe, the contribution of voids in the total volume increases, and that of the gravitationally collapsed regions decreases.

Wiltshire's theory of the differential running of watches in voids and walls, establishes that after the moment of recombination, some 300,000 years after the Big Bang, the imaginary clocks, located in different regions of the Universe, started to differ increasingly, because in regions with high matter density, gravity makes watches run slower, and in voids, relatively faster. Wiltshire revived one of the implications of general relativity, long overlooked by astrophysicists, though already explained by Einstein himself, who said: "*Let us examine the rate of a unit clock, which is*

---

[191] David Wiltshire, "Cosmic Clocks, Cosmic Variance and Cosmic Averages", in: *New Journal of Physics* (2007): 80.

arranged to be at rest in a static gravitational field...: the clock goes more slowly if set up in the neighbourhood of ponderable masses."[192]

In 2007, Wiltshire proposed his *timescape* model of differential clock rates in an inhomogeneous universe.[193] This model is capable of explaining the apparent acceleration of the expansion of the Universe and other phenomena that have motivated other cosmologists to speculate about dark energy to explain them. Wiltshire distinguishes three different clocks, i.e. slow running clocks in gravitationally dense and collapsed regions that measure time τ_w (w for walls), rapid clocks in the voids with time τ_v (v for voids), and a global–average clock with time t. These clocks yield three differential clock ratios, e.g. $\gamma = \frac{\gamma_w}{\gamma_v} = \frac{d\tau_v}{d\tau_w} \approx 1.5$, also called the *lapse function*[194]; $\gamma_v = \frac{dt}{d\tau_v} = 0.92$; and $\gamma_w = \frac{dt}{d\tau_w} = 1.38$. Only at the beginning of the Universe, at the moment of recombination, the Universe was an almost perfectly smooth and homogeneous cloud of hydrogen and helium and, consequently, at that time, the different clocks yielded equal times, so that shortly after the Big Bang, $dt/d\tau_w = dt/d\tau_v = d\tau_v/d\tau_w = 1$.

The deceleration of the expansion velocity of the Universe is less in voids than in walls. Since our galaxy cluster is at the centre of a huge void, we observe a nearby expansion deceleration that is less than the global average. In general, the expansion deceleration in walls with time τ_w differs 5.5 *cm* per s^2 ($5.5 * 10^{-10} ms^{-2}$) from the deceleration in voids with time τ_v. This seems little, but the accumulated effect through the entire history of the Universe, since the Big Bang, is large. I shall come back to this point

---

[192] Albert Einstein, "The foundation of the general theory of relativity", in: *Annalen der Physik*, vol. 49 (1916), translated into English in: *The Collected Works of Albert Einstein*, vol. 6 (1989): 197–198.
[193] David Wiltshire, "Cosmic Clocks, Cosmic Variance and Cosmic Averages", in: *New Journal of Physics* (2007): 337–442.
[194] David Wiltshire, "Cosmological Equivalence Principle and the Weak Field Limit", in: *Physical Review D*, vol. 78 (2008): 8–9.

shortly. The fact that we measure this acceleration with our galaxy cluster wall clock, which runs slower than a global average clock, enhances this apparent acceleration even more.

The unequal deceleration in voids and walls implies that the Hubble parameter is not equal in different regions of the Universe. In order to establish the value of a Hubble constant, one must make measurements with different clocks, for example, one with the global average–time t and one with the proper time of the observer in our galaxy cluster τ_w. The value of the Hubble constant is different, when measured with the same global average clock measuring time t in walls H_w, voids H_v, or the Universe at large H_0. The values of the Hubble constant for the global average Universe are $H_w(t) = 34.9$, $H_v(t) = 52.4$ and $H_0(t) = 48.2 \, kms^{-1}Mpc^{-1}$, respectively. The present global average Hubble constant of the Universe at large H_0 also varies when measured by wall clocks, which run slower, and therefore register a higher expansion velocity; void clocks, which run faster, and so register a lower velocity; or global average clock, resulting in a global Hubble constant of $H_0(\tau_w) \cong 66.5$, $H_0(\tau_v) \cong 44.3$ and $H_0(t) \cong 48.2 \, kms^{-1}Mpc^{-1}$, respectively.

In a recent publication, the Wiltshire team has begun to map the Hubble flow anisotropy, starting with the different regions of the Local Group, the Local Void and the Great Attractor, taking into account the differential cosmic expansion rates in walls, voids and the Universe at large.[195] They restrict their *timescape–Szekeres* model to be consistent with the observed *CMB* anisotropy, which seems to be appropriate, since *both* anisotropies are supposed to have had their origin in the original, first fraction of a second Big Bang quantum fluctuations.

---

[195] Krzysztof Bolejko, Ahsan Nazer and David Wiltshire, "Differential Cosmic Expansion and the Hubble Flow Anisotropy", in: *Journal of Cosmology and Astroparticle Physics*, vol. 6 (2016): 35.

The following equation gives us the Hubble constant as measured with our terrestrial wall clock, as well as its rate of decrease over time:[196]

$$H_0(\tau_w) = \frac{dt}{d\tau_w} H_0(t) - \frac{d\left(\frac{dt}{d\tau_w}\right)}{dt}.$$

In order to establish its value, one must make measurements with two clocks, that is the one with the global average time t and the other one, with the proper time τ_w of the observer in a galaxy cluster.

The cosmological redshift z, as determined by wall observers, is related to the redshift determined by volume average co–moving observers $\bar{z}$, by the following transformation:

$$1 + z = \frac{dt/d\tau_w}{dt_0/d\tau_{w_0}}(1 + \bar{z}).$$

Section 2.2.3 The backreaction

In order to define the past and present–day parameters as a function of the global average time, Buchert's backreaction formalism is needed to average the values of parameters measured with clocks in walls and voids, respectively.

Since I have referred to the term 'backreaction', it is necessary to say a few words about its meaning. The Universe is a collection of regions with high matter density (walls), where local clocks run slowly, and regions with low matter density (voids), where clocks run faster. We can make this inhomogeneity disappear through a process of averaging also known as smoothing, but that does not take away the fact that the values of cosmological parameters in walls and voids differ among themselves and from the global average. The curvature of space is not homogeneous. How

[196] David Wiltshire, "Exact Solution to the Averaging Problem in Cosmology", arXiv:0709.0732 (2007): equation 8; and David Wiltshire, "Cosmic Clocks, Cosmic Variance and Cosmic Averages", in: *New Journal of Physics* (2007): 25 (equation 42).

can we obtain the global–average values of the cosmological parameters? Obviously, that could be done in theory by obtaining a weighted average of their values in different regions of the Universe, in its voids and walls. There is a complication, however, since the values of these parameters evolve and change with time, so that their magnitude is not constant, neither on the local scale, nor on the large scale. Consequently, we have two options, the first one of which would be to obtain the average of the original values of the parameter in different regions at the beginning of the Universe, and then see how this average evolves. The second option would be to let the parameter evolve with time in different regions of the Universe, and obtain a present time average of these independently evolving values at this stage of the evolution of the Universe.

Normally, the operation of averaging (smoothing) and that of evolving in time (deriving over time) are *commutative*, so that the same result is obtained, independently of the order in which these two operations are executed, as can be seen in the following example ($\partial = \frac{d}{dt}$):

1) $\partial \langle t^3 + 2t^2 - 3t + 6 \rangle = \partial \left(\frac{1}{4}t^3 + \frac{1}{2}t^2 - \frac{3}{4}t + 1.5 \right) = \frac{3}{4}t^2 + t - \frac{3}{4}$.

2) $\langle \partial t^3 + \partial 2t^2 - \partial 3t + \partial 6 \rangle = \langle 3t^2 + 4t - 3 + 0 \rangle = \frac{3}{4}t^2 + t - \frac{3}{4}$.

The problem with general relativity is that the operations of averaging and deriving over time of its tensor equations are NOT commutative operations. It is not the same to let an average matter distribution and its corresponding spatial geometry evolve in time, or let the matter distributions of different regions and their corresponding spatial geometries evolve in time and then average the final results. Cosmologists tend to first obtain a hypothetical, original average matter–energy distribution and its corresponding geometry, and use Einstein's equations to obtain the homogeneous geometry that results from the evolution in time of this average. Actually, the proper procedure would be to first resolve Einstein's equations for the different geometries of the different regions of the Universe, then let these results evolve in time, and then average the final results.

The first one to draw attention to the fact that the operations of averaging and resolving Einstein's equations are not commutative, was George Ellis, in 1984:[197] *"Thus, a significant problem at the foundation of cosmology is to provide suitable definitions of averaged manifolds[198]... of metric ($G_{\mu\nu}$) and stress–tensor ($T_{\mu\nu}$) averaging and smoothing procedures, and to show these have appropriate properties."*[199] The backreaction constitutes the difference between the average of different densities that has evolved in time, and the average of different densities after they have evolved in time. Ellis obtained Einstein's tensor, integrating the backreaction term $P_{\mu\nu}$, but without being able to give an estimate of its solution:

$$G_{\mu\nu} = R_{\mu\nu} - \frac{1}{2}g_{\mu\nu}R = -\frac{8\pi G}{c^4}T_{\mu\nu} + P_{\mu\nu}.$$

Roustam Zalaletnidov has obtained the exact solution to the averaging problem of Einstein's tensor equations.[200]

Regrettably, during fifteen years, Ellis' warnings were not taken into account by many cosmologists in the construction of their models. There was a general tendency to estimate the values of the global cosmological parameters at the present time, and then project them back to the origins of the Universe, in simplified models, where Newtonian gravitational dynamics were assumed to be valid at so–called non–relativistic velocities,

---

[197] George Ellis, "Relativistic cosmology: its nature, aims and problems", in: Bruno Bertotti *et al.*, eds, *General Relativity and Gravitation* (1984) 215–288.

[198] 'Manifolds' are multiples of different space–time regions of the Universe.

[199] George Ellis, "Relativistic cosmology: its nature, aims and problems", in: Bruno Bertotti *et al.*, eds., *General Relativity and Gravitation* (1984) 231.

[200] See Roustam Zalaletnidov, "Averaging out the Einstein's Equations", in: *General Relativity and Gravitation*, vol. 24 (1992): 1015–1031; and Roustam Zalaletnidov, "Averaging Problem in General Relativity, Macroscopic Gravity and Using Einstein's Equations in Cosmology", in: *Bulletin of the Astronomical Society of India* (1997): 401–416.

and the Universe was assumed to be homogeneous, from beginning to end. Often these assumptions were not even consciously made.

However, the problems became more acute at the end of the 1990s, when the apparent acceleration of the expansion of the Universe was discovered by Kirshner, Knop, Perlmutter, Riess and Schmidt. Only in the case that the local expansion rates were equal to the global–average expansion rate, as would be the case in a homogeneous and isotropic Universe, the magnitude of this backreaction term would be zero ($Q = 0$), but, as we shall now see, this assumption proves to be invalid. Not following the proper order of operations yields erroneous results: *"the geometry which arises from the time evolution of an initial average of the matter distribution does not generally coincide, at a later time, with the average geometry of the full inhomogeneous matter distribution evolved via Einstein's equations".*[201]

Thomas Buchert, a German astrophysicist working in France, followed up on Ellis's suggestions. In the case of expanding, spherical, *inhomogeneous* volumes θ, the operations of averaging and evolving in time are NOT commutative, so, in that case, $d_t \langle \theta \rangle_D \neq \langle d_t \theta \rangle_D$, and as a result, a backreaction Q_D is generated, which represents the difference between the present average of the domains that evolved separately in time $\langle d_t \theta \rangle_D$, and the evolution of the original average of these domains $d_t \langle \theta \rangle_D$. Buchert did not give empirical estimates of this difference. Following this idea, Edward Kolb, Sabino Matarrese and Antonio Riotto suggested that Ω_{Q_D} might be equal to Ω_Λ,[202] making speculations on dark energy Λ superfluous.

As we saw in Math box 1.2, omega (Ω) is the ratio of the density ρ of the Universe and its critical density ρ', which, if exceeded, would cause the

---

[201] David Wiltshire, "Exact Solution to the Averaging Problem in Cosmology", arXiv:0709.0732.

[202] Edward Kolb, Sabino Matarrese & Antonio Riotto, "On Cosmic Acceleration without Dark Energy", in: *New Journal of Physics* (2006): 322–346.

Universe to start collapsing. Its empirical value determines the final destiny of the Universe, as we may appreciate in Table 2.2.

Table 2.2 Friedmann–Lemaître model of the Universe

Universe	Density ρ	Omega $\Omega = \rho/\rho'$	Constant k	Expansion rate	Final destiny
Closed	$\rho > \rho'$	$\Omega > 1$	$k = +1$	decreases, and then negative	collapse
Flat	$\rho = \rho'$	$\Omega = 1$	$k = 0$	decreases to zero in $t \to \infty$	eternal expansion
Open	$\rho < \rho'$	$\Omega < 1$	$k = -1$	always positive	eternal expansion

Omega's total sum value is determined by its components, which are the matter–energy density of the Universe (Ω_M), the dark energy component (Ω_Λ), and/or the component due to the backreaction (Ω_Q). Kolb's, Matarrese's and Riotto's conjecture was that the standard cosmological ΛCDM model, which has $\Omega = \Omega_M + \Omega_\Lambda$, could be replaced with another one, which has $\Omega = \Omega_M + \Omega_Q$. It is at this point in the recent history of cosmology that Wiltshire appears on the scene. He demonstrates two important things:

1. The value of Ω_Q is very small and, besides, negative, so as to make it impossible for it to be a candidate to replace Ω_Λ.

2. In a cosmological model that drops the supposition of homogeneity, and makes serious use of Einstein's general relativity, both on the theoretical and the practical–mathematical level, the apparent, recent acceleration of the expansion velocity of the Universe can be explained without any need for speculations on the repulsion produced by dark energy.

In Math box 2.5, proof is delivered that corroborates the first point. Then, in Section 2.2.4, with Math boxes 2.6 and 2.7, I prove the second point.

Math box 2.5 The backreaction according to Buchert and Wiltshire

The term D indicates a certain space–time domain of the Universe; H_D is the Hubble constant in that domain; θ_D is the expansion of that domain's volume; Q_D represents the difference between the present average of the domains that evolved separately in time $\langle \frac{d}{dt}\theta \rangle_D$, and the evolution of the average of the original quantities $\frac{d}{dt}\langle \theta \rangle_D$; σ represents the shear, which is the distortion of elements of a flow caused by the interaction with the matter surrounding it. Buchert has the following backreaction[203]:

(1) $Q_D = \frac{2}{3}\left(\frac{d}{dt}\langle \theta \rangle_D - \langle \frac{d}{dt}\theta \rangle_D \right) - 2\langle \sigma^2 \rangle_D = \frac{2}{3}\left(\langle \theta^2 \rangle_D - \langle \theta \rangle_D{}^2 \right) - 2\langle \sigma^2 \rangle_D.$

Giving a zero value to the shear ($\sigma = 0$), Buchert obtains:

(2) $Q_D = \frac{2}{3}\left(\langle \theta^2 \rangle_D - \langle \theta \rangle_D{}^2 \right).$

Wiltshire further developed these equations, in the context of his own relativistic model, to obtain an empirical value of Q_D:[204]

(3) $\langle \theta^2 \rangle_D = 9 f_w H_w{}^2 + 9 f_v H_v{}^2.$

(4) $\langle \theta \rangle_D{}^2 = 9 f_w{}^2 H_w{}^2 + 9 f_v{}^2 H_v{}^2 + 18 f_w f_v H_w H_v.$

By definition, the volume of walls f_w in the Universe equals its total volume minus the volume of voids f_v:

---

[203] Thomas Buchert, "On Average Properties of Inhomogeneous Cosmologies", arXiv:gr-qc/00010556 (2000): 306–321; and Thomas Buchert, "Averaging Inhomogeneous Newtonian Cosmologies", in: *Astronomy and Astrophysics*, vol. 320 (1997): 1–7.
[204] David Wiltshire, "Cosmic Clocks, Cosmic Variance and Cosmic Averages", in: *New Journal of Physics* (2007): 21.

(5) $f_w = 1 - f_v$.

Combining (2), (3), (4) and (5), we obtain:

(6) $Q_D = \left(6(1 - f_v)H_w{}^2 + 6f_vH_v{}^2 - 6(1 - f_v)^2H_w{}^2 - 6f_v{}^2H_v{}^2 - 12(1 - f_v)f_vH_wH_v\right)$.

(7) $Q_D = \left(6f_v(1 - f_v)H_w{}^2 + 6f_v(1 - f_v)H_v{}^2 - 6f_v(1 - f_v)2H_wH_v\right)$.

(8) $Q_D = 6f_v(1 - f_v)\left(H_v{}^2 + H_w{}^2 - 2H_vH_w\right)$.

(9) $Q_D = 6f_v(1 - f_v)(H_v - H_w)^2$.

Since $f_v = 1 - f_w = 0.759$; $H_v(t) = 52.4$ and $H_w(t) = 34.9$, it follows that:

(10) $Q \cong 336$.

The value of Q_D gives us the value of Omega due to the backreaction Ω_Q:

(11) $\bar{\Omega}_Q = -\dfrac{Q_D}{6*\bar{H}_0{}^2}$.

The global average Hubble constant for the Universe as a whole, measured with a global average clock, $\bar{H}_0(t)$ is:

(12) $\bar{H}_0(t) = 48.2$.

From (10), (11) and (12) we obtain:

(13) $\bar{\Omega}_Q \cong -\dfrac{336}{6*48.2^2} \cong -0.024$.

I will now reformulate the equation of Ω_Q in terms of the different Hubble constants for walls $H_w(t)$, voids $H_v(t)$ and the Universe at large $\bar{H}_0(t)$, as measured with the global average clock t, in order to have a better look at the evolution in time of the backreaction Q and the value of the Omega component Ω_Q that depends on it. As we saw,

(14) $\gamma_v = \dfrac{dt}{d\tau_v} = 0.92$ and

(15) $\gamma_w = \dfrac{dt}{d\tau_w} = 1.38,$[205]

and the lapse function is:

(16) $\gamma = \dfrac{\gamma_w}{\gamma_v} = \dfrac{d\tau_v}{d\tau_w} \cong 1.5.$

Also, by definition:

(17) $\bar{H}_0(t) = \gamma_w H_w = \gamma_v H_v.$

From (12), (15) and (17), we obtain:

(18) $\bar{H}_0(t) = \dfrac{dt}{d\tau_w} H_w = 1.38 * 34.9 = 48.2.$

Also, by definition, the Hubble constant for walls is the recession velocity in walls divided by distance:

(19) $H_w(t) = \dfrac{da_w}{dt} \dfrac{1}{a_w}.$

By definition:

(20) $\bar{H}_0 = \dfrac{dt}{d\tau_w} H_w.$

From (19) and (20), we obtain:

(21) $\dfrac{dt}{d\tau_w} = \dfrac{\bar{H}_0 a_w dt}{da_w}.$

Also, by definition, the Hubble constant for the Universe at large is a weighted average of the Hubble constants for walls and voids:

---

[205] Ben Leith, Cindy Ng & David Wiltshire, "Gravitational Energy as Dark Energy: Concordance of Cosmological Tests", in: *Astrophysical Journal Letters*, vol. 672 (2008): L94; and David Wiltshire, "Cosmological Equivalence Principle and Weak Field Limit", in: *Physical Review D*, vol. 78 (2008): 9.

(22) $\bar{H}_0 = f_w H_w + f_v H_v = 0.241 * 34.9 + 0.759 * 52.4 = 48.2.$

From (21) and (22), we obtain:

(23) $\dfrac{dt}{d\tau_w} = \dfrac{(f_w H_w + f_v H_v)a_w dt}{da_w} = \dfrac{f_w H_w a_w dt +}{da_w} + \dfrac{f_v H_v a_w dt}{da_w}.$

From (19) we obtain:

(24) $\dfrac{a_w dt}{da_w} = \dfrac{1}{H_w},$

and from (23) and (24), and remembering that $f_w = 1 - f_v$, we obtain:

(25) $\dfrac{dt}{d\tau_w} = f_w + f_v \dfrac{H_v}{H_w} = (1 - f_v) + f_v \dfrac{H_v}{H_w} = 1 + f_v \left(\dfrac{H_v}{H_w} - 1 \right).$

We multiply both terms with $(H_w/H_v)/(H_w/H_v)$:

(26) $\dfrac{dt}{d\tau_w} = \dfrac{H_w/H_v + f_v(1 - H_w/H_v)}{H_w/H_v}.$

From (18) and (26), we obtain:

(27) $\bar{H}_0 = \dfrac{H_w/H_v + f_v(1 - H_w/H_v)}{H_w/H_v} H_w.$

From (27), we derive:

(28) $II_w = \dfrac{\bar{H}_0 (H_w/H_v)}{H_w/H_v + f_v(1 - H_w/H_v)}.$

Since, by definition:

(29) $H_v = H_w/(H_w/H_v),$

we obtain from (28) and (29):

(30) $H_v = \dfrac{\bar{H}_0}{(H_w/H_v) + f_v(1 - H_w/H_v)}.$

With (9), (28) and (30) we obtain the backreaction in terms of the three Hubble constants, as measured with a global–average clock:

$$(31) \quad Q_D = 6f_v(1 - f_v)\frac{\bar{H}_0{}^2(1-H_w/H_v)^2}{[H_w/H_v +f_v(1-H_w/H_v)]^2} \; .$$

Let us define:

$$(32) \quad h_r(t) = (H_w/H_v).$$

From (14), (15), (17) and (32), we obtain:

$$(33) \quad h_r(t) = \frac{\gamma_v}{\gamma_w} = \frac{1}{\gamma} = 0.6667.$$

We can simplify the equation (31) with the help of equation (32):

$$(34) \quad Q_D = 6f_v(1 - f_v)\frac{\bar{H}_0{}^2(1-h_r)^2}{[h_r +f_v(1-h_r)]^2} \cong 336.[206]$$

Now let us remember the definition of Omega due to the backreaction:

$$(11) \quad \bar{\Omega}_Q = -\frac{Q_D}{6\bar{H}_0{}^2} \; .$$

From (11) and (34), we obtain:

$$(35) \quad \bar{\Omega}_Q = -f_v(1 - f_v)\frac{(1-h_r)^2}{[h_r +f_v(1-h_r)]^2} = -0.024.[207]$$

The purpose of this exercise is to show the evolution of the backreaction in time, as observed in Graph 2.4. The present, local time value of $\bar{\Omega}_Q$ is at the left–hand side of the graph, with $\bar{\Omega}_Q = -0.024$ and $z = 0$.

It can be inferred from Graph 2.4 that there are two points in time when Q_D and $\bar{\Omega}_Q$ are zero. These zero values are outside the scope of the graph, much higher up than $\bar{\Omega}_Q = -0.024$, and at the left of $z = 0$ and the right of $z = 6$. One point in time when the backreaction Q_D and Ω_Q were zero, were the first 300,000 years of the history of the Universe, when $h_r = 1$,

---

[206] David Wiltshire, "Cosmic Clocks, Cosmic Variance and Cosmic Averages", in: *New Journal of Physics*, vol. 9 (2007): equation 31.
[207] *Ibidem*: equation 10.

so that $1 - h_r = 0$ (see equations (35) and (36), respectively).

A value of $h_r = 1$ means that the wall clocks, void clocks and global average clocks had not yet differentiated and registered the same time,

$$h_r = \frac{H_w}{H_v} = \frac{\gamma_v}{\gamma_w} = \frac{1}{\gamma} = 1.$$

Graph 2.4 The value Ω_Q of as a function of redshift z[208]

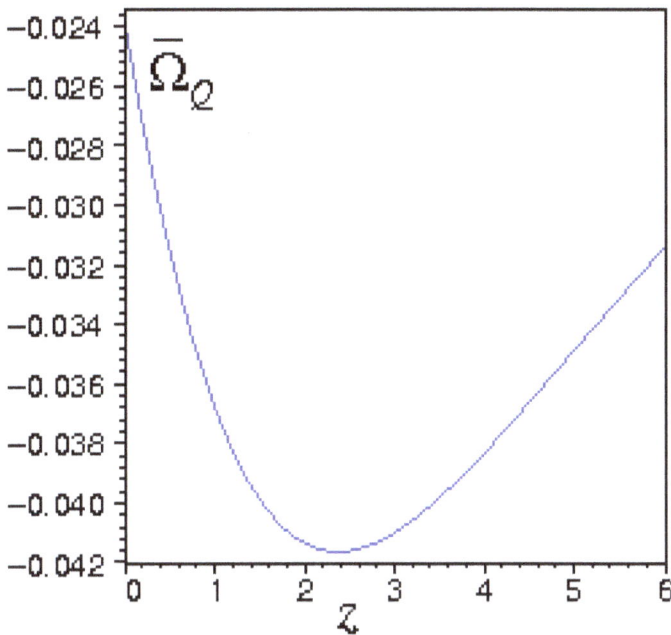

The other point in time when $Q_D = \overline{\Omega}_Q = 0$, will be when the Universe is much older and $f_v = 1$, so that $(1 - f_v) = 0$. This would be the case, because the Universe will end up as an immense, almost infinite, void (see Chapter 6).

---

[208] Courtesy of David Wiltshire, in a private e–mail to the author, dated May 5, 2009.

Section 2.2.4 Wiltshire's *timescape* model

Taking into account not only the backreaction, but also the consequences of differential clock rates in the Universe for $\overline{\Omega}_M$, we obtain the following values for Omega in Wiltshire's relativistic model of an inhomogeneous Universe, which he presents as the *timescape* model (see Math box 2.6):

$$\overline{\Omega}_M = 0.125; \quad \overline{\Omega}_Q = -0.0241; \quad \overline{\Omega}_k = 0.8991.$$

Math box 2.6 Omega in Wiltshire's relativistic model

We remember from chapter 1 that the matter–energy density is

(1) $\rho = 3H^2/8\pi G + 3kc^2/8\pi GR^2$.

The critical density ρ' is obtained when $k = 0$:

(2) $\rho' = 3H^2/8\pi G$.

The constant Ω is the ratio of the density of the Universe and its critical density:

(3) $\Omega = \dfrac{\rho}{\rho'} = 1 + \dfrac{kc^2}{H_0^2(t)R^2}$,

where k is the curvature parameter, with $k = +1$ standing for a closed universe, $k = -1$ for an open one, and $k = 0$ for a flat one.

In the ΛCDM model, the equation governing the expansion of the Universe contains the different contributions to the energy–matter density:

(4) $\Omega_M + \Omega_\Lambda + \Omega_k = \dfrac{8\pi G\rho_M}{3H_0^2(t)} + \dfrac{\Lambda c^2}{3H_0^2(t)} + \dfrac{-kc^2}{H_0^2(t)R^2} = 1$,

where R is the radius of the Universe and Λ is the cosmological constant. What is the value of Ω_M? We know G and $H_0(t)$, but we do not know the average energy–matter density of the Universe ρ_M, so an estimate has to be made, based on observations. In 1997, Dekel, Burnstein and White gave a very broad estimate of $0.3 < \Omega_M < 1.2$, leaving open the question whether the Universe is open, closed or flat, and whether dark energy exists or not.[209] Three years later, Bahcall made an estimate, based on the relationship between mass and galaxy cluster luminosity, of $\Omega_M = 0.25$.[210] A year later, Peacock and his team of 27 astrophysicists used data on the anisotropy of the *CMBR* and the redshift of 200,000 galaxies of the Sloan Digital Sky Survey (=*SDSS*) and gave an estimate of $\Omega_M = 0.30$.[211] Finally, in 2004, Max Tegmark and his team of 36 astrophysicists, together with the *SDSS* collaborators, used the *SDSS* redshift data, the Wilkinson Microwave Anisotropy Probe (=*WMAP*) data and supernova type *SN1a* data, and obtained six possible results, lying between $0.251^{+0.036}_{-0.027} < \Omega_M < 0.334^{+0.027}_{-0.024}$.[212] The wide range of results is due to the fact that the estimates are model–dependent, that is, they depend on whether one introduces certain parameters in the model, or not. The authors express a certain preference for the result that supposes $k_{mx} = 0.15\,\dfrac{h}{Mpc}$, that leads to $\Omega_M = 0.297^{+0.038}_{-0.032}$.

The value of $\Omega_M \approx 0.3$ represents the consensus of today's adherents to the ΛCDM model. With respect to the model–dependency of these results, the authors express the following warning:

"The cosmology community has now established the existence of dark matter, dark energy and near–scale invariant seed fluctuations. Yet we do not know why they exist or the physics responsible for generating them. Indeed, it is striking that standard model physics fails to explain any of the four ingredients of the cosmic matter budget: it gives too small

[209] Avishai Dekel, David Burnstein & Simon White, "Measuring Omega", in: Neil Turok ed., *Critical Dialogues in Cosmology* (1997): 175–192.

[210] Neta Bahcall, "Clusters and Cosmology", in: *Physics Reports*, vol. 33 (2000): 233–244

[211] John Peacock *et al.*, "A Measurement of the Cosmological Mass Density from Clustering in the 2nd Galaxy Redshift Survey", in: *Nature*, vol. 410, number 6825 (2001): 169–173.

[212] Max Tegmark, "Cosmological parameters from SDSS and WMAP", in: *Physical Review D*, vol. 69 (2004): 19 (Table 7). There are many other estimates in this survey, depending on whether certain parameters are introduced in the model, or not. The *WMAP* team has $\Omega_M = 0.26 \pm 0.05$.

CP–violation to explain baryogenesis, does not produce dark matter particles, does not produce dark energy at the observed level and fails to explain the small yet non–zero neutrino masses."[213]

In retrospect, this warning was quite warranted, since, as a matter of fact, dark matter and dark energy are myths, as I argue in this chapter. In Wiltshire's *timescape* model, where dark energy is being disposed of, the global average omega $\bar{\Omega}$ is the sum of different components:

(5) $\bar{\Omega}_M + \bar{\Omega}_Q + \bar{\Omega}_k = 1 \rightarrow \gamma_w^{-3}(\Omega_M) + \dfrac{-Q}{6H_0{}^2(t)} + \bar{\Omega}_k = 1.$

Wiltshire's matter density $\bar{\Omega}_M$ is a function of the Ω_M value in the ΛCDM model, corrected for the differential clock rates $\dfrac{dt}{d\tau_w}$ in walls and the global–average Universe. Since Wiltshire takes Ω_M to be $\Omega_M = 0.33$, it follows that:

(6) $\bar{\Omega}_M = \gamma_w^{-3}(\Omega_M) = \left(\dfrac{dt}{d\tau_w}\right)^{-3} \Omega_M = 1.38^{-3}(0.33) = 0.125.$

The contribution of the backreaction to Omega is, as we saw (equation (36) of Math box 2.5):

(7) $\bar{\Omega}_Q = \dfrac{-336}{6(48.2)^2} = -0.0241.$

So, the Omega for the total matter–energy density in the *TS* model is:

(8) $\bar{\Omega}_M + \bar{\Omega}_Q = 0.125 - 0.0241 = 0.1009.$

This means that $\bar{\Omega}_M + \bar{\Omega}_Q < 1$, and that our Universe is open ($k = -1$).

---

[213] Max Tegmark, "Cosmological Parameters from SDSS and WMAP", in: *Physical Review D*, vol. 69 (2004): 21. For dark energy not being produced at the observed level, see also Steven Weinberg, "The Cosmological Constant Problem", in: *Review of Modern Physics*, vol. 61 (1989): 6–23

We are now ready to understand that the *apparent* acceleration of the expansion velocity of the Universe is actually an optical illusion, derived from the fact that our observations are made with wall clocks, mistakenly taken for global–average clocks. Why do I speak, like Wiltshire, of *apparent* acceleration? Is the acceleration not *real*? Yes, and no. The fundamental relativistic principle that guides us is the following one, formulated by Wiltshire: "*Systematically different results will be obtained when averages are referred to different clocks.*"[214]

If we observe a supernova, that moves away from us, and is located at the other end of the immense void that surrounds our galaxy cluster, the velocity and redshift of its light, passing through this large void, and reaching an observer in a strong gravitational field, will be higher when measured by the observer's wall clock than by the clock in the void, or a global–average clock. The clock in the denser region runs more slowly than the one in the void. For that reason, the deceleration measured with the wall clock will be different from the one measured with the clock in the void, or a global–average clock. When measured with the clock in the void, or the global–average clock, the supernova will appear to move away from us at a slower rate, and its light will appear to have a smaller redshift, than when measured with the wall clock, which is the one we use, because of the differential clock rates.

A terrestrial observer, measuring the redshift of the supernova's light with his own slower running wall clock, will observe an acceleration, whereas an observer located in the void surrounding us, will measure, with his faster running clock, a deceleration: "*it is quite possible to obtain regimes in which the wall observers measure apparent acceleration, $q < 0$, even though void observers do not*".[215] Void observers would measure a normal

[214] David Wiltshire, "Cosmic Clocks, Cosmic Variance and Cosmic Averages", in: *New Journal of Physics*, vol. 9 (2007): 27.
[215] *Ibidem*: 29.

deceleration of the expansion velocity, yielding a positive deceleration parameter $q > 0$.

Besides, the volume of the less rapidly decelerating voids increases dramatically, as compared to the walls. Today, the parameter of the global deceleration, as observed by wall observers like Kirshner, Knop, Perlmutter, Riess and Schmidt, has a negative value of $q = -0.04280$ (an apparent acceleration). This same parameter, in the proper time of the global–average observer has a value of $q = +0.01533$ (a deceleration),[216] as I explain in Math box 2.7. This is why Einstein's theory is called general *relativity*: time and space are *relative*. The history of the Universe has three periods:

1) a brief period that was dominated by energy;
2) a long period that is now reaching its end, dominated by matter; and
3) a third epoch that is now beginning, dominated by voids.

Right now, we live in a transition period between the matter dominated epochs and the following one dominated by ever larger volumes of voids. It is precisely in such a transition period that we may observe an apparent acceleration (see Math box 2.7):

"Depending upon parameter values, it is possible for wall observers to register an apparent acceleration with the deceleration parameter taking values of q < 0. Backreaction and the rate of decrease of $\frac{dt}{d\tau_w}$ are largest in an epoch during which the universe appears to undergo a void–dominance transition, or equivalently a transition in which spatial curvature Ω_k becomes significant. The reason for apparent acceleration at such an epoch [is that] in the transition epoch the volume of the less rapidly decelerating regions increases dramatically, giving rise to apparent acceleration in the volume–average."[217]

---

[216] David Wiltshire, "Cosmic Clocks, Cosmic Variance and Cosmic Averages", in: *New Journal of Physics*, vol. 9 (2007): 34.
[217] *Ibidem*: 30.

Math box 2.7 The deceleration parameter in the relativistic model

Wiltshire obtains the following value for the deceleration parameter, when measured with wall clock time τ_w:

(1) $q(\tau_w) = \dfrac{-(1-f_v)(8f_v{}^3+39f_v{}^2-12f_v-8)}{(4+f_v+4f_v{}^2)^2}$.[218]

Since $f_v = 0.759$, we see an *apparent* expansion acceleration:

(2) $q(\tau_w) \cong -0.042785$.

Now the relationship between $\bar{\Omega}_M$ and Ω_M is the following:

(3) $\bar{\Omega}_M = \gamma_w{}^{-3}(\Omega_M)$.

In equation 4 of Math box 2.6, we saw Ω_M has the following value:

(4) $\bar{\Omega}_M = 0.125$.

Wiltshire obtains the following value for the *real* deceleration parameter as measured with the global–average clock time t:

(5) $q(t) = \frac{1}{2}\bar{\Omega}_M + 2\bar{\Omega}_Q$.[219]

Since $\dfrac{dt}{d\tau_w} = 1.38$; $\Omega_M \cong 0.33$; $Q \cong 336$, $H_0(t) \cong 48.2$, and the density derived from the backreaction is $\bar{\Omega}_Q = -0.24$, we obtain:

(6) $q(t) \cong 0.0145$.

The fact that the deceleration parameter $q(t)$, when measured with a global–average clock, has a positive value (equation 6), means that the expansion velocity of the Universe is decelerating, as might normally be

---

[218] David Wiltshire, "Exact Solution to the Averaging Problem in Cosmology", arXiv0709.0732 (2007): equation 26.
[219] David Wiltshire, David Wiltshire, "Cosmic Clocks, Cosmic Variance and Cosmic Averages", in: *New Journal of Physics*, vol. 9 (2007): equation 61.

expected after a big explosion like the Big Bang. The erroneous generalization of the negative value of the deceleration parameter, as measured with our terrestrial wall clock (equation 2), as if it were the global–average deceleration parameter, has its origin in not taking seriously Einstein's general relativity: time is relative.

Section 2.2.5 Comparing the ΛCDM and the *timescape* models

The speculation about dark energy is not necessary, and things can be explained with orthodox, general relativity. This conclusion does not mean that the ΛCDM model does not fit the data. Both the ΛCDM model and the *timescape–Szekeres* model fit the data. In order to compare the fitness of either model, we must first get a bit more familiar with Bayesian probability calculus, named after Thomas Bayes (1701–1761) (see Math box 2.8).

Math box 2.8. The fundamentals of Bayesian probability

Bayesian evidence $E(M)$ in favour of some cosmological model M is defined as the probability P that certain empirical data D are observed in a sample, in the case that this model M would be the one that corresponds to the physical reality of the Universe:

(1) $E(M) = P(D_M)$.

and the Bayes factor B_{ij} is the rate of the Bayesian evidence for both models M_i and M_j :

(2) $B_{ij} = \dfrac{M_i}{M_j}$.

If the M_i model has a greater probability of being a good fit for the data than the M_j model, $B_{ij} > 1$.

One subsequently compares each model with a base model under the null hypothesis M_0, which is the hypothesis that a particular model, like M_i or M_j, does NOT explain certain phenomena.

Then we draw the natural logarithm of the Bayes factor, in order to compare the respective models M_i and M_j with the base model M_0:[220]

Table 2.3 Bayesian probability calculus

$\left|\ln B_{ij}\right|$	$B_{ij} = \dfrac{P(M_i)}{P(M_0)}$	Probability of significant difference	Strength of evidence
0	1	$p=0$	No evidence
0.139762	1.15	$0<p<0.75$	Almost no evidence
<1	<2.719	$p<0.75$	Inconclusive
1	≈ 2.719	$p=0.75$	Weak evidence
2.5	≈ 12	$p=0.923$	Moderate evidence
5	≈ 150	$p=0.993$	Strong evidence

This inductive method is analogous to the χ^2, where the observed distribution is compared to the expected distribution under the null hypothesis, and a decision is made whether this difference is statistically significant.

Elgaroy and Multamäki analysed the Riess–Kirshner and the Astier supernovae samples, comparing both with a base model of a flat universe with a constant deceleration factor, and concluded there is no evidence for a transition of deceleration to acceleration:

"[T]he best model in both cases has q(z) constant. It therefore seems fair to conclude that there is no significant evidence in the present supernovae data for a transition from deceleration to acceleration, and claims to the

---

[220] R. Trotta, "Bayes in the sky: Bayesian Inference and Model Selection in Cosmology", arXiv:0803.4089, p. 14.

contrary are most likely an artefact of the parameterization used in the fit of the data... It is at the moment not possible to say anything about when, or indeed if the Universe went from deceleration to acceleration."[221]

Shapiro and Turner, analysing the Riess–Kirshner supernovae sample (the 'Gold set'), and the Knop–Perlmutter and Astier samples reached similar conclusions: *"the present Supernovae 1a data cannot rule out the possibility that the universe has actually been decelerating for the past 3 Gyr [=three thousand million years] (i.e. since z = 3.0)".*[222]

Responding to an inaccurate criticism of Kwan, Francis and Lewis,[223] Smale and Wiltshire showed that one must take into account the different fitter used in different datasets. The SALT/SALT II fitters *"provide Bayesian evidence to favour the spatially flat ΛCDM model over the TS model"* (as is the case in four supernova samples: the Union sample, the Constitution sample, the Salt–2 sample, and the Union–2 sample), but if the MLCS2k2 fitter is used (as is the case in four other samples: the Riess–07 sample, the MCLS–17 sample, the MLCS–31 sample and the SDSS–II sample), then *"Bayesian evidence favours the TS model over the spatially flat ΛCDM model"*, so that, *"basically, both models are a very good fit".*[224] In both cases the Bayesian evidence is so slight that it is inconclusive and the primary question is the fitter, not the fitness of any model. Ishak and Sussman, using χ^2, compared another relativistic and inhomogeneous model, with $\Omega_\Lambda = 0$ (the Szekeres model[225]), with the homogeneous Λ*CDM*

---

[221] Øystein Elgaroy & Tuomas Multamäki, "Bayesian Analysis of Friedmannless Cosmologies", arXiv:astroph/ 0603053, pp. 5–6.

[222] Charles Shapiro & Michael Turner, "What do we Really Know about Cosmic Acceleration?", in: *Astrophysical Journal* vol. 649 (2006): 566.

[223] Juliana Kwan, Matthew Francis & Geraint Lewis, "Fractal Bubble Cosmology: A Concordant Cosmological Model?", arXiv:0902.4249 (2009): 2.

[224] Peter Smale & David Witshire, "Supernova Tests of the *Timescape* Cosmology", arXiv:1009.5855v1, p. 19.

[225] See P. Szekeres, "A Class of Inhomogeneous Cosmological Models", in: *Communications in Mathematical Physics*, vol. 41 (1975): 55–64.

model, that has $\Omega_\Lambda = 0.73$. The observational data of 94 Supernovae type 1a fitted both models, with $\chi^2 = 112$ and $\chi^2 = 105$, respectively, yielding no significant differences (see graph 2.5)

Graph 2.5 Both the ΛCDM and *timescape* models fit supernovae data[226]

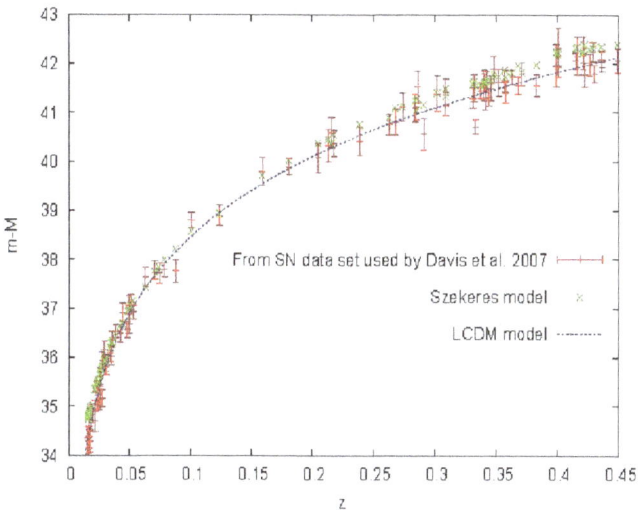

From SN data set used by Davis et al. 2007
Szekeres model
LCDM model

---

[226] Mustapha Ishak, Roberto Sussman *et al.*, "Dark Energy or Apparent Accceleration Due to a Relativistic Cosmological Model More Complex than FLRW?", arXiv:0708.2943 (2008): 5. Supernova fits for the Szekeres model (green crosses) and ΛCDM (blue curve) models. The 94 Supernova (up to $1 + z = 1.449$) data are from Tamara Davis *et al.*, "Scrutinizing Exotic Cosmological Models using ESSENCE Supernova Data", in: *Astrophysical Journal,* vol. 662 (2007): 716-725; Michael Wood–Vasey *et al.*, "Observational Constraints on the Nature of the Dark Energy: First Cosmological Results from the ESSENCE Supernova Survey", in: *Astrophysical Journal*, vol. 666 (2007): 694–715; and Adam Riess *et al.,* "Type Ia Supernova Discoveries at $z < 1$ From the Hubble Space Telescope: Evidence for Past Deceleration and Constraints on Dark Energy Evolution", *The Astrophysical Journal*, vol. 607 (2004): 665-738. The Szekeres model fits the data with a $\chi^2 = 112$. This is close to the $\chi^2 = 105$ of the ΛCDM model.

Then there is the Bayesian evidence resulting from comparing the luminosity distance of 140 Supernovae that the Riess–07, Union and Constitution samples have in common, in the case of the ΛCDM and the *timescape* models. The likelihood that one model is a better fit than the other, in these three samples, is between 15% and 20%, with $lnB = 0.14$, $lnB = 0.14$ and $lnB = 0.17$, respectively: *"the models are statistically indistinguishable for the 140 SNe 1a regardless of the fitter used"* and *"both models are a very good fit"*.[227]

If we have two explanations for the same phenomenon, in this case the apparent recent acceleration of the expansion of the Universe, and one is a model based on orthodox general relativity, and the other one a model based on speculations about something that has never been observed or proven to exist in replicable experiments, i.e. dark energy, we find ourselves under the obligation, applying Ockham's razor, to take the orthodox model of general relativity without dark energy as the valid one.

Perhaps, the reader is wondering how two models, one with $\Omega_\Lambda = 0$, and the other with $\Omega_\Lambda = 0.7$ can both fit the data? As a matter of fact, this is not so strange as it seems at first sight. We should take into account that when computer software is modelled in conformity with ΛCDM, it integrates the assumption of $\Omega_\Lambda \cong 0.7$ in the model as a pre-condition. So, the model does not *prove*, but rather *supposes*, that the observed recent acceleration of the expansion velocity is caused by the repulsion of some mysterious dark energy overcoming the gravitational attraction. These kinds of assumptions, which form part of a computer model, are sometimes called 'equations of state'. This means that the simulations of the computer model do not corroborate the ΛCDM model, they only reveal that the model works, meaning that the empirical data are compatible with the model.

---

[227] Peter Smale & David Witshire, "Supernova tests of the *timescape* cosmology", arXiv:1009.5855v1, pp. 12, 18.

The whole process is rather tautological. It is much like the Ptolemaic cosmology explaining the orbits of the planets around the Earth. Again, and again, the Ptolemaic model was shown to be compatible with the observations of these planetary orbits. That is why, for fourteen centuries, the academic community strongly believed in the model. Nobody really knew what was the mysterious mechanism that made the planets follow these orbits of circles on circles. In the same way, modern astrophysicists do not know what dark energy is, it just appears that something out there makes the recent expansion velocity of supernovae apparently accelerate and they call it dark energy. In both these instances, nothing is really explained, but something mysterious is invented or supposed to be there in order to fill the gap in our knowledge.

It so happened, however, that Newton could explain Kepler's data about elliptical orbits of planets around the Sun, as being caused by the gravitational force between the Sun and each planet, the same force that is familiar to everybody from observing how an apple falls from a tree. In the same way, today, the speculation about dark energy can be abandoned since the same observations have now been explained by orthodox general relativity, a theory that has been corroborated by the facts, again and again. This does not mean that planets, seen from Earth, do not appear to follow orbits of circles on circles, neither does this mean that the expansion velocity of the Universe, as observed from Earth, with our wall clocks, does not appear to be recently accelerating. The observations stay in place, but scientific explanations of these observations derived from orthodox physics replace the speculative pseudo–explanations.

CHAPTER 3

FACT: THE FINE TUNING OF PHYSICAL CONSTANTS

Over the course of the twentieth century, the scientific community reached an understanding of the fine tuning of our Universe. Fine tuning is defined as the set of initial conditions and specific values of certain universal constants in the physical laws that made possible the emergence of stars and complex life on Earth. As far as I know, Paul Davies, a theoretical physicist from England, born in 1946, was the first one to use the term 'fine tuning':

> "*[T]he structures of many of the familiar systems observed in nature is determined by a relatively small number of universal constants. Had these constants taken different numerical values from those observed, then these systems would differ correspondingly in their structure. What is especially interesting is that, in many cases, only a modest alteration of values would result in a drastic restructuring of the system concerned. Evidently the particular world organization that we perceive is possible only because of some delicate 'fine tuning' of these values.*"[228]

Roger Penrose claims that our Universe is a very special one in a set of an almost infinite number of possible universes: "*The Creator's pin has to find a tiny box, just one part in $(10^{10})^{123}$ of the entire phase-space volume, in order to create a universe with as special a Big Bang as that we actually [find]*", [229] as we see in Figure 3.1.

---

[228] Paul Davies, *The Accidental Universe* (1982): 60.
[229] Roger Penrose, *The Road to Reality* (2004): 730.

Figure 3.1 A very special universe[230]

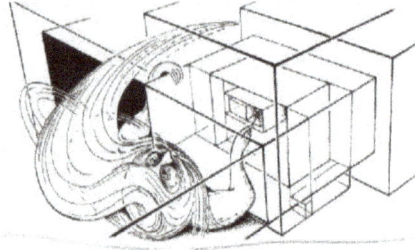

Luke Barnes recently analysed the fine tuning of 15 fundamental and five derived physical constants, necessary for the emergence of intelligent life – and thoroughly refutes Victor Stenger's denial of this fine tuning –.[231] In this chapter, I analyse six cases, showing that the precise value of a single constant is a *sine qua non* condition for the existence of various complex structures and, in the summary, I argue that the precise values of many constants and initial conditions must be simultaneously satisfied to make it possible for intelligent life to emerge, as in our Universe:

1. The value of Ω in the Big Bang and star and galaxy formation.
2. The production and conservation of protons in the Big Bang.
3. The beginning of nuclear fusion in the stars.
4. The fusion of carbon, oxygen and other elements in the stars.
5. The existence of planets and stable, planetary orbits.
6. The existence of stable atoms and complex molecules.
7. Summary.

---

[230] Adapted from: Roger Penrose, *The Road to Reality* (2004): 730 and Roger Penrose, *The Emperor's New Mind* (1991): 343. The image of the creator does not imply a statement of faith in a divine creation on the part of Penrose.

[231] Luke Barnes, "The Fine-Tuning of the Universe for Intelligent Life", in: *Publications of the Astronomical Society of Australia*, vol. 29 (2012): 529-564, criticising Victor Stenger, *The Fallacy of Fine-Tuning. Why the Universe is Not Designed for Us* (2011).

Section 3.1 The value of Ω in the *Big Bang* and galaxy formation

There is a delicate balance between the two extremes of a continuously expanding Universe and a gravitational collapse that determines if stars and galaxies emerge over time, and this depends on the constant Ω, as one can see in Table 3.1.

Table 3.1 Friedmann-Lemaître model of the Universe

Universe	Density ρ	Omega $\Omega = \rho/\rho'$	Constant k	Expansion rate	Final destiny
Closed	$\rho > \rho'$	$\Omega > 1$	$k = +1$	decreases, and then negative	Collapse
Flat	$\rho = \rho'$	$\Omega = 1$	$k = 0$	decreases to zero in $t \to \infty$	Eternal expansion
Open	$\rho < \rho'$	$\Omega < 1$	$k = -1$	always positive	Eternal expansion

The constant Ω is the ratio of the observed density of the Universe ρ to the critical density ρ', i.e. $\Omega = \rho/\rho'$. If the observed density exceeds the critical density, there would be a gravitational collapse; if the opposite, there would be an everlasting expansion. In the limit, when R tends to infinity and ρ to zero, $v_{ex}{}^2$ tends to $-kc^2$ (see Math box 3.1). So, there are three gravitational scenarios in space-time (see Graph 3.1).

The possibility of complex life in the Universe depends on a very important initial condition, specifically, the value of Ω. For the Universe to not collapse on itself, nor expand too quickly, Ω must have initially been equal to or slightly less than unity. Too rapid an expansion would have hindered the collapse of hydrogen and helium clouds, and consequently, the formation of galaxies and stars. Conversely, too slow an expansion would have resulted in a premature collapse of the new-born Universe, before stars and galaxies could be formed.[232]

Graph 3.1 Three Friedmann-Lemaître models according to value Ω_M

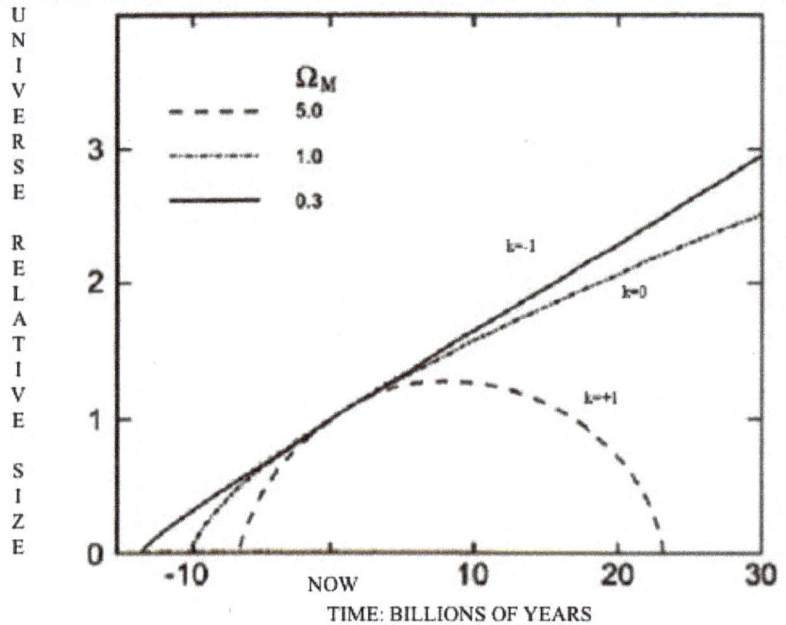

This does not mean that the value of Ω is exactly one *today*. However, although today Ω is thought to be $\Omega \cong 0.3$ (in the ΛCDM model), or $\Omega \cong 0.125$ (in Wiltshire's time-scape model), this does not exclude fine tuning from the *Big Bang*:

"*Ω may not be exactly one, but it is now at least 0.3. At first sight, this may not seem to indicate fine tuning. However, it implies that Ω was very close indeed to unity at the beginning of the Universe.*"[233]

---

[232] Guillermo Gonzalez & Jay Richards, *The Privileged Planet* (2004): 197. The graph is from *ibidem*: 185.

[233] Martin Rees, *Just Six Numbers* (2000): 97–99.

Math box 3.1 The critical density of the Universe

According to Newton, the escape velocity v_{esc} is the velocity an object requires to escape from a gravitational field of another object with mass M and radius R. Mathematically, it is defined as:

(1) $v_{esc} = \sqrt{\dfrac{2GM}{R}} \implies v_{esc}^2 = \dfrac{2GM}{R}$,

where G is the gravitational constant from Newton's universal law of gravitation.

We now define the expansion velocity v_{ex} as the escape velocity v_{esc} with a constant χ, such that $\chi > 0$, when the expansion velocity v_{ex} is bigger than the escape velocity v_{esc}; $\chi < 0$, when the expansion velocity v_{ex} is smaller than the escape velocity v_{esc}; or $\chi = 0$, when the expansion velocity v_{ex} is equal to the escape velocity v_{esc}:

(2) $v_{ex}^2 = v_{esc}^2 + \chi = \dfrac{2GM}{R} + \chi$.

The mass M of a spherical object is equal to the product of the volume of the sphere ($V = \frac{4}{3}\pi R^3$, where R is the radius of the sphere) and its density is ρ:

(3) $M = \frac{4}{3}\pi R^3 \rho$.

Combining (2) and (3), we obtain:

(4) $v_{ex}^2 = \frac{8}{3}\pi G R^2 \rho + \chi$.

Albert Einstein published his treatise on general relativity in 1917. In 1922 and 1927, respectively, Alexander Friedmann and Georges Lemaître, one independently of the other, used Einstein's general relativity to transform the Newtonian escape and expansion velocities and apply them to the Universe, and they obtained:

(5) $\chi = -kc^2$,

where k is the Friedmann-Lemaître-Robertson-Walker constant of the curvature of the Universe and c, the velocity of light. Combining (4) and (5) we obtain:

(6) $v_{ex}^2 = \frac{8}{3}\pi G R^2 \rho - kc^2$,

such that if $k = +1$, the Universe is closed, and it collapses before its radius R reaches infinity; if $k = 0$, the Universe is flat and it will expand forever, with the expansion velocity reaching zero when its radius R reaches infinity and its density ρ, zero ($v_{ex} \to 0$, when $R \to \infty$ and $\rho \to 0$); and if $k = -1$, the Universe is open and it will expand forever, with the expansion velocity being constant when R and t reach infinity.

I now introduce the Hubble constant. The Hubble constant is the ratio of the recession velocity of a galaxy (which is equal to the expansion rate of the Universe v_{ex} at that point in the Universe) and its distance d from Earth:

(7) $H = v_{ex}/d \Longrightarrow v_{ex} = Hd$.

Suppose that we are looking at a galaxy with a high recession velocity, at the limit of the observable Universe. Thus, in this case, the distance d is equal to the radius R of the observable Universe. So, equation (7) becomes:

(8) $H = v_{ex}/R \Longrightarrow v_{ex} = HR \Longrightarrow v_{ex}^2 = H^2 R^2$.

From (6) and (8), we deduce the density of the Universe ρ is:

(9) $\rho = 3H^2/8\pi G + 3kc^2/8\pi G R^2$.

The critical density ρ' is obtained when $k = 0$:

(10) $\rho' = 3H^2/8\pi G$.

Omega is the ratio of the density of the Universe and its critical density:

(11) $\Omega = \frac{\rho}{\rho'}$.

Thus, combining equations (9), (10) and (11), we get Ω as a function of the constant k, the speed of light c, the radius of the Universe R and the Hubble constant H:

(12) $\Omega = \dfrac{\rho}{\rho'} = 1 + \dfrac{kc^2}{H^2 R^2}$.

Equation (12) is equation (366) from my treatise on general relativity in my Spanish language e-book on the Universe.[234] The reader may not immediately grasp how to reach today's value of $\Omega \approx 0.3$ or $\Omega \approx 0.124$ starting with $\Omega \cong 1$ in the Big Bang. Let us look at a numerical example. Let us give the critical density the value of $\rho' = 1$ and suppose the Big Bang begins with a density with an empirical value of almost 1:

(13) $\rho = (1 - 1/10^3) = 999/1{,}000 \cong (1 - 1/10^{2.76}) = 998/1000 \cong 1$.

Therefore, shortly after the Big Bang, after $n = 1$ time units have passed, Ω has a value of 1:

(14) $\Omega_{t=1} = [(1 - 1/10^3)/1]^1 \cong [(1 - 1/10^{2.76})/1]^1 \cong 1$.

The initial density is very high but it decreases with time because of the expansion of the Universe, so that, after $n = 1{,}200$ time units have passed, Omega, which started as $\Omega_{t=1} \cong 1$, is reduced to:

(15) $\Omega_{t=1{,}200} = \left[\left(1 - 1/10^3\right)/1\right]^{1{,}200} \cong 0.3$, or

(16) $\Omega_{t=1{,}200} = \left[\left(1 - 1/10^{2.76}\right)/1\right]^{1{,}200} \cong 0.124$.

Graph 3.2 illustrates how the possibility of the evolution of complex structures and life strongly depend on the initial value of Omega:

---

[234] John Auping, "La construcción de la geodésica y el tensor de Einstein", en: *El Origen y la Evolución del Universo* (2016): 631–696.

Graph 3.2 The delicate balance between the expansion of the Universe and its gravitational collapse[235]

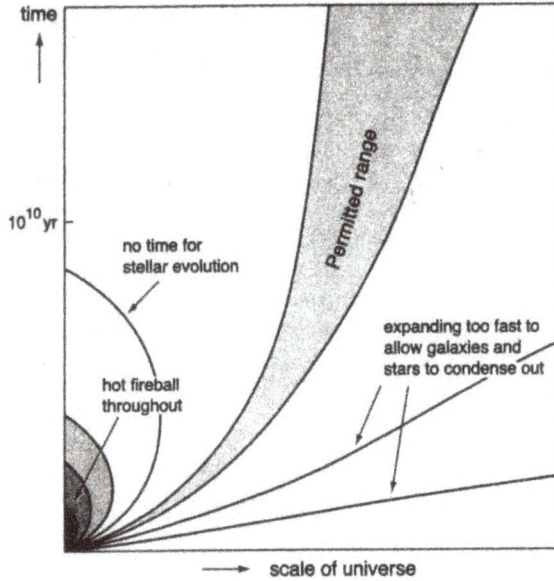

Martin Rees, a British astrophysicist born in 1942, concludes that, at the beginning of the Universe, Ω had a very finely tuned value, only slightly differing from unity, such that stars and galaxies could form later on:

"*[I]t looks surprising that our universe was initiated with a very finely tuned impetus, almost exactly enough to balance the decelerating tendency of gravity… the required precision is astonishing: at one second after the Big Bang, Ω cannot have differed from unity by more than one part in a million billion (one in 10^{15}) in order that the universe should now, after ten billion years, be still expanding and with a value of Ω that has certainly not departed wildly from unity.*"[236]

---

[235] Adapted from Martin Rees, *Just Six Numbers* (2000): 98.
[236] *Ibidem*: 99.

Section 3.2 The production and conservation of protons in the Big Bang

The basic observable building blocks of matter are protons, which have a positive charge, causing the nucleus of an atom to have a positive charge, which in turn means that an equal number of electrons with negative charge orbit this nucleus, so that the atom has a net charge of zero. It so happens that in the Universe, neutrons, with zero charge, are unstable nucleons and the protons, with charge +1, are the stable ones. If it were not for a type of *fine tuning,* the protons would have been unstable and there would not be stable atoms.

Previously,[237] we saw that from 10^{-10} to one second of the evolution of the Universe, matter was arranged as *protons and neutrons.* Very energetic photons can produce protons and anti-protons and neutrons and anti-neutrons, which are then annihilated, generating gamma rays.[238] However, cosmologists agree that after the mutual destruction of matter and anti-matter, a surplus of matter remained, in the form of protons and neutrons. Due to the weak nuclear force, neutrons, which have an average life of just over 10 minutes, decay into protons and electrons, or it is also possible for neutrons and positrons to combine to produce protons; and vice-versa, protons produce neutrons and positrons, or protons combine with electrons to produce neutrons.[239] While the temperature was high enough, the transformation of neutrons into protons and vice-versa was possible, despite the rapid expansion of the Universe, and thermodynamic equilibrium existed. But when the temperature dropped, the equilibrium was broken and the relative abundance of both particles was fixed at the values they had at the prevailing temperature when equilibrium was broken.[240] Below this

---

[237] In Section 2 of Chapter 1.

[238] Jonathan Allday, *Quarks Leptons and the Big Bang* (2002): 244–266.

[239] See Section 2 of Chapter 1. In these transformations, electron neutrinos v_e and electron anti-neutrinos $\bar{v}_e$ make up for missing mass-energy and surplus mass-energy.

[240] Paul Davies, *The Accidental Universe* (1983): 62.

temperature, neutrons decay to generate protons, but not vice-versa, so in the Big Bang, neutrons would have ceased to exist if the process of nuclear fusion had not started, which conserved a certain number of neutrons in helium nuclei.

Math box 3.2 Two cases of fine tuning: the ratio of the weak force to the gravitational force and the difference in proton and neutron mass

The critical temperature for the quantity of neutrons and protons to be fixed is T_F

(1) $k_B T_F \approx G^{1/6} g_w^{-2/3} \hbar^{11/6} c^{7/6}$ [note 241],

where the Boltzmann constant $k_B \cong 1.38 * 10^{-23} JK^{-1}$. There are two coincidences that come into play here. Firstly, the Carr-Rees coincidence, *"a fundamental relationship between the weak nuclear force [g_w] and the gravitational force [G]"*, that is to say, *"a delicate coincidence between the gravitational and weak interactions"*,[242] where α_w is the weak interaction constant:

(2) $\left(\frac{Gm_e^2}{\hbar c}\right)^{1/4} \approx \left(\frac{g_w m_e^2 c}{\hbar^3}\right) \equiv \alpha_w \cong 6.47 * 10^{-12} \implies$

(3) $g_w = \left(\frac{Gm_e^2}{\hbar c}\right)^{1/4} * \left(\frac{\hbar^3}{m_e^2 c}\right) \equiv G^{1/4} m_e^{-3/2} \hbar^{11/4} c^{-5/4}.$

If we substitute the value of g_w from equation (3) into equation (1), we obtain:

(4) $k_B T_F \cong m_e c^2.$

---

[241] Paul Davies, *The Accidental Universe* (1983): 63.

[242] Bernard Carr & Martin Rees, *Nature,* vol. 278 (1979) quoted by Paul Davies, *The Accidental Universe* (1983): 63. John Barrow & Frank Tipler, in: *The Anthropic Cosmological Principle* (1986): 399, faithful to their standardized annotation, fix $\hbar = 1$ and $c = 1$ and obtain $G_F m_e^2 \approx (Gm_e^2)^{1/4}$, which should not mislead us.

The second inexplicable 'Goldilocks' coincidence (meaning neither too little nor too much) is the fact that the mass difference between the neutron and proton Δm is a bit more than the electron mass m_e:[243]

(5) $\Delta m - m_e = 1.396 MeV - 0.511 MeV = 0.883 MeV \Rightarrow \Delta m -$

$m_e \geq 0 \Rightarrow$

(6) $\Delta m \geq m_e$.

If we substitute (6) into (4), we obtain:

(7) $k_B T_F \approx (\Delta m)c^2 \Rightarrow \frac{(\Delta m)c^2}{kT_F} \approx 1.$

On the other hand, we have the Boltzmann factor, which determines the relative abundance of different particle species when their quantities are fixed at the values they had when thermodynamic equilibrium was broken at temperature T:[244]

(8) $\frac{N_{p2}}{N_{p1}} = e^{-(m_2 - m_1)c^2/k_B T} = \frac{N_{p2}}{N_{p1}} = e^{-(\Delta m)c^2/k_B T}$,

where e is a mathematical constant, also known as Euler's number, with an approximate value of $e \cong 2.71828$. Substituting (7) into (8), we obtain the following approximation:

(9) $\frac{N_{neutrons}}{N_{protons}} \approx e^{-1} \approx 0.37.$

There are two coincidences that come into play at the level of the critical temperature fixing the relative abundance of protons and neutrons. These are, firstly, a special relationship between the weak nuclear force and the gravitational force, and secondly, the fact that the difference in mass

---

[243] Paul Davies, *The Accidental Universe* (1983): 63.
[244] *Ibidem*: 33.

between the proton and neutron, Δm, is a bit more than the electron mass m_e. These coincidences give us the theoretical relative abundance of neutrons and protons of approximately 27% and 73% (since (27/73=0.37, see Math box 3.2). Empirically, the relative abundances are slightly different: 13% and 87%, respectively. However, the exact values are not as important here as the fact that the two coincidences mentioned were the cause of many more protons remaining than neutrons. If the ratio of the abundance of protons to that of neutrons had, for example, been $e^{-0.1} \approx 0.9$,[245] the number of protons and neutrons would have been almost equal, with disastrous consequences, as we shall see below.

After the formation of protons and neutrons, in the history of the emerging Universe,[246] comes the process of nuclear fusion. From minute 3 to minute 20, approximately, when the Universe was transformed into an enormous hydrogen bomb, with a radius of almost a light year, helium nuclei were formed along with a small amount of deuterium and lithium. Fred Hoyle noted that the generation of helium and hydrogen in relatively abundant proportions depends strongly on the finely tuned density of matter-radiation present in the Big Bang:

"*If the density is too low, the resulting protons do not combine with the remaining neutrons, and very little helium is formed. On the other hand, if the density is too high, there is a complete combination of neutrons and protons, and with the further combination of the resulting deuterium into helium, very little hydrogen remains as time increases.*"[247]

The temperature had dropped sufficiently and the matter-energy was sufficiently dense so that the deuterium nuclei could remain intact, and the

---

[245] If $k_B T_F > \Delta mc^2$, for example, $(\Delta mc^2/k_B T_F) \approx 0.1$.
[246] See Section 1.2 of Chapter 1.
[247] Fred Hoyle & Roger Tayler, "The mystery of the cosmic helium abundance", in: *Nature*, vol. 203 (1964): 1108.

nuclear reactions that had already begun a little earlier, generating unstable nuclei, now produced stable nuclei of helium (and some isotopes of this and other light elements such as lithium), with a balance of 13% neutrons and 87% protons.[248] Manuel and Antonio Peimbert calculated the exact abundance of primordial helium (two protons and two neutrons) in the Universe to be 24.77% of the total elemental mass[249], leaving a hydrogen (one proton) proportion of 75.23%. The main result of these fusions is the existence of helium $^4He^2$ and the survival of neutrons; *"Once bound in this way [in stable nuclei], the strong interaction between the protons and neutrons stabilizes the neutrons preventing them from decaying."* [250]

However, *"if the Boltzmann factor were close to unity, there would be little hydrogen left over [in the Universe]".*[251] We would have been left with a universe with clouds of pure helium, which would have precluded the formation of solar systems with planets. This is because the fusion processes of hydrogen to helium are relatively slow, but those in which helium is transformed into carbon and oxygen are relatively fast. Helium stars would burn their fuel a hundred times more quickly than hydrogen stars with comparable mass, and would not last long enough *"to encourage the gradual evolution of biological life-forms in planetary systems"*[252] and, *"without hydrogen there would be no organic material and no water"*,[253] essential for life.

---

[248] Steven Weinberg, *The First Three Minutes of the Universe* (1977): 98.

[249] Manuel Peimbert, Valentina Luridiana & Antonio Peimbert, "Revised Primordial Helium Abundance Based on New Atomic Data", in: "*Astrophysical Journal*, vol. 666 (2007): 636–646.

[250] Jonathan Allday, *Quarks Leptons and the Big Bang* (2002): 263.

[251] Paul Davies, *The Accidental Universe* (1983): 64.

[252] John Barrow & Frank Tipler, *The Anthropic Cosmological Principle* (1986): 399, referred to by Guillermo Gonzalez & Jay Richards in *The Privileged Planet* (2004): 202.

[253] Paul Davies, *The Accidental Universe* (1983): 65, reproduced in John Barrow & Frank Tipler, *The Anthropic Cosmological Principle* (1986): 399.

If the difference between the proton mass and the neutron mass were less than the electron mass, the result would be a universe of pure neutrons. The difference in mass between a neutron and a proton is one thousandth of the mass of the proton. Now, if this difference in mass had been "*only a third of this value, then free neutrons would be unable to decay into protons, because they would not have enough mass to produce the required electron*",[254] so the beta decay of a neutron into a proton and an electron would not have been possible. According to Rees, "*if the neutron mass were only 0.998 of its actual value (that is if the u quark [the up quark] were slightly heavier than the d quark [the down quark]) then free protons would decay into neutrons by positron emission: $p \rightarrow n + e^+ + \nu_e$*"[255] or $p + e^- \rightarrow n + \bar{\nu}_e$.[256]

In both cases, the unstable nucleon would be the proton and the structures of the Universe would decay into structures of neutrons by the annihilation of protons and electrons: "*This would lead to a World in which stars and planets could not exist*", because "*without electrostatic forces to support them, solid bodies would collapse rapidly into neutron stars (if smaller than about $3M_\odot$ [=3 solar masses]) or black holes*".[257] The consequence of all this would have been, according to Barrow and Tipler, in their book on the anthropic cosmological principle, that "*no atoms would ever have formed and we would not be here to know it*".[258] The physical-mathematical details can be found in Math box 3.3.

---

[254] Paul Davies, *The Accidental Universe* (1983): 65.

[255] *Ibidem*: 65 and John Barrow & Frank Tipler, *The Anthropic Cosmological Principle* (1986): 400.

[256] John Barrow & Frank Tipler, *The Anthropic Cosmological Principle* (1986): 400.

[257] *Ibidem*: 400.

[258] *Ibidem*: 400.

Math box 3.3 The mass difference between the proton and neutron and its relationship with the electron mass

We saw (in Math box 3.2) that the mass difference between the proton and neutron Δm is a bit more than the electron mass m_e:[259]

(1) $\Delta m - m_e = 1.396 MeV - 0.511 MeV = 0.883 MeV \Longrightarrow \Delta m - m_e \geq 0 \Longrightarrow$

(2) $\Delta m \geq m_e$.

In the case of $\Delta m < m_e$, there would be no way to produce helium, and the result would be a universe of pure neutrons. The difference in mass Δm between a neutron m_n and a proton m_p is one thousandth of the mass of the proton.

(3) $\dfrac{\Delta m}{m_p} = \dfrac{m_n - m_p}{m_p} = \dfrac{939.6 - 938.3}{938.3} \cong 0.0014.$

However, if the difference in mass Δm were only one third of its empirical value, free neutrons could not decay into protons, because they would not have sufficient mass to produce the required electron, given that a difference of 0.465 *MeV* does not cover the 0.511 *MeV* required by the electron. So, the following decay would be impossible:

(4) $n \nrightarrow p + e^- + \bar{v}_e$.

and, instead of (29), there would be the reverse decay:

(5) $p + e^- \longrightarrow n + \bar{v}_e$.

If the mass of the neutron were even less, for example, 98.8% of its empirical value, the neutron would have less mass than the proton, preventing the production of electrons from happening, as in equation (6). The unstable nucleon would be the proton, with the following decay:

---

[259] Paul Davies, *The Accidental Universe* (1983): 63, reproduced in John Barrow & Frank Tipler, *The Anthropic Cosmological Principle* (1986): 400.

(6) $p \rightarrow n + e^+ + \nu_e.$

In both cases, that is (5) and (6), the structures of the Universe would decay into neutron structures through pe^- annihilation.[260] Neither protons, nor electrons would exist, and therefore, neither would atoms or life on Earth.

Section 3.3 The beginning of nuclear fusion in stars

Previously,[261] we analysed the ratio of the gravitational force to the original kinetic energy of the Big Bang. Owing to the fact that the gravitational force and the momentum of the original explosion of the *Big Bang* were finely tuned, clouds of hydrogen and helium were formed in the Universe that began to collapse to form stars. What will happen to these stars? How long will they live? Will there be fusion of heavier elements? This depends on the interaction of three forces, specifically, the gravitational force, the electromagnetic force and the strong nuclear force. There are two possibilities. The first possibility is that the stellar mass does not exceed a critical limit, so that the pressure and the temperature in the interior of the star are not sufficient to overcome the electromagnetic repulsion between protons, nor to start nuclear fusion, which is only triggered at very short distances between protons. The star ends its life as a brown dwarf or as a planet such as Jupiter.

The second possibility is that the stellar mass exceeds a critical limit. If the stellar mass exceeds this critical limit, gravity will do its job and supply enough pressure, density and temperature in the sphere to overcome the electromagnetic repulsion between protons. Once gravity has overcome the electromagnetic repulsion of the protons, there will be enough protons that collide with sufficient velocity to trigger the process of nuclear fusion.

---

[260] John Barrow & Frank Tipler, *The Anthropic Cosmological Principle* (1986): 400.
[261] In Section 1.3.1 of Chapter 1.

When there are many high velocity collisions of protons and neutrons, from time to time two nuclei of deuterium, which consists of one proton plus one neutron, are combined to form a helium nucleus, that has two protons and two neutrons. If the probability that two protons are combined – transforming one of the protons into a neutron in the process – is $p = 10^{-n}$ and if there are 10^n high velocity proton collisions per second per cubic centimetre, it follows that on average every second, in every cubic centimetre, a helium atom is produced.

What would happen if the gravitational force were not so weak? Let us suppose that it were a million times stronger and that the ratio of the electromagnetic force and the gravitational force were not 10^{36} but, for example, 10^{30}. In this case, objects would not have to be so big for the gravitational force to be able to overcome the electromagnetic repulsion between protons. The fusion processes would be triggered in much smaller stars with a lifetime a million times shorter. A typical star would run out of fuel in 10,000 years, before organic evolution had even started![262] If we assume that biological evolution takes time, as was the case on Earth, it follows that in the case of a slightly stronger gravitational force, there would be no complex life as we now know it.

Conversely, with a weaker gravitational force, complex structures of long duration would certainly be possible. But in the case of a gravitational force a million times weaker, the Sun would have to be almost a million times larger in volume to trigger the process of nuclear fusion that provides energy to planets such as the Earth, and this planet would have to be almost a million times larger in volume to hold on to its atmosphere. The larger volume would offset the lower gravitational force. However, there would be insurmountable obstacles to achieving such a volume. A star with a volume a million times greater than the Sun, would be the result of

---

[262] Martin Rees, *Just Six Numbers* (2000): 34.

the gravitational collapse of a number of hydrogen atoms a million times more than those of the Sun. But this implies that a gravitational force a million times weaker would have to occasion the collapse of a million times more atoms. The chance of so many hydrogen atoms being so close so as to produce their gravitational collapse is so small that this would be a nearly impossible feat and it would only occur in a few exceptional cases.

In conclusion, the beginning of nuclear fusion in stars depends on the fine tuning of the interaction of three fundamental forces, that is, the gravitational, the electromagnetic, and the strong nuclear force. Perhaps the reader might object that a fine tuning of the ratio of the gravitational and electromagnetic force with a precision of $\pm 10^6$ is not very fine. In fact, it is, because $\pm 10^6/10^{36} = \pm 1/10^{30}$. Therefore, to achieve this ratio, the gravitational force is fine tuned to a precision of one in a million million million million million, so that nuclear fusion can take place in stars.

Math box 3.4 The elephant, the beetle and the ant

In Math box 1.11, we saw that nuclear fusion can begin in a sphere of hydrogen gas, if it contains a minimum of 10^{54} hydrogen atoms. This number, 10^{54}, comes close to $N \approx 10^{60}$ that appears in the 'number coincidences' that Bondi reported in his book *Cosmology* of 1959. Rees comments that "*the 'coincidence' that Dirac and Bondi discussed does not in itself now cause puzzlement: there is really just one very large number in physics: it is the reciprocal of the 'gravitational fine structure constant' α_G^{-1}*".[263] The constant α_G is the squared ratio of the proton mass m_{pr} and the Planck mass m_{pl}: $\alpha_G = \left(m_{pr}/m_{pl}\right)^2$, and $m_{pr} \approx 10^{-20}m_{pl}$, so it follows that the value Rees refers to is:

---

[263] Martin Rees, *Just Six Numbers* (2000): 30–35 and "Numerical coincidences and 'tuning' in cosmology", arXiv:astro-ph/0401424v1

(1) $1/\alpha_G \approx \left(m_{pl}/m_p\right)^2 \approx (10^{20})^2 \cong 10^{40}$.

The ratio of electromagnetic and gravitational constants is about equal to the reciprocal of the gravitational fine structure constant $1/\alpha_G$:

(2) $\alpha_C/\alpha_G \cong 10^{36} \approx 1/\alpha_G \cong 10^{40}$.

Only through a sufficiently great mass can gravity produce enough kinetic energy to overcome the electromagnetic repulsion between protons. This looks like another Bondi coincidence: electromagnetism is 10^{36} times stronger than gravity and we need some $\sim\!10^{36}$ mass units (Planck masses), that is $\sim\!10^{56}$ protons, to start nuclear fusion. But this is like the story of the elephant (the start of nuclear fusion), the dung beetle (a Planck mass) and the leafcutter ant (a proton). An elephant can carry 9,000 kilos, i.e. 375 times the weight the dung beetle can, and the latter can lift up 24 kilos, 960 times the weight the ant can, so that 360,000 ants are needed to lift up the weight an elephant can.

Brandon Carter comments with good reason that this is not a case of the 'weak anthropic principle' but can be perfectly explained in terms of the masses and forces already fixed,[264] as was later suggested also by Rees.

Section 3.4 The fusion of carbon, oxygen and other elements in stars

Once the nuclear fusion that transforms hydrogen into helium has started, additional conditions are required for producing elements heavier than helium. This point brings us to the best-known topic of all the literature on fine tuning, i.e. the production of carbon and oxygen in stars. I shall first explain the nuclear fusion process that produces carbon and oxygen in the stars (Section 4.1) and then the fine tuning in this process (Section 4.2).

---

[264] Brandon Carter, "Large Number coincidences and the Anthropic Principle in Cosmology", in: Malcolm Longair, ed., *Confrontation of Cosmological Theories with Observational Data* (1974): 292.

Section 3.4.1 The fine tuning needed for producing carbon and oxygen

The importance of hydrogen, carbon and oxygen in the biochemistry of life is fairly well known.[265] These are the three most common elements in living organisms. Two molecules of utmost importance for life are derived from them: firstly, the carbon dioxide used by plants to produce oxygen, plants being necessary as the first link in the animal food chain, and secondly, water, which due to its special properties is of paramount significance for the evolution and continuity of life on Earth.[266] Four of the six carbon electrons may be shared with other elements; a carbon atom has a valence of 4 and consequently can bond to another four atoms, which gives it a unique significance as a cornerstone in the construction of organic molecules.[267] Helium does not make up organic molecules, but it is an essential link in the fusion chain, from hydrogen to carbon and oxygen in the stars.

Helium ($^4He^2$) is formed from hydrogen ($^1H^1$) through the mediation of deuterium ($^2H^1$). The first reaction, where two protons combine to form a deuterium nucleus, has a very low probability. A proton in the centre of the Sun takes on average 10^{10} years to collide with another proton and form deuterium. Two protons do not form a diproton ($^2H^2$), because the nuclear force is not strong enough to overcome the electromagnetic repulsion between two protons. Only if one proton, by collision with another, is transformed into a neutron, will the two form a deuterium nucleus – one proton plus one neutron. Then, two deuterium nuclei can combine to form a helium nucleus – two protons plus two neutrons – because the two neutrons in the helium nucleus provide two units of strong nuclear force, without adding anything to the electrical repulsion in the nucleus, which enables the joint nuclear force of four nucleons to overcome the electrical repulsion of two protons.

---

[265] John Barrow & Frank Tipler, *The Anthropic Cosmological Principle* (1986): 510–575.
[266] *Ibidem*: 524–541.
[267] See Section 7.3 of Chapter 7 of this book.

If the nuclear force were a bit stronger than it is, it would have been able to overcome the electromagnetic repulsion of two protons to form a diproton ($^2H^2$). In this case, in the Big Bang, all of the protons would have combined to form diprotons in the first few minutes of the Big Bang. With no hydrogen left to be transformed into deuterium, and then into helium, nor helium to be transformed into carbon and oxygen and other heavier elements, nor hydrogen and oxygen to produce water molecules, planets with heavy elements would not have emerged, neither would water or life.[268] We can mark the range of this fine tuning with precision. If the strong nuclear constant a_s had a value ranging between 3.4% and 3.7% more than its empirical one, the strong nuclear force would overcome the electromagnetic repulsion of two protons and combine them into a single nucleus, producing an event known as the 'diproton disaster':

"The existence of deuterium and the non-existence of the diproton therefore hinge precariously on the precise strength of the nuclear force. If the strong interaction were a little stronger, the diproton would be a stable bound state with catastrophic consequences: all the hydrogen in the Universe would have been burnt to $^2H^2$ during the early stages of the Big Bang and no hydrogen compounds or long-lived stable stars would exist today. If the diproton existed we would not."[269]

And vice-versa: if the strong nuclear force had a value between 9% and 11% less than its empirical value, a proton could not stick to a neutron,[270] and if this deuterium (= neutron plus proton) were missing, the main link in the chain that transforms hydrogen into helium in the stars would be

---

[268] Martin Rees, *Just Six Numbers* (1999): 55.

[269] John Barrow & Frank Tipler, *The Anthropic Cosmological Principle* (1986): 322.

[270] Martin Rees, *Just Six Numbers* (1999): 55; John Barrow & Frank Tipler, *The Anthropic Cosmological Principle* (1986): 322.

missing. Thus, we would have a universe made out of pure hydrogen with no chemistry or nuclear fusion.

In summary, if the strong nuclear force were a little bit stronger than it in fact is, we would have a universe made out of pure diprotons. And if it were a little weaker, we would have a universe of pure hydrogen. The fact that the strong nuclear force is located in a narrow region, between this upper and lower limit, is an example of *fine tuning,* as can be seen in Graph 3.3.[271] In addition, if $\alpha_s \leq 0.3\alpha^{1/2}$, the carbon nucleus would be unstable and there would be no life based on carbon and oxygen.[272]

Graph 3.3 The limits imposed on α_s and α by the fact we are here [273]

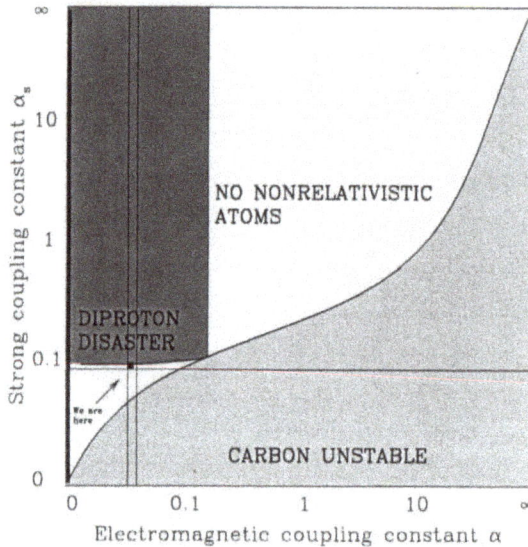

[271] Max Tegmark, "Is 'the Theory of Everything' merely the Ultimate Ensemble Theory", in: *Annals of Physics*, vol.270 (1998): 14–16; John Barrow, *The Anthropic Cosmological Principle* (1986): 320–322; John Barrow, *The constants of Nature* (2002): 168.

[272] Max Tegmark, "Is 'the Theory of Everything' merely the Ultimate Ensemble Theory", in: *Annals of Physics*, vol. 270 (1998): 15.

[273] *Ibidem*:16; John Barrow, *The Constants of Nature* (2002): 168.

In Section 1.3.3 and Math box 1.15 of Chapter 1, we saw that a remarkable feat of fine tuning is also required for the production of carbon and oxygen in the stars. Carbon and oxygen, two elements essential for life, are produced in the core of stars by nuclear fusion processes. Carbon is 'lucky' twice over. Firstly, because there is an excited state of carbon that has a resonance in the fusion of beryllium and helium, with the result that it exponentially accelerates the fusion process. Due to the speed of the fusion, carbon can be generated before the end of beryllium's extremely short lifetime.

The second time that carbon is 'lucky', is due to a lack of resonance in the fusion of carbon and helium, which sufficiently slows down the rate of nuclear fusion, so that not all the carbon is transformed into oxygen. It is only due to the fine tuning of the strong nuclear force, which has a decisive impact on the value of the reaction rate in nuclear fusion processes, that both carbon and oxygen formed in abundant quantities, subsequently enabling the evolution of life. In *The Road to Reality*, Penrose comments that this circumstance is a clear case of fine tuning: "*It is remarkable that the constants of Nature are so adjusted that such an energy level should be in just the right place, so life, as we know it, could come about*".[274]

When Hoyle discovered all this, no one was yet able to calculate this fine tuning that enables the production of carbon and oxygen. But in 1998, Oberhummer, Pichler and Scótó produced a brief study of nuclear astrophysics, demonstrating the extreme dependence of the value of the nuclear fusion rate on the nuclear force factor (see Table 3.2).

Obviously, the increase in the nuclear force factor p in Table 3.2, does not adversely affect the amount of carbon produced, but it would affect, as we have already seen, the production of oxygen. The nuclear force factor must be very finely tuned for both carbon and oxygen to exist.

---

[274] Roger Penrose, *The Road to Reality* (2005): 759.

Table 3.2 The triple alpha nuclear reaction rate at a temperature of 10^8 Kelvin as a function of the nuclear force factor p [275]

Effective nucleon-nucleon interaction	MN Model	V1 Model	V2 Model
Force factor p	f(p) [276]	f(p)	f(p)
1.002	422	337	64.4
1.001	20.2	11.4	7.9
1.000	1.0	1.0	1.0
0.999	0.05	0.09	0.13
0.998	0.003	0.008	0.02

In the year 2000, the same three nuclear physicists published their conclusions on the subject in more detail, in the journal *Science*.[277] The nuclear fusion processes depend on two fundamental forces in nature, firstly, the strong nuclear force that brings together the baryons in the nuclei, when they collide under high temperature and high density, overcoming the electromagnetic repulsion that the protons experience due to their positive charge, and, secondly, the electromagnetic force which has to

---

[275] Heinz Oberhummer, Rudolf Pichler & Attila Csótó, "The Triple-Alpha Process and Its Anthropic Significance", in: Nikos Prantzos, Elisabeth Vangioni-Flam & Michel Cassé, eds, *Nuclei in the Cosmos V. Proceedings of the International Symposium on Nuclear Astrophysics* (1998): 119–122.

[276] The increase or reduction of the triple-alpha reaction is given by $f(p) = \frac{r_{3\alpha}(p)}{r_{3\alpha}} \approx$ $\exp\left(\frac{\varepsilon - \varepsilon(p)}{k_B T}\right)$; ε is the energy of the resonance and $r_{3\alpha} = 3^{3/2} N_\alpha^3 \left(\frac{2\pi \hbar^2}{M_\alpha k_B T}\right)^3 \frac{\Gamma_\gamma}{\hbar} \exp\left(-\frac{\varepsilon}{k_B T}\right)$ where M_α is the mass density of α, N_α its number density, T the temperature of the plasma of the star, and $\omega\gamma \cong \Gamma_\gamma$ is the strength of the resonance.

[277] Heinz Oberhummer, Attila Csótó & Helmut Schlattl, "Stellar Production Rates of carbon and its abundance in the Universe", in: *Science*, vol. 289 (2000): 88–90.

be overcome. They made their calculations for stars with masses 20 times, 5 times and 1.3 times that of the Sun. Their conclusion is worth quoting:

"We conclude that a change of more than 0.5% in the strength of the strong interaction or more than 4% change in the strength of the Coulomb [electromagnetic] force would destroy either nearly all carbon or all oxygen in every star. This implies that irrespective of stellar evolution the contribution of each star to the abundance of carbon or oxygen in the interstellar material would be negligible. Therefore, for the above cases the creation of carbon-based life in our universe would be strongly disfavoured (...) Therefore, the results of this work are relevant [to] the anthropic cosmological principle."[278]

Graph 3.4 The existence of carbon and oxygen depends on the fine tuning of the strong nuclear force and the electromagnetic force[279]

STRONG NUCLEAR FORCE VARIATION

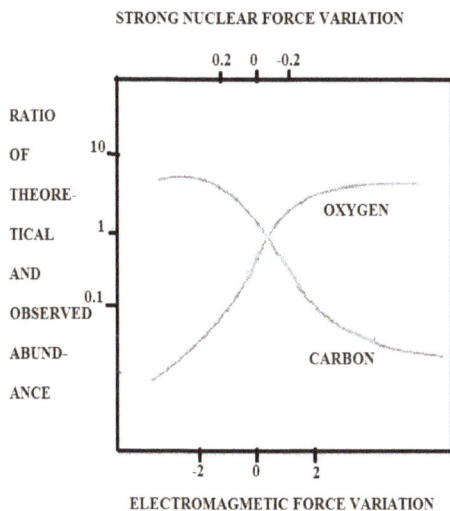

RATIO OF THEORE- TICAL AND OBSERVED ABUND- ANCE

OXYGEN

CARBON

ELECTROMAGMETIC FORCE VARIATION

---

[278] Heinz Oberhummer, Attila Csótó & Helmut Schlattl, *ibidem*: 90.
[279] Figure from Heinz Oberhummer Attila Csoto & Helmut Schlattl, in: *Science*, vol. 289 (2000), adapted by John Barrow, *The Constants of Nature* (2002):155 and by me.

The anthropic principle, referred to by authors like Barrow, Davies, Oberhummer and others, exists in two forms. In its 'weak' form it is no more than a statement of the principle of causality, that is, that natural laws (the cause) must be such as to have enabled the emergence of life (the effect) about 14 billion years after the Big Bang, since we are here analysing them. In its 'strong' form, the anthropic principle is a metaphysical statement and asserts that the natural laws of the Universe are finely tuned by some cause so as to make possible the emergence of human life.[280] I will analyse these questions in more detail in another chapter of this book.[281]

At a conference in New York in 1999, Steven Weinberg expressed doubts about the fine tuning of the Universe.[282] His argument against fine tuning has two steps: he first *reduces* the phenomenon of fine tuning to the particular case of the fine tuning present in the production of an excited state of the carbon nucleus and then, he *denies* the fine tuning in this triple-alpha chain. Let us look at both of these steps. Regarding the first step, fine tuning is much more extensive than Weinberg admits when he reduces it to the case of production of carbon in stars. In this chapter, I analyse no less than seven different cases of fine tuning. With regard to the second step, i.e. his denial of fine tuning in the production of an excited state of carbon, Weinberg refers to two studies, those of Hong, 1999, and Livio, 1989,[283] who argue that carbon has the required energy level to be 'naturally' created from the encounter of helium and beryllium 8.

---

[280] Both principles are discussed by Davies in chapter 5 of *The Accidental Universe* (1984) and by Barrow and Tipler in *The Anthropic Cosmological Principle* (1987).

[281] See Chapter 6 of this book.

[282] Steven Weinberg, "A Universe with no Designer", in: *Annals of the New York Academy of Sciences*, vol. 950; and in: Miller, James, ed., *Cosmic Questions* (2001): 169–174.

[283] Suk-ho Hong & Suk-yoon Lee, "Alpha Chain structure in ^{12}C", in: *Journal of Korean Physics* (1999): 46–48. Mario Livio, D. Hollowell & J. Truram "The anthropic significance of the existence of an excited state of ^{12}C", in: *Nature* (1989): 281–284.

Although they were published before Weinberg's talk at the 1999 conference, he makes no reference to two other articles by Pichler, Oberhummer and others, 1997,[284] and Fedorov and Jensen, 1996,[285] who criticize the studies of Hong and Livio. These authors refute the statements of Hong and Livio and prove that carbon is generated, effectively, by the resonance in the triple-alpha process that was first discovered by Hoyle. Oberhummer notes that the values for the rest mass energies of beryllium 8 and carbon 12 used by Hong and Lee are very different from their empirical values, which makes their model highly questionable.[286] Weinberg also made no reference to the first article by Oberhummer, Pichler and Scótó, published one year before his 1999 conference.

In his 2009 book on cosmology,[287] there is not a single reference to the phenomenon of fine tuning, nor to authors such as Fedorov, Jensen, Oberhummer, Pichler, or Scótó, who do prove the existence of fine tuning. This is a blatant case of cherry picking in favour of his hypothesis.

Section 3.5. The existence of planets and stable planetary orbits

Before looking at the conditions for the stability of planetary orbits, it is first necessary to look at the conditions for the very existence of planets. According to a classic study by Brandon Carter, the division of the stars in the main sequence from blue giants to red dwarfs depends critically on the interrelationship between the gravitational coupling constant, the

---

[284] Rudolf Pichler, Heinz Oberhummer, Attila Csótó & S. Moszkowski, "Three alpha structures in ^{12}C", in: *Nuclear Physics A*, vol. 618 (1997): 55–64.

[285] D. Fedorov & A. Jensen, "The Three-Body Continuum Coulomb Problem and the 3α Structure of ^{12}C", in: *Physics Letters B*, vol. 389 (1996): 631–636.

[286] Guillermo Gonzalez & Jay Richards, *The Privileged Planet* (2004): 392, note 8. Heinz Oberhummer *et al.*, "The Triple-Alpha Process and Its Anthropic Significance", in: Nikos Prantzos & Sotoris Harissopoulis, eds, *Nuclei in the Cosmos V. Proceedings of the International Symposium on Nuclear Astrophysics* (1998).

[287] Steven Weinberg, *Cosmology* (2009).

electromagnetic coupling constant and the proton and electron mass ratio. Carter states that "*this condition – by a remarkable coincidence – is only just satisfied*" and if the gravitational coupling constant were a bit stronger, or the electromagnetic coupling constant a little weaker, "*then the main sequence would consist entirely of radiative blue giants*", which "*would be incompatible with the formation of planets and as such, of observers*".[288] Solar systems with planets are to the upper left of the Sun on the main sequence, approaching blue giants, in the Hertzsprung-Russell diagram, named after Ejnar Hertzsprung (1873–1967) and Henry Russell (1877–1957), who classified the stars according to absolute magnitude, luminosity and colour.[289] So, minor variations in the gravitational and electromagnetic coupling constants would result in the existence of very few solar systems with planets, adversely affecting the probability of the emergence of complex life in the Universe (see Math box 3.5).

Math box 3.5 Fine tuning on the main sequence

According to a classic study by Brandon Carter, the division of the stars on the main sequence from blue giants to red dwarfs depends critically on the interrelationship of the gravitational coupling constant α_G, the electromagnetic coupling constant α and the mass ratio β:

(1) $\sqrt{\alpha_G} \geq \alpha^6 \beta^2$.

Inserting the values of these constants into this equation, we obtain:

(2) $7.75 * 10^{-20} \geq 4.48 * 10^{-20}$,

---

[288] Brandon Carter, "Large Number Coincidences and the Anthropic Principle in Cosmology", in: Malcolm Longair, ed., *Confrontation of Cosmological Theories with Observational Data* (1974): 296–297. Carter uses the symbols $m_p^2 \equiv (m_p/m_{pl})^2$, e^2 and m_e/m_p to indicate α_G, α and β, respectively. With the symbols used by Carter, the equation looks like this: $m_p \geq e^{12}(m_e/m_p)^2$.

[289] The reader can find and might want to consult the *Hertzsprung-Russell diagram* on the internet (Wikipedia) or in one of the many textbooks on astronomy that reproduce it.

so that this condition – by a remarkable coincidence – is just satisfied. If the gravitational coupling constant were stronger, or the electromagnetic coupling constant a little weaker, we would have:

(3) $\sqrt{\alpha_G} \gg \alpha^6 \beta^2$.

so that the main sequence would be entirely made up of radiative blue giants, which is incompatible with the emergence of observers like us.

We will now have a look at the question of the stability of planetary orbits. In the *Collected Scientific Papers* of Paul Ehrenfest (1880–1933), an Austrian-Dutch theoretical physicist, the editor, Martin Klein, included a translation of his 1920 article *Which roles does the three-dimensionality of space play in the basic laws of physics?* Ehrenfest demonstrates that stable planetary orbits can only exist in a universe with three spatial dimensions. I summarised this article, together with a summary of Newton's analysis of planetary orbits as determined by his gravitational laws, in another book:[290]

"There is a characteristic difference between two or three [spatial] dimensions, on the one hand, and a number of dimensions greater than three, on the other hand, with regard to the stability of the circular orbit. While in R_3 the orbit remains finite when a small perturbation of not too great an energy occurs... in the case of R_4, R_5, R_6, etc., the circular orbits, although they are of course still possible, are disrupted at the slightest perturbation, steering the planet into a downwards spiral towards the central body or outwards into infinity."[291]

---

[290] Paul Ehrenfest, "Which roles does the three–dimensionality of space play in the basic laws of physics?", in: *Collected Scientific Papers* (1959). See also John Auping, "La geometría analítica de la elipse"; "Las leyes de Kepler y Newton en un solo sistema axiomático"; and "Las órbitas planetarias estables según Ehrenfest", in: *Origen y Evolución del Universo*, e-book (2016): 531-565.
[291] Paul Ehrenfest, *ibidem*: 440–441.

Ehrenfest proves that in a universe of four or more spatial dimensions, stable, closed orbits cannot exist, but are possible in a universe of three spatial dimensions (see Table 3.3). He derives the following equation of radial velocity for the theoretical set of universes with n spatial dimensions: $\frac{dr}{dt} = \pm\frac{1}{r}\sqrt{Ar^2 + Br^{4-n} - C^2}$, where $A = \frac{2E}{m}$; $B = \frac{2GMm}{(n-2)}$; and $C = \frac{L_k^2}{m^2}$. The total energy is $E = \frac{G^2M^2m^2(e^2-1)}{2L_k^2}$. Only in three-dimensional space, elliptical orbits, as observed by Kepler, are possible, since in such orbits the eccentricity has values of $0 < e < 1$, so that total energy E is always negative, yielding two values of r where $\frac{dr}{dt}$ is zero, i.e. at the maximum and minimum distance of the planet's elliptical orbit from the star, where $Ar^2 + Br^{4-n} = C^2$; and giving positive ($\frac{dr}{dt} > 0$) or negative values ($\frac{dr}{dt} < 0$) of the radial velocity between these zero values, such that $Ar^2 + Br^{4-n} - C^2 > 0$, so that $\pm\frac{1}{r}\sqrt{Ar^2 + Br^{4-n} - C^2}$ is a real number.

Table 3.3 Summary of some of Ehrenfest's conclusions

Number of space dimensions	There are two positive values of r where radial velocity is zero, and between these points it is positive or negative	Closed (elliptic) and stable orbits are possible	Falling toward the star or moving toward infinity is possible
$n = 3$	Yes	Yes	Yes
$n \geq 4$	No	No	Yes

Figure 3.2, by Tegmark, illustrates the impossibility of stable planetary orbits in a universe of four or more spatial dimensions. The grey part contains the trajectories of objects gravitationally attracted by another object and shows there are only two possibilities: the lower mass body either falls into the body of larger mass or is launched into infinity. No elliptical orbits!

Figure 3.2 Two body problem in universe of $n \geq 4$ spatial dimensions

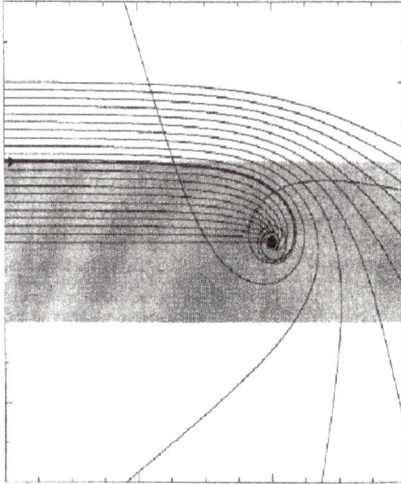

Tegmark extended Ehrenfest's analysis, by varying the number of both spatial and temporal dimensions from 0 to 5, which gives $6 * 6 = 36$ combinations and draws the conclusion that *homo sapiens* is only viable in a universe with only one time dimension and three spatial dimensions.[292]

Section 3.6 The existence of stable atoms and complex molecules

In Section 3.2, we saw the required conditions for the existence of stable protons and electrons, which are necessary for the existence of atoms. I will now discuss the other conditions required for atoms, which generally consist of protons, neutrons, and electrons, to be stable: the ratios of the neutron, proton and electron masses, also crucial for stable protons, and the ratios of the strong nuclear force, the weak nuclear force and the electromagnetic force; as well as the number of spatial dimensions in the Universe.

---

[292] Max Tegmark, "Is the 'Theory of Everything' merely the ultimate Ensemble Theory," in: *Annals of Physics*, vol. 270 (1998): 16–21. The figure is from *ibidem*: 17.

We can extend this analysis of universes with $N \geq 4$ spatial dimensions, from the macro-world of solar systems to the micro-world of atoms. Recent studies by Tangherlini[293] and by Gurevich & Mostepanenko[294] solve the Schrödinger equations for the hydrogen atom (one proton and one electron), for the general case of $n \geq 4$ spatial dimensions, and conclude that in such cases not only stable macro-structures such as planets orbiting stars are impossible, but that stable micro-orbits of electrons around a nucleus are not possible either. In the words of Gurevich and Mostepanenko:

"In spaces with $n \geq 4$ [spatial dimensions] electrons must fall on the nuclei and therefore the atomic structure of matter does not exist... The atomic matter and therefore life are possible only in 3-dimensional space."[295]

In the words of Barrow and Tipler, *"stable atoms, chemistry and life can only exist in $N < 4$ spatial dimensions"* and, *"there are no stable bound orbits for $N > 3$".*[296]

Not only is the number of spatial dimensions important for the existence of stable atoms, but also the fine tuning of the values of certain constants, as we shall now see. For the existence of complex atoms and molecules, the gravitational force is no longer important, as it is so weak that on small scales it almost does not count. However, on microscopic scales there are three dimensionless numbers[297] that are very important:

---

[293] Frank Tangherlini, "Atoms in Higher Dimensions", in: *Nuovo Cimento,* vol.27 (1963): 636.

[294] L. Gurevich & V. Mostepanenko, "On the existence of atoms in n-dimensional space", in: *Physics Letters A*, vol. 35 (1971): 201–202.

[295] L. Gurevich & V. Mostepanenko, "On the existence of atoms in n-dimensional space", in: *Physics Letters A*, vol. 35 (1971): 202.

[296] John Barrow & Frank Tipler, *The Anthropic Cosmological Principle* (1986): 265.

[297] The various units of the different variables of the equations cancel each other out.

1) The fine structure constant is $\alpha_C = \dfrac{e^2}{(2\,\varepsilon_0 hc)} = 1/137$, where $e = 1.6022 * 10^{-19}C$ is the electron charge; $\varepsilon_0 = 8.854 * 10^{-12}N^{-1}m^{-2}C^2$ is the vacuum permittivity; $C \cong 3 * 10^9 esu$ is the electrical charge; $h = 6.6260755 * 10^{-34}Js$ is the Planck constant; $J = Nm = kgm^2s^{-2}$ is the unity of work or energy; and $c \cong 3 * 10^8 ms^{-1}$ is the velocity of light. This constant α_C refers to the strength of the electromagnetic interaction between elementary charged particles.

2) The strong nuclear force constant is $\alpha_S = \dfrac{g_S^2}{\hbar c} = 0.1182\pm0.0027$, where $\hbar = \dfrac{h}{2\pi}$ is a version of the Planck constant; $g_S = 6.1152 * 10^{-14}(Jm)^{1/2}$ is the strong nuclear force, operating at a range of $10^{-15}m$, which is the diameter of an average atom's nucleus.

3) The ratio of the electron mass to the proton mass is $\beta = 1/1836.$[298] The masses of these particles are: electron mass, $m_e = 0.511MeV/c^2$; neutron mass, $m_n = 939.6MeV/c^2$; and proton mass, $m_p = 938.3MeV/c^2$.

The small value of the dimensionless constant β ensures that the nucleus has a stable location, as Tegmark explains: "*In a stable ordered structure, for example a chromosome, the typical fluctuation in the location of a nucleus relative to the inter-atomic spacing is $\beta^{1/4}$, so for such a structure to remain stable over long time scales, one must have $\beta^{1/4} << 1$.*"[299]

I will now analyse the importance of the ratio of the strong nuclear and electromagnetic forces for the stability of the atom. In the nuclear fusion of the triple-alpha process, where carbon is produced from three helium

---

[298] $\beta = m_e/m_p = 1/1836$.

[299] Max Tegmark, "Is the 'Theory of Everything' merely the ultimate Ensemble Theory", in: *Annals of Physics*, vol. 270 (1998): 15. See also John Barrow & Frank Tipler, *The Anthropic Cosmological Principle* (1986): 304.

atoms, and in the fusion of one helium atom and one carbon atom to generate oxygen, the nuclei in the fusion process have greater rest mass energy than the nucleus resulting from the fusion, and part of the energy from the original rest mass is released in the form of radiation energy.[300] In general, from hydrogen to iron, the nuclear fusion releases rest mass energy and, from iron to uranium, the nuclear fission releases it. From hydrogen to iron, the nucleus resulting from the fusion has a lower rest mass energy than the sum of the rest mass energies of the nuclei fused together: the difference in the rest mass energies is released as radiation energy. From uranium to iron, it is the reverse: the resulting nuclei from the fission have a rest mass energy inferior to the rest mass energy of the split nucleus. The surplus rest mass is released as radiation energy.

Now, the stability of nuclei already generated by nuclear fusion is a function of the exact value of the strong nuclear force and the electromagnetic force. The strong nuclear force causes protons and neutrons to stick together in a nucleus, forming a sphere of nucleons. But this force only operates at very short distances, such that in large nuclei, the nucleons at opposite ends no longer attract each other. This causes the electrical repulsive force between protons to start to deform the nucleus, transforming the perfect sphere into an oval shape with eccentricity e. The relocation of nucleons in the nucleus is possible because the nucleus is not a solid mass point, but is more like a flexible drop of water.[301] At a certain point, the electrical repulsive force overcomes the strong nuclear force, and the nucleus disintegrates into two parts. Given this interaction of opposite forces in the nucleus, this instability in large nuclei starts when (number of

---

[300] According to the famous formula $E = mc^2$.

[301] The comparison with the water drop is by Niels Bohr, "Neutron Capture and Constitution", in: *Nature*, vol. 137 (1936): 344–348; Hans Bethe and R. Bacher, "Nuclear Physics A. Stationary States of Nuclei", in: *Reviews of Modern Physics* vol. 8 (1936): 82–229; and Carl Friedrich von Weizsäcker, *Die Atomkerne, Grundlagen und Anwendungen ihrer Theorie* (1937).

protons)^2 / neutrons is greater than 49.   For example, uranium has the atomic number 92 and an atomic weight of 238, i.e. it contains 92 protons and 146 neutrons, so that $(92^2/146) = 58$ and $58 > 49$, so uranium is unstable. The details of this point are explained in Math box 3.6.

Math box 3.6 The stability of a nucleus

In what follows, E_N is the rest mass energy of the nucleus; ΔE_N is the mass quantity converted into radiation energy; Z is the number of protons and A the number of neutrons:

(1) helium fusion $\rightarrow$ carbon: $\Delta E_N = 3E_N(Z/3, A/3) - E_N(Z, A) > 0$.

(2) uranium fission: $\Delta E_N = E_N(Z, A) - 2E_N(Z/2, A/2) > 0$.

The boundary between stable and unstable nuclei is calculated as follows. In the following equation, there are two dimensionless constants:

(3) $a_s = 17.313$ and $a_c = 0.702$.[302]

(4) $\Delta E_N = \frac{e^2}{5}\left(a_c Z^2 A^{-1/3} - 2a_s A^{2/3}\right)$.

In large nuclei, the boundary between stability and instability occurs when $E_N = 0$ and if this boundary is exceeded, the nucleus disintegrates:

(5) $\Delta E_N = \frac{e^2}{5}\left(a_c Z^2 A^{-1/3} - 2a_s A^{2/3}\right) = 0 \Longrightarrow \frac{Z^2}{A} = \frac{2a_s}{a_c} \cong 49$,

so that the nucleus is unstable when

(6) $\frac{Z^2}{A} > 49$.

If we want to generalise this formula to other conceivable universes with different values of a_s and a, we would have an unstable nucleus if:

---

[302] John Barrow & Frank Tipler, *The Anthropic Cosmological Principle* (1986): 325.

(7) $\quad \dfrac{Z^2}{A} > 49 \left(\dfrac{\alpha_s}{10^{-1}}\right) \left(\dfrac{1/137}{\alpha}\right).$[303]

Given that in this Universe $\alpha_s \cong 10^{-1}$ and $\alpha \cong 1/137$, equation (7) becomes (6).

A small variation in the value of the electromagnetic and strong nuclear forces would cause unstable nuclei with atomic numbers much lower than uranium in the periodic table. If the strong nuclear force were reduced by 50%, iron would be unstable (see graph 3.5). In general, if the electromagnetic force were a bit stronger, or the strong nuclear force a bit weaker, or both, nuclei crucial to biology such as carbon and oxygen could not exist (Z is the atomic number):

"Thus, if the electromagnetic interaction were stronger (increased α_C) or the strong interaction a little weaker (decreased α_S), or both, then biologically essential nuclei like carbon would not exist in Nature. For example, if the electron charge were increased by a factor ~3 no nuclei with Z > 5 would exist and no living organisms would be possible. The existence of carbon-based organisms hinges upon a 'coincidence' regarding the relative strengths of the strong and electric forces."[304]

Graph 3.5 illustrates how three phenomena, i.e. the electromagnetic repulsion between protons, the strong nuclear force that glue neutrons and protons together in the nucleus, and the size of the nucleus, determine the stability or instability of the nucleus. The curves are the collection of combinations of atomic size – measured in the vertical axis – and the theoretical value of the strong nuclear force constant, as a function of its empirical value – measured in the horizontal axis – leaving the

---

[303] John Barrow & Frank Tipler, *The Anthropic Cosmological Principle* (1986): 326.
[304] *Ibidem*: 326.

electromagnetic force constant without variation. The upper curve represents a weaker electromagnetic force, the lower curve a stronger one.

Graph 3.5 The stability of atomic nuclei depends on the values of the strong nuclear force, and the fine-structure constant[305]

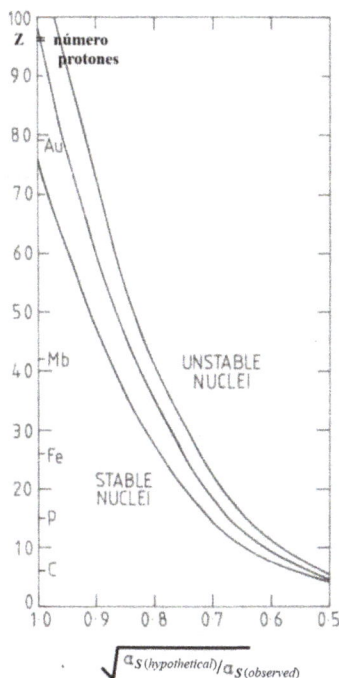

Section 3.7 Summary: calculating the fine tuning of the Universe

There are other limits imposed on the variation of the constants by the fact that we are here, for example, various chemical properties that enable high fidelity DNA replication.[306] In Section 3.2, we also saw that if the weak nuclear force were weaker, all the hydrogen would have been converted into

---

[305] Graph by Paul Davies, "The variation of the coupling constants", in: *Journal of Physics*, vol. 5, 1972, p. 1300, reproduced in John Barrow & Frank Tipler, *The Anthropic Cosmological Principle* (2002): 326.

[306] John Barrow & Frank Tipler, *The Anthropic Cosmological Principle* (1986): chapter 4.

helium in the first fifteen minutes of the Big Bang, and without hydrogen, we would also not have molecules based on hydrogen, carbon and oxygen, which are the elements required for life as we know it.[307]

By way of a mathematical summary, I take the analysis done by Smolin as a starting point, and then correct and expand on it. Smolin makes an overall estimate of the degree of specialness of our Universe, with, among other things, stars that live for billions of years, with nuclear fusion to emerge, a necessary prior condition for the production of heavy elements, which, in turn, are a necessary prior condition for life to emerge on Earth.[308] The analysis is carried out in Math box 3.7.

Math box 3.7 The accumulated *fine tuning* of the Universe

Smolin states that for the fine tuning of the Universe, the value of three basic constants must be set, specifically *the Planck constant h*, the *speed of light c* and *the gravitational constant G*, that between the three of them define the Planck mass, and, additionally, set the mass of four relatively stable particles, namely, the proton, neutron, electron and neutrino, as multiples of the Planck mass:

(1) Planck constant $h = 6.6260755 * 10^{-34} \, Js.$

(2) Gravitational constant $G = 6.673 * 10^{-11} kg^{-1} m^3 s^{-2}.$

(3) Speed of light $c = 299{,}792{,}458 \, ms^{-1} \cong 3 * 10^8 m/s.$

(4) Planck mass $1 \, m_{planck} = 2.1 * 10^{-8} \, kg \equiv hcG^{-1/2}.$

(5) Neutrino mass $m_\nu \approx 5 * 10^{-35} \, kg = 2.38 * 10^{-27} \, m_{pl}.$

(6) Proton mass $m_p = 1.67262 \; 10^{-27} \, kg = 7.965 * 10^{-20} \, m_{pl}.$

---

[307] Max Tegmark, "Is 'the theory of everything' merely the ultimate ensemble theory", in: *Annals of Physics*, vol. 270 (1998): 16.

[308] Lee Smolin, *The Life of the Cosmos* (1997): 325–326.

(7) Neutron mass $m_n = 1.6749 * 10^{-27} \, kg = 7.976 * 10^{-20} \, m_{pl}$.

(8) Electron mass $m_e \cong 9.11 * 10^{-31} kg = 4.338 * 10^{-23} m_{pl}$.

For stars that live for billions of years to emerge in a universe, the ratios of the neutrino, proton, neutron and electron masses must have the values that have been empirically confirmed. The probability that the observed ratio of the proton to the Planck mass exists by chance would be $p = 1/10^{19}$. To form stable nuclei, the mass of the electron must be a minimum fraction of the mass of the proton, for example, 1/1836; the mass of the neutrino must be almost zero; and the mass of the neutron must be slightly greater than that of the proton, for example, 1.00138. Each of these last three things has a probability of approximately $p = 1/10^{22}$. The cumulative probability of these four ratios is $p = 1/10^{85}$. Smolin also introduces a cosmological constant that must be adjusted with a precision of $p = 1/10^{60}$. Accumulating these probabilities, we obtain $p = 1/10^{145}$.

Up to now, we have only considered the mass and the gravitational constant. In order to have stable atoms and the successful fusion of heavy elements in stars, the *relative* values of the four basic forces must also be set, specifically the gravitational force, the strong nuclear force and its carriers, namely the gluon and π meson; the weak nuclear force and its carriers, namely bosons W^+, W^- and Z_0; and the electromagnetic force and its carrier, the photon. Let us first look at the values of the dimensionless constants of the main forces, with two versions of the weak nuclear force constant:

(9) Gravitational force[309] $\alpha_G \approx (m_p/m_{pl})^2 \approx 6.34392 * 10^{-39}$, infinite range;

---

[309] The (speculative) carrier of the gravitational force is the graviton, with mass=0 and spin=2

(10) Electromagnetic force[310] $\alpha_C = e^2/(2\varepsilon_0 hc) = 1/137.036$,[311] infinite range;

(11) Strong nuclear force[312] $\alpha_S(M_{Z^0}) \approx \frac{g_S^2}{\hbar c} \cong 0.1182$,[313] range 10^{-15}m;[314]

(12) Weak nuclear force[315] $\alpha_W = g_W m_e^2 c/\hbar^3 \cong 6.4692 * 10^{-12}$, range 10^{-18}m;[316]

(13) Weak nuclear force[317] $\alpha_W = g_W m_p^2 c/\hbar^3 \cong 2.1816 * 10^{-5}$, range 10^{-18}m.

The ratios of the dimensionless constants are the following:

(14) $\alpha_S/\alpha_c = 16.1934$.

(15) $\alpha_S/\alpha_W = 5.418 * 10^3$.

---

[310] The carrier of the electromagnetic force is the photon γ with mass=0 and spin=1.

[311] Jean-Philippe Uzan, "The fundamental constants and their variation: observational and theoretical status", in: *Reviews of Modern Physics*, vol. 75 (2003): 405; John Barrow, in: *The Anthropic Cosmological Principle* (1986): 293 and in: *The Constants of Nature* (2002): 46, 86, where stating that $\alpha = \frac{e^2}{\hbar c} = \frac{2\pi e^2}{hc}$, is incorrect.

[312] The carrier of the strong nuclear force, between quarks is the gluon; between nucleons, the π meson.

[313] Siegfried Bethke, "α_s at Zinnowitz 2004", hep-ex/0407021v1; the average of the measurements is $\alpha_s \approx 0.12$. Michael Schmelling, "Status of the Strong Coupling Constant", hep-ex/9701002v1, 1996, estimated $a_S(M_z) = 0.118 \pm 0.003$. At higher energies its value increases, to $\alpha_S(Q^2 = 100(GeV/c)^2) \approx 0.16$, according to Bogdan Povh *et al.*, *Particles and Nuclei* (2002): 109.

[314] Diameter of a medium sized atomic nucleus.

[315] Definition by Bernard Carr & Martin Rees, "The Anthropic Principle and the Structure of the Physical World", in: *Nature,* vol. 278 (April 1979): 611. The carriers of the weak nuclear force are the bosons W^+, W^-, Z_0 with $m > 80GeV$ and spin=1.

[316] 0.1% of a proton diameter.

[317] Definition by Paul Davies, *The Accidental Universe* (1983): 21. The difference with Carr & Rees is explained by the difference in mass of the electron and the proton: $10^{-11} * (1/\beta)^2 = 3.37 * 10^{-5} \approx 10^{-5}$.

(16) $\alpha_S/\alpha_G = 1.8632 * 10^{37}$.

(17) $\alpha_C/\alpha_G = 1.1506 * 10^{36}$.

(18) $\alpha_C/\alpha_W = 3.345 * 10^2$.

(19) $\alpha_W/\alpha_G = 3.439 * 10^{33}$.

In addition, the respective ranges of these forces must be set. Four basic forces yield three ratios relative to each other. Of these three, Smolin only sets two, specifically, the ratios of the weak nuclear force and the electromagnetic force to the strong nuclear force, which he estimates at $p = 1/10^2$ each: "*Taking the strong nuclear interaction as the measure, the weak and electromagnetic interaction are each about one part in 100*".[318] The cumulative probability is now:

(20) $p = 10^{-145} * 10^{-2} * 10^{-2} = 10^{-149}$.

Now, we need to set the ranges in which these forces operate. In the case of the electromagnetic and gravitational forces, it is the radius of the entire Universe. In the case of the weak nuclear force and the strong nuclear force, it is the radius of a nucleus. The ratio of the radius of the Universe to the radius of the nucleus is about 10^{40}, so the cumulative precision of the fine tuning should be, according to Smolin:

(21) $p = 10^{-149} * 10^{-40} * 10^{-40} = 10^{-229}$.

Smolin concludes: "*We reach the conclusion that the probability for the world to have turned out as ours, with stars lasting billions of years, and thus with nuclear and atomic physics more or less like ours – were the parameters of the standard model picked randomly – is at most one part in* 10^{229}."[319]

---

[318] Lee Smolin, *The Life of the Cosmos* (1998): 325
[319] *Ibidem*: 325.

This estimate of Smolin can be somewhat improved, by means of some clarifications. The probability for a universe to have all the characteristics that, as a matter of fact, our universe displays is $p \approx 3.75 * 10^{-267}$, as is shown in Math box 3.8:

Math box 3.8 Refining Smolin's estimates

We can refine Smolin's estimates somewhat:

1) Firstly, the cosmological constant is a myth rendered useless by general relativity, as I argued in Section 2 of Chapter 2, and is therefore not part of the fine tuning of physical constants. This reduces the accumulated precision of fine tuning to $p = 1/10^{169}$.

2) On the other hand, the ratio of the strong nuclear force to the weak one is not $1/10^2$, but, approximately $1/10^6$, which increases the precision of the fine tuning to $p = 10^{-173}$.

3) If the ratio of the radius of the Universe and a medium sized nucleus is 10^{40}, as Smolin estimates, the ratio of the radius of the Universe to the range of the weak nuclear force is a bit higher, that is 10^{43}, because the range of the weak nuclear force is a bit less than that of the strong nuclear force.[320] This increases the total probability of the fine tuning to $p = 10^{-176}$.

4) On the other hand, although Smolin sets the value of the gravitational force, he does not set the ratio of the gravitational force to the strong nuclear force, which is $1/6*10^{39}$. This gives a cumulative fine tuning of $p \approx (1/6) * 10^{-215}$.

5) Smolin does not take into account the fine tuning of the momentum (kinetic energy) of the initial expansion of the Big Bang, which Rees

---

[320] In the case of the weak nuclear force, it is 10^{-18} m, which is 0.1 percent of a proton diameter; in the case of the strong nuclear force, it is 10^{-15} metres, i.e. the radius of a medium-sized nucleus.

estimates to be $1/10^{15}$. This raises the degree of cumulative fine tuning to $1/6*10^{-230}$.

6) We must also take into account the number of spatial dimensions necessary for closed, stable planetary orbits. With regard to the possible maximum number of spatial dimensions, there really is no limit, but I will take the number 11 that is used in some superstring theories, so the probability that three spatial dimensions come out is $p = 1/11 = 0.091$. This raises the degree of fine tuning to $p = 0.015 * 10^{-230}$.

7) We must take into account the probability that the Universe is sufficiently large that, in some solar system, the required initial conditions are met for the emergence of complex life. Here there are only two possibilities, specifically, that the size of the Universe *is* or *is not* large enough for the emergence of complex life, i.e. $p = 0.5$. The cumulative probability of all of these cases of fine tuning is $p = 0.75 * 10^{-232}$.

8) Smolin does not take into account the fine tuning of the neutrino mass, possibly thinking that it makes no difference if this mass is zero or almost zero. The neutrino mass is $5 * 10^{-35}$ kg. Davies points out that a slightly more massive neutrino would lead to an early collapse of the Universe and to "*severe disruption of the galactic structure*".[321] On the other hand, a less massive neutrino would have led to an accelerated expansion of the Universe, preventing the formation of stars and galaxies because, although the neutrino has very little mass, its density is very high ($N_v = 10^9/m^3$), which implies that "*the accumulated neutrino mass could outweigh all the stars*".[322] The cumulative probability of fine tuning, including the neutrino mass, is $p = 3.75 * 10^{-267}$.

---

[321] Paul Davies, *The Accidental Universe* (1983): 62.
[322] *Ibidem*: 61.

Supernovae also impose limits on the neutrino mass. Although stars are transparent to neutrinos, the implosion that precedes the supernova is so powerful and compresses the interior of the star to such a degree that it significantly slows down the flow of neutrinos, which then generate the necessary pressure for the periphery to explode. In this way, heavy elements created in the star prior to the supernova are dispersed into space. Thus, carbon, oxygen and iron in our solar system were dispersed into space by a supernova that gave life to the Sun and the planets. Therefore, "*without supernovae, Earth-like planets would not exist*".[323] If the neutrino had zero mass, it would have no capacity for pressure. If it were more massive, it would become part of the black hole that is left when a star ends its life.

The margin of error allowed in the fine tuning is unlikely to exceed 10%. This gives us a narrow margin for the cumulative fine tuning of the Universe in accordance with the emergence of stars of long duration, fusion of heavy elements and intelligent life:

(1) $p \approx 3.375 * 10^{-267} <$ fine tuning $< p \approx 4.125 * 10^{-267}$.

This result, $p \approx 3.375 * 10^{-267} <$ fine tuning $< p \approx 4.125 * 10^{-267}$ (equation 1 of Math box 3.8) concludes our overview of the fine tuning of physical constants in the laws of nature which determined the evolution of the Universe, that started with the Big Bang, and made possible the evolution of stars with a long life cycle and nuclear fusion of elements heavier than helium, and the emergence of solar systems and people like us to discover all this.

---

[323] Paul Davies, *The Accidental Universe* (1983): 68

CHAPTER 4

MYTH: MODERN THEORIES OF THE MULTIVERSE

When faced with the undeniable evidence of fine tuning, many cosmologists opt for a set of theories which have as a common denominator the so-called multiverse. These theories propose a mechanism of unlimited multiplication of universes, in space or in time, and of variation of the physical laws and constants in these universes, such that the probability of universes emerging with physical laws and constants and initial conditions favourable to the emergence of complex life, is close to one. These theories of the multiverse can be grouped into six categories:

1. Hoyle's theory of creation fields in the observable Universe.
2. The theories of black holes as origins of universes:
2.1 Wheeler's theory of the Big Crunch.
2.2 Smolin's theory of the multiverse through black holes.
3. Barrow's theory of the variation of physical constants.
4. The Guth-Linde theories of inflation and the multiverse.
5. Superstrings and Susskind's theory of the multiverse.
6. Conclusions on multiverse theories.

We will look at each theory, specifically, the theories of Hoyle (Section 4.1), Wheeler (4.2.1), Smolin (4.2.2), Barrow (4.3), Guth-Linde (4.4) and Susskind (4.5).

Section 4.1 Hoyle's theory of creation fields in the observable Universe

Fred Hoyle (1915–2001) coined the term 'Big Bang' on BBC radio's *Third Programme* broadcast, on March 28, 1949. It was his way of expressing his contempt for the theory. Somehow, though, it became accepted as the mainstream scientific denomination of the origin of the Universe. Hoyle was a fierce critic of the Friedmann-Lemaître model, and when looking for an alternative, he came up with his own theory, which he called the Steady State model.

Hoyle worked with two colleagues, Thomas Gold (1920–2004) and Hermann Bondi (1919–2005), both born in Vienna, who had both escaped from Nazi-occupied Austria. After the war, in September 1945, they went to see a film in which a character recounts a nightmare to a group of people who tell him their own nightmares. He then awakes and visits his neighbours to tell them what happened. They each recount their nightmares and the character awakes again, realising it was a just dream, and so on. The film gave Gold the idea that there could be situations that are both stable and dynamic at the same time, just like a flowing river. This led to the idea of a continuously expanding eternal universe in which the growing spaces between galaxies are filled with new spontaneously formed atoms in what Hoyle called a 'creation field', before forming gas clouds that condense into new stars and galaxies, which also move away from each other and form new creation fields, in turn giving rise to the formation of new atoms, and so on. Hoyle recalls the idea in his reminiscences of 1946–1947:

"In a sense, the steady-state theory may be said to have begun on the night that Bondi, Gold and I patronised one of the cinemas in Cambridge. The picture, if I remember rightly, was called Dead of Night. *It was a sequence of four ghost stories, seemingly disconnected as told by the several characters in the film, but with the interesting property that the end of the fourth story connected unexpectedly with the beginning of the first, thereby*

setting up the potential for a never-ending cycle. When the three of us returned that evening to Bondi's rooms in Trinity College, Gold suddenly said: 'What if the Universe is like that?'... One tends to think of unchanging situations as being necessarily static. What the ghost-story film did sharply for all three of us was to remove this wrong notion. One can have unchanging situations that are dynamic, as for example, a smoothly flowing river. The universe had to be dynamic, since Hubble's red-shift law proved it to be so... From this position, it did not take us long to see that there would need to be a continuous creation of matter."[324]

Hoyle pointed out that such a universe would be eternal: "*The present model has both an infinite future and an infinite past.*"[325] It follows that certain galaxies (which Hoyle referred to as nebulae) would escape from the observable Universe and new galaxies would emerge in the spaces created by the expansion of the Universe, suggesting the conservation not only of mass but also of the density of mass in the observable Universe. One problem was that nobody knew how the physics of this 'creation field' worked. Hoyle argued that it did not entail the creation of whole stars and galaxies, merely the gradual creation of new atoms at a rate of "*one atom every century in a volume equal to the Empire State Building*".[326]

The creation of atoms out of nothing in Hoyle's Steady State model implied the metaphysical concept of *creatio ex nihilo* (the creation of something out of nothing), albeit without explicitly acknowledging the fact. As I shall explain in Chapter 8 in more detail, in doing science, we make the metaphysical supposition that anything that is out there to see, must have a cause. Hoyle's concept of *creatio ex nihilo* of anything, regardless of its size or volume, differs from this metaphysical supposition. Although quantum

---

[324] Cited in Malcolm Longair, *The Cosmic Century* (2006): 324.
[325] Fred Hoyle, "A New Model for the Expanding Universe", in: *Monthly Notices of the Royal Astronomical Society*. vol. 108 (1948): 381.
[326] Cited in Simon Singh, *Big Bang. The Origin of the Universe* (2004): 347.

physics provides for particles and anti-particles popping up in space-time, these are immediately annihilated, leaving just radiation. This process is different from that proposed by Hoyle as it is not *creatio ex nihilo*, given that space-time abounds in gravitational, electromagnetic and quantum fields, responsible for the production of these particles. I will come back to this point in Section 8.1 of Chapter 8, when discussing Krauss' proposal.

The trio published their results in two articles in 1948: a first, more philosophical article authored by Bondi and Gold,[327] and a second, more mathematical article written by Hoyle.[328] An observable consequence of Hoyle's model, that could be corroborated or refuted by facts, was that in such an eternal universe, *distant and nearby galaxies would look the same* because in this model the new galaxies emerge in both distant and nearby creation fields. Bondi and Gold acknowledged this hypothesis, whose corroboration would make or break the model:

"We take the perfect cosmological principle to imply that no feature of the universe is subject to any consistent change, and no observer hence capable of any unique definition of a universal time. This will be satisfied only if the ages of galaxies in any sufficiently large volume follow a certain statistical distribution...Furthermore, the age distribution of galaxies in any volume will be independent of the time of observation, and it will hence be the same for distant galaxies as for near ones, although in the former case the light has taken long to reach us."[329]

By solving the problem of the production of heavier elements, particularly carbon, as analysed in Section 1.3 of Chapter 1, Hoyle helped both models

---

[327] Hermann Bondi & Thomas Gold, "The Steady State Theory of the Expanding Universe", in: *Monthly Notices of the Royal Astronomical Society*, vol. 108 (1948): 252–270.

[328] Fred Hoyle, "A New Model for the Expanding Universe", in: *Monthly Notices of the Royal Astronomical Society*, vol. 108 (1948): 372–382.

[329] Hermann Bondi & Thomas Gold, "The Steady State Theory of the Expanding Universe", in: *Monthly Notices of the Royal Astronomical Society*, vol. 108 (1948): 257.

(Big Bang and Steady State), since up to that point neither had been able to explain the mystery of the production of heavier elements, most notably carbon. Yet there was still no decisive empirical evidence to rule out either of the theories or both of them.

Hoyle's theory not only envisages the continuous creation of new matter-energy in the Universe, but it also contemplates variations in the physical constants of natural laws in different space-time regions of the observable Universe. Hoyle himself was confounded by the fine tuning of the ratios of the electromagnetic, gravitational and strong nuclear forces, and the phenomenon of nuclear resonance, which I analysed in Section 1.3 of Chapter 1 and Section 3.4 of Chapter 3, that enabled the nuclear fusion of carbon and oxygen in sufficient quantities for life:

"I do not believe that any scientist who examined the evidence would fail to draw the inference that the laws of nuclear physics have been deliberately designed with regard to the consequences they produce inside the stars. If this is so, then my apparently random quirks have become part of a deep-laid scheme. If not, then we are back again at a monstrous sequence of accidents."[330]

Since anything to do with a deliberate design by a creator god was completely unacceptable to Hoyle, he resorted to the theory of 'creation fields' that generate new matter-energy and the variation of the value of physical constants in these different fields. We happen to be in one such field, where, by chance, the physical constants have the right values to make possible the production of the right levels of carbon and oxygen:

"The curious placing of the levels in C^{12} and O^{16} need no longer have the appearance of astonishing accidents. It could simply be that since creatures like ourselves depend on a balance between carbon and

---

[330] Cited in John Barrow, *Constants of Nature* (2002): 157.

oxygen, we can exist only in the portions of the universe where these levels happen to be correctly placed. In other places, the level of O^{16} might be a little higher, so that the addition of alpha-particles to C^{12} was highly resonant [and] creatures like ourselves could not exist."[331]

The Steady State theory envisages an eternal Universe, where new space-time regions are continuously created which function as 'creation fields', the existence of which Hoyle conjectured without providing any empirical proof. This statement is logically compatible with the other that the constants in the natural laws vary in different space-time regions.

Let us now go back a bit in time in order to understand the empirical evidence which served decisively to put to the test both the *Big Bang* and the Steady State models. Let us remember that Alpher and Gamow had argued that not enough helium was produced in stars to explain its abundance in the Universe. However, this abundance could be explained by the nuclear fusion reactions in the first 20 minutes after the Big Bang (see Section 1.2 of Chapter 1). Some scientists were not convinced, though, suspecting that Gamow and Alpher had adjusted their mathematics to coincide with the real volumes of helium in the Universe. Gamow and Alpher made use of the first computers, recently built, to confirm their calculations of the relative abundance of hydrogen and helium.

The suspicions nevertheless persisted, fuelled by the fact that Alpher was a student who had not yet graduated and Gamow a popular cosmology writer for amateurs who liked to play the clown. He had once claimed that God lived nine light years from the Earth because when war broke out between Russia and Japan in 1905 and the Russians prayed in their churches, it was not until 1923 that Japan was hit by a powerful earthquake. An atheist himself, Gamow also cited Pope Pius XII's comments in favour of the Big Bang model to irritate some of his colleagues. In light of these

---

[331] Cited in John Barrow, *Constants of Nature* (2002): 156.

jokes, Alpher was angry that Gamow's buffoonery had prevented people from taking their research seriously:

"Because he injected a considerable amount of humour into his presentations, he was frequently not taken seriously by too many of his fellow scientists. His not being taken seriously is something that rubbed off on the two of us as his colleagues."[332]

Since the validity of objective theories is independent of the subjective process by which scientists accept or reject them, the question of which theory was true, Big Bang or Steady State, was still undecided. Since 1948, there had been two objective theories, both of which predicted observable phenomena. They could not both be true: either both were false or one of them was true, the other one being false. To make this decision, empirical evidence was needed. Both theories had in common that the Universe was expanding, so that the light from receding galaxies would be redshifted, so this fact in itself would not discriminate between the two theories. The Big Bang theory implied that nuclear fusion shortly after the Big Bang would result in a distribution of hydrogen and helium in a proportion of 10:1 (by the number of atoms) or 3:1 (by atomic weight). This prediction had been verified (see Section 1.2 of Chapter 1), but too many doubts remained, meaning it did not yet decisively distinguish between the two theories.

Gamow, Alpher and Herman's theory predicted that a relic of the light from the original explosion (*CMBR*) should be observable throughout the Universe, with a wavelength in the range of microwaves and a temperature below 15 degrees Kelvin. Hoyle, Bondi and Gold's theory predicted that distant galaxies and nearby galaxies all would have the same average age because new galaxies are continuously being formed everywhere in the Universe, so that young galaxies would be homogeneously distributed throughout the observable Universe in their different stages of evolution.

---

[332] Cited in Simon Singh, *Big Bang. The Origin of the Universe* (2004): 335.

The corroboration or refutation of these two predictions would make it possible to decide which theory was true and which one was false.

In 1928, AT&T had initiated a transatlantic radiotelephone service and had commissioned Bell Laboratories in Crawford Hill, New Jersey, to detect natural sources of radio waves in order to neutralise them in its telephone service. Karl Jansky (1905–1950), a US electronics engineer who worked at Bell Labs, discovered two sources of 'noise': the first was near and far-away storms; the second was an extremely weak radiation that repeated every 23 hours and 56 minutes, just short of a day. Jansky mentioned the phenomenon to Melvin Skellett, who explained that the four missing minutes could be accounted for by the fact that the sidereal day is 23 hours and 56 minutes. Every year, the Earth spins 365¼ times on its axis while also orbiting the Sun. This means that, with respect to the stars, Earth actually rotates 366¼ times on its axis. Since there are 8,766 hours in a year, a sidereal day lasts 8,766 / 366¼ = 23 hours and 56 minutes. This enabled Jansky to find the source of these radio waves in the centre of our galaxy and he published his results in 1933.

Using radio interception equipment abandoned after the Second World War, two English astrophysicists, Sir Martin Ryle (1918–1984), who had worked with radars during the war, and Antony Hewish (born 1924) improved the precision of radio telescopes in 1946 by pointing various telescopes at single sources of radio signals, so as to increase the accuracy of radio source catalogues, known as 1C, 2C, 3C, 4C and detect strong but normally invisible radio sources. Both men were awarded the 1974 Physics Nobel Prize for their discovery. Focusing a visible light telescope on the Cygnus A radio source, Walter Baade (1893-1960) found the signals were not coming from stars but from galaxies, now known as 'radio galaxies'.[333]

---

[333] Walter Baade & Rudolph Minkowski, "Identification of the radio sources in Casssiopeia, Cygnus A, and Puppis A", in: *Astrophysical Journal*, vol. 119 (1954): 206–214.

By 1961, Ryle had catalogued 5,000 radio galaxies and made a discovery that was confirmed in the Southern skies by a team of radio astronomers in Sydney, Australia: radio galaxies occurred more frequently at greater distances. Since the Steady State model predicted that different types of galaxies were distributed homogeneously and isotopically, not just in space but also in time, the fact that certain radio galaxies were far away and not close by, falsified the Steady State model, and Ryle was well aware of this:[334] *"This is a most remarkable and important result, but if we accept the conclusion that most of the radio stars are external to the Galaxy, and this conclusion seems hard to avoid, then there seems no way in which the observations can be explained in terms of a Steady State theory."*[335]

Hoyle refused to accept these results and repeatedly challenged the observation technique and interpretation of the data. Besides, in the 1960s, Fred Hoyle and Roger Tayler (1929–1997) obtained more accurate results concerning the production of primordial hydrogen and helium in the Big Bang (75% and 25% by mass, respectively), a result that confirmed the Big Bang model, but they avoided mentioning this fact explicitly.[336] Bondi and Gold also refused to accept that their theory had been refuted by the facts. They had previously announced that they were so convinced of the validity of their model that they would continue to believe it, even in the face of evidence to the contrary: *"we regard the principle as of such fundamental importance that we shall be willing if necessary to reject theoretical extrapolations from experimental results if they conflict with the perfect cosmological principle even if the theories concerned are generally*

---

[334] Martin Ryle, "Radio Stars and their Cosmological Significance" in: *The Observatory*, vol. 75 (1955): 137–147.
[335] Cited in Malcolm Longair, *The Cosmic Century* (2006): 326.
[336] Fred Hoyle & Roger Tayler, "The Mystery of the Cosmic Helium Abundance", in: *Nature*, vol. 203 (1964): 1108–1110.

accepted." [337] An incredible confession that goes right in the face of Popper's philosophy of science, which requires scientists to abandon theories that are refuted by facts in the real physical world (see Chapter 7).

Further evidence, obtained using radio astronomy, also proved decisive. In 1963, Maarten Schmidt, a Dutch–US astronomer born in 1929, discovered the first quasi-stellar object, or quasar, at a distance of over one billion light years from our galaxy (3C 273, or radio source 273 in Ryle's third catalogue), with a red shift so large that it implied it was receding at 16% of the speed of light (48,000 km/s). Other quasars subsequently discovered were also extremely far away. The fact that quasars are younger objects showed that the Universe had evolved over time. From this point, many cosmologists who had remained sceptical of the Big Bang model began to accept it, including Dennis Sciama (1926–1999), a British-Italian astrophysicist and one of the founders of modern cosmology, who stated that these quasar observations were *"the most decisive evidence so far obtained against the Steady State model of the Universe."*[338]

By now, the Steady State theory had been thoroughly refuted by the facts, but corroboration of the Cosmic Microwave Background Radiation (*CMBR*) hypothesis was still pendent and necessary to validate the Big Bang model. The subsequent discovery of the *CMBR*, predicted in 1948 by Ralph Alpher and Robert Herman[339], and corroborated empirically by Arno Penzias and Robert Wilson in 1965[340] (see Section 1.2 of Chapter 1), definitely corroborated the Big Bang model.

---

[337] Hermann Bondi & Thomas Gold, "The Steady State Theory of the Expanding Universe", in: *Monthly Notices of the Royal Astronomical Society*, vol. 108 (1948): 255.
[338] Cited in Simon Singh, *Big Bang. The Origin of the Universe* (2004): 421.
[339] Ralph Alpher & Robert Herman, "Evolution of the Universe", in: *Nature*, vol. 162 (1948): 774–775.
[340] Arno Penzias and Robert Wilson, "A Measurement of Excess Antenna Temperature at 4080 MHz", in: *Astrophysical Journal*, vol. 142 (1965): 419–421.

Section 4.2 Black holes as origins of universes

John Wheeler (1911–2008), a USA astrophysicist, speculated that our Universe began with the explosion of one black hole of infinite density, resulting from the collapse of a previous universe. Smolin conjectures that any black hole in whatever universe gives rise to a new universe. We shall pay attention to both these multiverse theories.

Many cosmologists talk about a singularity at the start of the Big Bang. Strictly speaking, a singularity is the core of a black hole, where all space-time and all matter-energy have collapsed to a point of infinite density. When black holes collide, they swallow each other up. In the case of a Big Crunch, this means that the end of everything is a supermassive black hole with a singularity at its centre. Penrose stresses the point that, even if we assume *"significant irregularities in the mass distribution of collapsing material..., in the situation of the global collapse of an entire universe..., singularities being a generic feature of gravitational collapse in classical general relativity... singularities are inevitable"*.[341]

The problem with proposals to locate the origin of the universe in a black hole is that they are in direct contradiction with the very low degree of gravitational entropy in the Big Bang, also conceived, by Penrose, as *"the extraordinary suppression of gravitational degrees of freedom in the* Big Bang"[342], which is quite the opposite of the maximum gravitational entropy existing in a supermassive black hole, as Penrose points out:

> *"[S]o long as the 2nd Law did indeed hold true at all times since the universe's inception, then the remote past must have been... constrained to be extremely highly organized macroscopically. The key to the 2nd Law, therefore, is the existence of an extraordinarily macroscopically*

---

[341] Roger Penrose, "Black holes and local irregularities", in: *Fashion, Faith and Fantasy* (2016): 230–241.
[342] Roger Penrose, *Fashion, Faith and Fantasy* (2016): 371.

organized initial state of the universe. But what was that state?... [T]his was the gigantic all-encompassing explosion known as the Big Bang! How can it be that such an unimaginably violent explosion actually represented an exceptionally low-entropy, incredibly macroscopically organized state?... [T]he very early universe was indeed extremely uniform, but with very slight irregularities in the density. Over the passage of time these density irregularities became gravitationally enhanced, consistent with a picture in which clumping of material gradually increased with time to produce stars, these being gathered in galaxies, with massive black holes in galactic centres, this clumping being ultimately driven by relentless gravitational influences. This indeed would have presented a vast entropy increase, illustrating that, when gravity is brought into the picture, the primordial fireball that is evidenced by the CMB must actually have been far from a maximum-entropy state... [W]ith gravity ultimately having a dominant presence, there is much gain the entropy obtained by moving away from the uniform distribution... Our existence, as we know it, depends on the low-entropy gravitational reservoir inherent in the initially uniform matter distribution."[343]

Strictly speaking, entropy is a concept belonging to thermodynamics. In Chapter 5, I shall present mathematical proof that the second law of thermodynamics does indeed apply to our Universe, being an adiabatic, thermodynamically closed, irreversible system, with its entropy, defined in the strict sense of the word, increasing with time. The term 'closed' in this case does not refer to the geometry of the Universe, which in Chapter 2 was proven to be open ($k < 0$), but to the fact that its total amount of matter-energy is constant ($dm/dt = 0$). What Penrose and others[344] argue,

[343] Roger Penrose, *Fashion, Faith and Fantasy* (2016): 249–250, 255–258.
[344] David Wallace, *Gravity, Entropy, and Cosmology: in Search of Clarity*, online, 2009.

however, is that gravitational and thermodynamic processes are intertwined, giving rise to the concept of gravitational entropy, so that the 'total' entropy at the beginning of the Universe was very low, not only because of the relatively low *thermodynamic* entropy at that time, but also because of the low *gravitational* entropy (see Chapters 5 and 6).

Section 4.2.1 Wheeler's Big Crunch theory

In 1973, John Wheeler, an American physicist (1911–2008), who initially coined the term 'black hole', published an essay titled *From Relativity to Mutability*.[345] Wheeler's theory conjectures there is an eternal succession of big bangs and big crunches. This succession of big bangs and big crunches supposes that our Universe is closed ($k > 0$), so that in the final collapse there will be a supermassive black hole at the end of each universe.

Wheeler also assumes that in each collapse everything is annihilated, not just matter and energy but also space-time and all information about physical laws and their constants. In addition, he assumes that when the phoenix rises from the ashes, the new universe has physical laws and constants that are different from the previous one. Wheeler's theory contemplates a variation with time of the physical constants in successive universes. Given that in each big bang the natural laws and constants are set differently, with initial conditions different from the previous universe, it follows from the law of large numbers that our Universe, so specifically tuned for the evolution of stars and life, had to randomly emerge at some time. Wheeler calls this phenomenon mutability:

"If the laws of conservation of particle number are transcended in black hole physics; if all dynamic laws are transcended in the collapse of the universe; if laws and constants of physics are first imprinted as initial

---

[345] John Wheeler, "From relativity to mutability", in: Jagdish Mehra ed., *The Physicist's Conception of Nature* (1973): 202–247.

conditions in the earliest phase of the Big Bang and erased in the final stage of gravitational collapse, then dimensionality [=the shape of these laws] itself can hardly be exempt from the universal mutability."[346]

There are problems with this theory. In the first place, although our Universe started with a Big Bang, there will be no Big Crunch. All of the evidence available to date indicates that our Universe is not closed, but rather flat or open, so it will continue to expand forever. Therefore, it will not collapse, nor is it stuck in such a cycle of big crunches and big bangs.

Secondly, in 2015, Stephen Hawking refuted Wheeler's no-hair theorem (which states that black holes are essentially bald, or featureless, except for their mass, angular momentum, and in the case of some black holes: electric charge) stating that when things are sucked up into a black hole, information does *not* get lost. His later paper described "*soft hair in terms of soft gravitons or photons on the black hole horizon, and shows that complete information about their quantum state is stored on a holographic plate at the future boundary of the horizon*".[347] When particles get sucked into a black hole, the information they carry, for example, about the wave function that describes their previous behaviour, and the laws of nature that caused it, is imprinted on a two-dimensional holograph on the event horizon, later to be ejected into the Universe by the Hawking radiation.

Section 4.2.2 Smolin's theory of the multiverse in black holes

Lee Smolin presents a theory of the multiverse in which universes are dispersed not in time, but in space. Each universe has different natural laws and constants. Smolin, like Wheeler, asserts that our Universe was born from a supermassive black hole that bounced back and exploded (the Big

---

[346] John Wheeler, in: Jagdish Mehra ed., *The Physicist's Conception of Nature* (1973): 241.
[347] Stephen Hawking, Malcolm Perry & Andrew Strominger, "Soft Hair on Black Holes", arxiv: 1601.00921v. (2016); in: *Physical Review Letters*, vol. 116, 231301 (2016): 1–9.

Bang) and that this black hole resulted from the collapse of a previous universe. In summary, "*the question about what happened 'before the Big Bang' in the event that quantum effects allow time to extend indefinitely into the past*", elicits the response that, beforehand, there was another universe that collapsed into a black hole and one can conjecture that "*it is possible that what is beyond the horizon of a black hole is the beginning of another universe?*"[348] This concept implies that there was no point $t = 0$, and the repeated expansions and collapses of the universe are everlasting.

Unlike Wheeler, Smolin conjectures that this happens to *all* black holes, including those that currently exist in our Universe, at the centre of galaxies for example, and not just the supermassive black hole that results from the collapse of an entire universe. So "*we live in a continuously growing community of 'universes', each of which is born from an explosion following the collapse of a star into a black hole*".[349]

My criticism of Smolin's theory is partly the same as for Wheeler's theory, that is to say, nobody knows of a physical theory that explains how physical laws and constants are being changed in every big bang resulting from black hole explosions. However, there is an additional criticism. Smolin's theory accepts that from the outside we cannot see what is happening inside a black hole, so his theory cannot be tested by observation. Smolin, however, is an advocate of Popper's logic of scientific discovery[350] and is quite aware of this danger. He maintains that his theory of the natural selection of universes in favour of universes with more black holes is *indirectly* falsifiable. He asserts that the large number of black holes in our Universe is an indication that we drew a winning ticket, that is, a universe finely tuned for life. He establishes a correlation between the abundance of black holes and the fine tuning of physical constants, arguing that the

---

[348] Lee Smolin, *The Life of the Cosmos* (1997): 87-88.
[349] *Ibidem*: 88.
[350] *Ibidem*: 76-77.

physical constants of our Universe favour the production of both abundant oxygen and carbon, necessary for life, and abundant black holes. However, Mario Livio correctly observes that nuclear fusion in stars depends on the very precise value of some physical constants, and the present value of the physical constants does *not* favour the production of black holes, since "*a reduction in the efficiency of nuclear reactions would result in more stars being unable to resist gravity's final pull and becoming black holes*".[351]

In general, although Smolin accepts falsifiability as the criterion that demarcates science and non-science, it is clear that his theory is not falsifiable, because the only observable universe is our own and with what criterion can we establish that in our universe there are relatively more or fewer black holes than in others? There is no way to make comparisons.

I conclude my review of Wheeler's and Smolin's proposals, turning to Penrose:

"*I have quite a lot of trouble with both the Wheeler and the Smolin proposals. In the first place, there is the extremely speculative nature of the key idea that some presently unknown physics can not only convert the space-time singularity of [gravitational] collapse into a 'bounce', but also slightly readjust the fundamental physical constants when this happens. I know of no justification from known physics to suggest such an extrapolation. But, to my mind, it is even more geometrically implausible that the highly irregular singularities that result from collapse can magically convert themselves into (or glue themselves to) the extraordinarily smooth and uniform Big Bang that each new universe would need of the kind that we are familiar with.*"[352]

---

[351] Mario Livio, *The Accelerating Universe* (2000): 188–189.
[352] Roger Penrose, *The Road to Reality* (2004): 761–762

Section 4.3 Barrow's theory of the variation of physical constants

John Barrow[353] and other authors have proposed that the most important physical constants vary in time within our own Universe. This hypothesis, being scientific, is falsifiable by evidence. It is to be noted that only observations in our observable Universe can falsify or corroborate our physical theories, for example, about the variability or invariability of physical constants.

In 1972, Freeman Dyson produced evidence that in all of the space-time regions of the observable Universe, the hypothesis of the variability of the fine-structure constant is excluded by observation, whereas the hypothesis of the variability of the gravitational constant is not yet testable within the range of the 1972 observational capabilities.[354] Dyson concluded:

"Of the five hypotheses A, B, C, D, E [that postulate a change in the fundamental constants] with which this discussion began, only D and E, the two which included a time-variation of the fine-structure constant, are yet excluded by observation... The two hypotheses, B and C, which include a time-variation of the gravitational constant, are consistent with present observational knowledge. Dirac's Hypothesis B is only barely consistent with the evidence from solar and stellar evolution and its tenability will be decisively tested by interplanetary ranging measurements within the next few years. There remain Dicke's Hypothesis C, with a much weaker time-variation of gravity than B, and the orthodox Hypothesis A in which all laboratory-measured constants are truly constant. A direct decision between A and C is not within the

---

[353] John Barrow & Frank Tipler, *The Anthropic Cosmological Principle* (1986): 255–257; John Barrow, *The Constants of Nature* (2002): 259–268; John Barrow & John Webb, "Inconstant Constants", in: *Scientific American*, vol. 16 (2006): 72–81

[354] Freeman Dyson, "The Fundamental Constants and Their Time Variation", in: Abdus Salam & E. Wigner, eds, *Aspects of Quantum Theory* (1972): 235–236

range of present observational capabilities [1972]. Probably the decision will be reached indirectly, by observations ... in the next ten years."[355]

At that time, Wheeler reached a similar conclusion: *"No change with time has ever been found in the fine structure constant [that determines the electromagnetic force], in the mass of any particle, or in any other constant of physics."*[356] In that same year, Wheeler and others debated the constancy of the gravitational constant, and concluded that this constant, defined by Isaac Newton and measured by Henry Cavendish (1731–1810), is indeed constant in general relativity, but in the majority of other metric theories it will vary from event to event in space-time, according to the distribution of matter in the Universe.[357] To be precise, the general theory of relativity is the most general metric theory that corroborates the hypothesis that physical laws and constants are invariant and independent of the coordinate system. Other, more recent works, whilst presenting theories that allow for a dynamic of variation in the physical constants, at the same time confirm the invariance of these constants in our observable Universe.[358]

A 2003 review by Jean-Philippe Uzan, a French astrophysicist, is of particular relevance.[359] He tests the hypothesis of variation for some fundamental constants, including the fine-structure constant α, the

---

[355] Freeman Dyson, "The Fundamental Constants and Their Time Variation", in: Abdus Salam & E. Wigner, eds., *Aspects of Quantum Theory* (1972): 235–236.

[356] John Wheeler, "From relativity to mutability", in: J. Mehra, *The Physicist's Conception of Nature* (1973): 202–247.

[357] Charles Misner, Kip Thorne & John Wheeler, "Is the gravitational constant constant?", in: *Gravitation* (1973): 1122–1226.

[358] Jean-Philippe Uzan, "The Fundamental Constants and their Variation: Observational and Theoretical Status", in: *Reviews of Modern Physics* (2003): 403–455; and Wendy Freedman & Michael Turner, "Measuring and Understanding the Universe", in: *Reviews of Modern Physics* (2003): 1433–1447.

[359] Jean-Philippe Uzan, "The Fundamental Constants and their Variation: Observational and Theoretical Status", in: *Reviews of Modern Physics* (2003): 403–455.

gravitational constant G, and the ratio β of the electron and proton masses. Uzan himself does not calculate the mean result from numerous studies of the hypothetical variation of some dimensionless constants. However, from his data we can determine the mean values for three constants, specifically, α, G, and β. By calculating these means, we can test the hypothesis of the variation of nature's constants in time or space. I present the results of this analysis below.

With respect to the variation of the fine-structure constant, $\Delta\alpha_{EM}/\alpha_{EM}$, Uzan presents results from 16 studies that measure its possible variation with time. Of these studies, 12 indicate zero variation, two, negative variation and two, positive variation. In all cases, the data are presented together with margins of error.

Math box 4.1 The variation of the fine-structure constant

By standardising Uzan's data, using the Wolfram *Mathematica* program, for a time period of a billion years and calculating the mean from the 16 studies, I obtain the following result:

(1) $+0.0278\% > \dfrac{\Delta\alpha_{EM}}{\alpha_{EM}} > -0.0278\%.$

The inevitable conclusion is that the hypothesis of variation of the fine-structure constant is refuted, given that its variation in 10^9 years, in our observable Universe, is zero, with an error margin of $\pm0.0278\%$, which for practical purposes means zero.

The first person to suggest that the gravitational constant G could have changed with time was Dirac.[360] Others followed. I will now test the

---

[360] Paul Dirac, "The Cosmological Constants", Letter in: *Nature*, vol.139 (1937): 323, reproduced in: Kenneth Lang & Owen Gingerich, eds, *A Source Book in Astronomy and Astrophysics, 1900–1975* (1979): 851–852.

hypothesis of the variation of *G*. Uzan reviewed 32 studies which aimed to measure the variation of this constant with time.

Math box 4.2 The variation of the gravitational constant

Once again standardising the time frame to 10^9 years, and calculating the mean from 32 empirical studies, using the Wolfram *Mathematica* program, I obtain the following result for the variation of the gravitational constant, within the margin of error:

(1) $+3.42\% > \Delta G/G > -3.34\%$.

This empirical evidence collated by Uzan corroborates the hypothesis, formulated by Davies, of a truly universal gravitational constant: "*The assertion that G is a universal constant is the claim that, wherever in the Universe, or at whatever point in history, one was to measure the force between two one-kilogram masses at one metre separation, then the result would always be* $6.7 * 10^{-11} N$."[361]

Let us now look at the hypothesis of the possible variation of *β*, that is, the ratio between the electron mass and proton mass. In Math box 4.3, it is shown that, given the minimal range of variation of *β*, the hypothesis of the variation of the constant *β* is also refuted, as is the case with the fine-structure constant and the gravitational constant.

Math box 4.3 The ratio of the electron and proton masses

Nine of the 10 studies reviewed by Uzan do not directly give the time period for the possible variation of this constant, but rather data on the redshift, due to the expansion of the Universe, of the objects studied. Kolb & Turner provide an equation for converting the redshift factor *z* to a time factor *t*:[362]

---

[361] Paul Davies, *The Accidental Universe* (1984): 10
[362] Equation from Edward Kolb & Michael Turner, *The Early Universe* (1990): 504

(1) $t = 2.0571 * 10^{17}(\Omega_0 h^2)^{-1/2}(1 + z)^{-3/2}$ s,
   with $\Omega_0 = 1$; $h = 6.626 * 10^{-34}$.

Applying this equation to the different values of z in the nine studies with redshift data that Uzan referred to, I obtain the following mean range of possible variation of β in 10^9 years, within the error margin:

(2) $3.934\% > \beta > -3.935\%$.

The obvious conclusion is that there is no empirical evidence whatsoever supporting the hypothesis of the variation of physical constants in our Universe. On the contrary, there exists overwhelming evidence falsifying this hypothesis. It is important to point out that the theory of the invariance of physical constants in space-time does not imply a metaphysical stance of deterministic causality. The problem of deterministic or indeterministic causality is presented elsewhere[363] and here I merely repeat the conclusion of that discussion, in the words of Popper:

"*We can admit that the world does not change in so far as certain universal laws remain invariant. But there are other important and interesting lawlike aspects – especially probabilistic propensities – that do change, depending upon the changing situation. Thus... there can be invariant laws and emergence of new and unpredictable things; for the system of invariant laws is not sufficiently complete and restrictive to prevent the emergence of new lawlike properties.*"[364]

---

[363] See Section 7.4 of Chapter 7 of this book.
[364] Karl Popper & John Eccles, *The Self and Its Brain* (1981): 25.

Section 4.4 The Guth-Linde theories of inflation and the multiverse

The multiverse theory of Andrei Linde presupposes Alan Guth's inflation theory.[365] In order to understand Linde's theory, I first briefly summarise Guth's inflation theory.

According to Guth, inflation is an exponential expansion of the Universe – much faster than the speed of light – during a fraction of its first second of existence. The Universe would have doubled in size every 10^{-34} second, during a fraction of the first second of the Universe.[366] At the end of this very brief inflationary period, this resulted in a total Universe with a diameter of 10^{19} light years and what later became an observable Universe with a diameter at that time of 3 metres, and today of $14.9*10^9$ light years.[367] *"Consequently, our visible universe must be a microscopic speck in [the] totality of the universe that is out there."*[368]

Following this inflation, many regions of the Universe did not come back into contact with each other. However, they are still homogeneous, because, according to Guth, *before* the inflation different 'regions' of the Universe *did* exchange energy and balance out differences in energy levels and temperatures, which would explain the homogeneity of the cosmic background radiation even after inflation and, in general, the horizon problem arising from the observed homogeneity of causally disconnected regions of space in the absence of a mechanism that created the same initial conditions all over the Universe.[369]

---

[365] Alan Guth, *The Inflationary Universe* (1998); Andrei Linde, *Inflation and Quantum Cosmology* (1990).

[366] Summary by Roger Penrose, *Fashion, Faith and Fantasy in the New Physics of the Universe* (2016): 295.

[367] Jonathan Allday, *Quarks, Leptons and the Big Bang* (2002): 256, 334–336.

[368] *Ibidem*: 336.

[369] George Smoot & Keay Davidson, *Wrinkles in Time* (1993):150–151, 176–184; Alan Guth, *The Inflationary Universe* (1998):180–186; Jonathan Allday, *Quarks, Leptons and the Big Bang* (2002): 334–343.

Where did the energy that powered this rapid expansion in the first second of the Big Bang come from? Guth speculates that this inflationary expansion of the Universe, in this fraction of the first second of the Universe, was driven by a negative gravitational force. In this fraction of a second, the gravitational force, instead of being attractive (compressing space-time and matter-energy) was repulsive due to the 'false vacuum' present at that time: *"the false vacuum actually leads to a strong gravitational repulsion"*.[370] Even though the Universe was rapidly expanding, the energy density remained constant, implying a *negative* pressure.

Guth's theory is rather bizarre, as he himself admits: *"the unusual concept of matter with a constant energy density has led us to the bizarre concept of negative pressure"*.[371]

In quantum field theory, *vacuum* refers to a state of space with the lowest energy potential possible.[372] A vacuum is called 'false' when it temporarily settles at a higher energy density level than the lowest one possible, making the false vacuum state unstable. Through quantum fluctuations or the creation of high-energy particles the false vacuum can transit to the true vacuum, which is stable. The high energy, massive particle that played that role is, according to Guth, the Higgs boson.

There is a problem with Guth's use of the Higgs boson. The Higgs boson was postulated by Peter Higgs, at the University of Edinburgh, in 1964. It is the quantum excitation of the Higgs field, a fundamental field of crucial importance in particle physics theory, and as such, before its discovery, it was the missing link in the Standard Model. According to Higgs, at the moment of electroweak symmetry breaking, when the electroweak and electromagnetic forces differentiate, the Higgs boson is instrumental in

---

[370] Alan Guth, *The Inflationary Universe* (1998): 173.

[371] *Ibidem*: 172.

[372] In the Big Bang space had a matter-energy density equivalent to 10^{80} grams per cubic centimetre.

giving mass to different particles: "*gauge vector mesons acquire mass if the symmetry to whose generators they are coupled breaks down spontaneously*".[373] Mesons are short-lived composite bosons made up of one quark and one anti-quark (I explain gauge symmetries and classification of particles shortly). The other bosons are the carriers of the fundamental forces of nature, for example, the photons and the gluons (see Table 4.1).

As late as 2006, the Higgs boson had not yet been detected.[374] But in 2008, the Large Hadron Collider at CERN, near Geneva, that enables protons to collide at nearly the speed of light, achieving very high energies, began to look for the Higgs boson and its existence has been confirmed in 2012.

Guth's particle, however, is not the Higgs boson, which has to do with the symmetry breaking of the electroweak force, at very high energy levels, of some 10^{14} *GeV,* whereas Guth's particle has other, very different characteristics. Says Penrose:

> "*In order to achieve this inflationary period, it is necessary to introduce a new scalar field φ into the menagerie of known (and conjectured) physical particle/fields. [T]his field φ is not taken to be directly related to any of the other known fields of physics, but is introduced solely in order to obtain an inflationary phase in the early universe. It is sometimes referred to as a 'Higgs' field, but it does not seem to be the 'ordinary' one, related to electroweak theory.*"[375]

John Hawley and Katherine Holcomb give a more detailed criticism of Guth's inflation theory.[376] I shall now summarize this criticism. The Higgs

---

[373] Peter Higgs, "Spontaneous Symmetry Breakdown without Massless Bosons", in: *Physical Review*, vol. 145 (1966): 1158.

[374] Particle Data Group, 2006, "Review of Particle Physics", in: *Journal of Physics G. Nuclear and Particle Physics*, vol. 33 (2006): 32.

[375] Roger Penrose, *The Road to Reality* (2004): 751.

[376] John Hawley & Katherine Holcomb, *Foundations of Modern Cosmology* (1998): 427–438.

boson, as is the case with any particle, is associated with a field, which in its turn is associated with a potential, and, just as is the case with any field, the Higgs field tends automatically to shift towards a state where the potential energy is minimum, just as mountains tend to erode away and form a smooth, flat plain, since this minimises the gravitational energy potential of the terrain. Before the moment of symmetry breaking, at very high temperatures, the Higgs field is supposed to have been stable in a rather precarious way (the way of the false vacuum), like the stability of a marble resting atop an inverted bowl, susceptible to any perturbation that would break this equilibrium. According to Guth, the energy density of the false vacuum acted like a cosmological constant, powering the inflation, which lasted only a fraction of a second. But according to Hawley and Holcomb, it turns out that the Higgs boson as we know it cannot have been responsible for inflation, because *"inflation occurred while the field was in the metastable false vacuum"*, and, contrary to Guth's speculation, *"nothing subsequently could eject the field from that metastable state"*, which would imply that *"the inflation never stopped"*.[377] The idea of inflation not ending, however, is contrary to Guth's idea that the inflation period was very short.

With the theoretical flexibility characteristic of verificationism (for the definition of this term see Chapter 7), the inflationists changed their model in order to rescue it from being refuted by the facts:

"The model was rescued by a change in the potential V. In this 'new inflation' scenario, inflation occurs not while the field is in the false vacuum, but during the transition from the false to the true vacuum. Since the potential plays the role of the cosmological constant, we can see that a slow decrease during inflation will provide a simple way for the inflation to come to an end after a period of time. In other words, the 'cosmological' constant changes slowly during the new inflation. To

---

[377] John Hawley & Katherine Holcomb, *Foundations of Modern Cosmology* (1998): 430.

ensure that 'enough' inflation occurs, the potential must be very flat so that the field carries out the transition very slowly; slowly, that is, in comparison to the characteristic rate of expansion at that time. If such an inflation occurred, it would have happened around 10^{-37} seconds after the Big Bang and would have required approximately 10^{-32} seconds to complete. During this cosmic eye blink, the scale factor would have been inflated by a dizzying $10^{40} - 10^{100}$ or even more. This more successful new inflationary theory was first proposed by Andrei Linde... New inflation... requires the existence, at the appropriate time in the history of the universe, of a particle with an extremely flat potential and a slow transition to the true vacuum. This generic particle has come to be known as the inflaton."[378]

Though Hawley and Holcomb do not refer explicitly to 'verificationism', their analysis reveals its presence in the history of inflation theory. To get around the problem that nothing could eject the Higgs field from the false vacuum, which would lead to eternal inflation, the inflationists changed the value of the potential V to make it possible for inflation to happen during the transition from the false to the true vacuum and then come to an end. Letting the potential, that plays the role of the cosmological constant in inflation theory, 'slowly' reduce to zero, they made it possible for inflation to last some time, before coming to an end. To liberate themselves from the constraints imposed by the very real properties of the Higgs boson, as conceived by Higgs, even before its discovery at the CERN accelerator, the inflationists invented the 'inflaton'. This procedure of 'saving' their theory by changing the value of some physical constants, like the potential V, and by inventing new physical objects, like the inflaton, and new physical forces, like gravitational repulsion, even though they have never been

---

[378] John Hawley & Katherine Holcomb, *Foundations of Modern Cosmology* (1998): 430–431.

encountered in reality or in any experiment, makes it impossible to refute it. It is the procedure of verificationism, not science.

Verificationism pops up also in other parts of inflation theory, for example, in the way the inflationists changed their theory to firstly predict a flat, then a closed, subsequently an open and finally, again, a flat universe, respectively.[379] Firstly, inflation explained the flatness of the Universe ($k = 0$, as any geometric irregularity that existed before inflation would have been 'stretched out' and therefore 'ironed out' to the point of disappearing after inflation. The assumption that geometric irregularities are 'ironed out' and disappear with a large-scale expansion, contains a "*fundamental error*",[380] according to Penrose, because, for example, fractal sets are never 'ironed out', however much they are stretched.

Later, there was a stage in cosmology when Hawking and Turok's proposal of a closed Universe ($k = +1$), without boundaries, was popular, and inflationists adapted their model to fit a closed Universe. Changing the values of some terms, inflation suddenly explained that the Universe was not flat, but closed.[381] This manner of 'adapting' a theory to the evidence that might otherwise refute it is the essence of verificationism. Subsequently, Hawking modified his theory, in view of the growing evidence for an open Universe ($k = -1$) and so did the inflationists. Now inflation explained the open Universe: "*In accordance with this trend in observations, inflation theorists began to provide inflationary models which now allowed $k \neq 0$, with $k < 0$ in fact.*"[382] But the situation changed again in 1998, because new evidence emerged, based on observations of supernovae that suggested acceleration in the expansion of the Universe. As we saw in Chapter 2, this apparent acceleration is an

---

[379] Roger Penrose, *The Road to Reality* (2004): 772, 1023.
[380] *Ibidem*: 756.
[381] *Ibidem*: 772.
[382] *Ibidem*: 1023.

optical illusion, arising from the erroneous notion of absolute time, and so, there is no need for a cosmological constant greater than zero. Ignoring all this, some astrophysicists postulated its existence ($\Lambda > 0$), implying that $\Omega_M + \Omega_\Lambda = 0.3 + 0.7 = 1$. Inflation theory adapted again, now predicting a flat universe, with $k = 0$. Penrose comments: "*Most inflationists appear to have reverted to k=0 as being a prediction of inflationary cosmology. I am not sure what Popper would have had to say about all this!*"[383]

Karl Popper, an important 20th century's philosopher of science, would have said that all this is verificationism (see Chapter 7). A theory capable of explaining $k = 0$, and $k = -1$, but also $k = +1$, and then again, $k = 0$, is not a scientific theory but a verificationist speculation. A theory so flexible as to predict all possible outcomes, cannot be refuted by any of them. The frontier, however, between scientific and non-scientific statements, according to Karl Popper, is the principle of falsifiability.[384] I agree with Roger Penrose, who maintains that the theory of inflation doesn't solve anything, because it does not predict anything.[385]

Let us now have a closer look at 'gravitational repulsion', the negative gravitational force that is the cornerstone of inflationary theory. Historically, the cosmological constant is a hypothetical physical force, that pops up as a *Deus ex machina* whenever astrophysicists think they observe something that in reality does not exist, and want to explain this non-existent phenomenon. When Einstein conceived the Universe to be a stable distribution of matter that did not collapse gravitationally, which is impossible (see Section 1.1.1 of Chapter 1), he introduced the cosmological constant Λ, which is a negative gravitational force, to solve the paradox. When Hubble published data on the redshift of galaxies that corroborated

---

[383] Roger Penrose, *The Road to Reality* (2004): 1023.
[384] Karl Popper, *The Logic of Scientific Discovery* (2005): 17–20, 57–73; also *Conjectures and Refutations* (1962): 36–40, and *passim*. For his philosophy of science, see Chapter 7.
[385] Roger Penrose, *The Road to Reality* (2004): 746–757, 772, 1020–1024.

the Friedmann-Lemaître model of an expanding Universe, Einstein publicly accepted that the facts refuted his model of a static universe, and supported the dynamic model of an expanding Universe, which did not need gravitational repulsion by a cosmological constant.[386]

The second time the cosmological constant popped up was when Adams, Perlmutter and Riess observed an apparent acceleration of the expansion of the Universe for which they received the 2011 Nobel Prize for Physics. As I explained in Section 2 of Chapter 2, this observation results from the fact that the watch of an observer, embedded in the strong gravitational field of a galaxy cluster, runs more slowly than the average watch of the Universe, that has many voids surrounded by walls of galaxy clusters, like a sponge. This means that a receding supernova that travels a certain distance, at the other side of a large void, makes that journey in less time, according to the observer's watch, compared to a watch mounted on the supernova, or according to the Universe's average watch, creating the perception of a recent acceleration of the expansion of the Universe.[387]

Guth introduced his cosmological constant to explain inflation. Again, this appears to be a case of the cosmological constant popping up to explain something that does not really exist. The difference, however, with the above-mentioned cases of a cosmological constant, is that the Guth-Linde theory does not predict observable consequences. Einstein's introduction of a cosmological constant predicted a stable universe; and the standard ΛCDM model's introduction of a cosmological constant predicted the recent acceleration of the expansion velocity of the universe. These predictions could be confronted with physical reality, and were refuted, or reinterpreted, respectively. This means that both were scientific theories proven to be wrong. The introduction of the cosmological constant in the inflation theory, however, does not predict any outcome that can refute it. Neither the cause

---

[386] See Section 1.1.3 of Chapter 1.
[387] I explained all this in detail in Section 2 of Chapter 2.

(gravitational repulsion), nor the supposed main effect (inflation during a fraction of the first second of the universe) are observable, making the theory irrefutable. And when inflation theory does make observable predictions, it successively predicts all of them, like the universe being open, closed or flat, whatever, so that none of them can refute the theory. Again, the theory is irrefutable. I agree with Michael Rowan-Robinson's comment that "*inflation does not in fact make any concrete predictions about the universe*" and "*there is no reason for believing in inflationary theory unless it makes testable predictions about the universe*".[388]

Why do I pay so much attention to the inflation theory in the context of fine tuning? I do so, because, according to Guth, his theory "*can explain why the Universe today is so incredibly flat and therefore resolve the fine tuning paradox pointed out by Bob Dicke*". [389] He said this when cosmologists thought the universe was flat, somehow relating its flatness to its fine tuning. As a matter of fact, inflation theory does not explain the fine tuning of the physical constants at all. It does not predict, for example, that the fine-structure constant is $\alpha_C = 1/137$, neither that the strong nuclear force constant $\alpha_S = 0.1182 \pm 0.0027$, nor that the ratio of the electron mass to the proton mass is $\beta = 1/1836$.

Some inflationists aim to explain fine tuning by thermalization, by which physical bodies reach thermal equilibrium through mutual interaction, which was prior to the inflation. However, according to Penrose,

"*[I]t is fundamentally misconceived to try to explain why the universe is special in any particular respect by appealing to a thermalization process. For, if the thermalization is actually doing anything (such as making temperatures in different regions more equal than they were before) then it represents a definite increase of entropy. Thus, the*

---

[388] Michael Rowan-Robinson, *The Nine Numbers of the Cosmos* (1999): 126.
[389] Alan Guth, *The Inflationary Universe* (1998): 179.

universe would have been even more special before the thermalization than after. This only serves to increase whatever difficulty we might have had previously in trying to come to terms with the initial extraordinarily special nature of the universe."[390]

Since thermalization appeared not to be fit, after all, to explain the fine tuning of the physical constants, the inflationists resorted to speculations about the multiverse. Shortly after Guth went public with his inflation theory, "*it was developed into hypotheses of multiple universes by... Andrei Linde and others and within a few years many-worlds cosmology was established as a small but thriving cottage industry*".[391] Guth and Linde extended the hypothesis of the inflationary Universe, postulating an eternal inflation, also known as the theory of chaotic inflation.[392] The false vacuum decays, says Guth, because it has a half-life of some 10^{-35} seconds. While in these 10^{-35} seconds half of the false vacuum decays, the other half inflates, with an expansion rate much faster than the rate of decay. That is, the volume of the non-decaying part of the false vacuum inflates until it reaches a volume much greater than the volume of the part of the vacuum that decays. So, in spite of the decay, the region of false vacuum grows: "*the false vacuum region grows forever: once inflation begins, it never ends*".[393]

How is this latter statement compatible with the concept of a very short period of inflation in our Universe? Guth and Linde answer that the part of the vacuum which decayed, in 10^{-35} seconds, is where the Big Bang happened and our Universe was born. Inflation in the part that decayed

---

[390] Roger Penrose, *The Road to Reality* (2004): 755.

[391] Helge Kragh, *Conceptions of Cosmos* (2007): 235.

[392] Alan Guth, *The Inflationary Universe* (1998): 245–252; Andrei Linde, *Inflation and Quantum Cosmology,* (1990): 18–25. See also, Andrei Linde, "Inflation, Quantum Cosmology and the Anthropic Principle", arXiv.org/pdf/hep-th/0211048.

[393] Alan Guth, *The Inflationary Universe* (1998): 246.

lasted a very short time. The other half, not yet decayed, divides itself in two: one which decays (giving rise to another universe) and another which inflates, later dividing into two parts: one which inflates and another which decays (giving rise to a third universe), etc. This process gives rise to an infinite number of universes.

Linde introduced the concept of '*mutation*'.[394] Each Universe has its own natural laws and constants: "*The universe becomes divided into many different exponentially large domains... In some of these mini-universes... physics is quite different from our own.*"[395]

Since the number of universes is infinite, it follows that, by chance, in some universes the physical laws and constants are such as to allow the evolution of long-lived stars and intelligent life. Linde feels that in this way he has explained the fine tuning of physical constants in our visible Universe.

Now it is time for a critique of the eternal inflation theory. Firstly, Linde's statement about an eternal Universe without a point $t = 0$ is refuted by the fact that in an ever-expanding Universe, eternal "*inflation does not seem to avoid the problem of the initial singularity (although it does move it back into an indefinite past)*".[396]

Secondly, the inflationists postulate that, after the brief inflationary period, the universe was in a smoothed-out low-entropy state, so implicitly they affirm that the universe went from a very low entropy state (prior to inflation) to a very low entropy state (after inflation), which is in contradiction with the second law of thermodynamics, about the increase of entropy in our Universe, as Penrose correctly argues:

---

[394] Andrei Linde, *Inflation and Quantum Cosmology*, (1990): 25–28.
[395] *Ibidem*: 26, with my emphasis.
[396] Arvind Borde and Alexander Vilenkin, "Eternal Inflation and the Initial Singularity", in: *Physical Review Letters*, vol. 72 (1994): 3307.

"Our existence as we know it depends on the low-entropy gravitational reservoir inherent in the initial uniform matter distribution. This leads us to consider a remarkable – indeed fantastical – thing about the Big Bang. It is not merely the mystery of its very occurrence, but that it was an event of extraordinarily low entropy... [T]he inflationary argument [is] aimed at showing that a smoothed-out universe would inevitably result from the inflationary process. Suppose that it is indeed true that the inflationary processes will almost invariably lead to a smoothed-out expanding universe after inflation has finished. This concept fundamentally conflicts with the 2nd Law... Let us reverse the direction of time from such a macroscopic state – but with generically perturbed sub microscopic ingredients – and let the reversed-time dynamical evolution (with equations still allowing for the possibility of inflation, with φ-field, etc) take over. This must lead us somewhere, but now with entropy increasing in the collapsing direction. Where it leads us would be generally some very complicated high-entropy black-hole congealed state."[397]

Thirdly, the theory of eternal inflation is a new version of the old Steady State model by Hoyle, Bondi and Gold.[398] Both models aim to explain the fine tuning of our Universe as the outcome of a random distribution of variable values of physical constants, across an infinite number of regions in our Universe (Hoyle, Bondi and Gold) or an infinite succession of universes in time (Guth and Linde). The main difference between the inflationary multiverse model and the Steady State model, is that the latter was a scientific theory, because it could be tested with observations of our Universe, and, as a matter of fact, it was refuted, whilst the former model is not refutable, because the other universes that are continuously created are out of reach of our observation. Guth and Linde have created a theory that is

---

[397] Roger Penrose, *Fashion, Faith and Fantasy* (2016): 258, 304.
[398] See Sextion 4.1 of this Chapter.

irrefutable, because it cannot be compared with observations of reality. Guth openly admits this: *"we can never expect to test the predictions of inflation for the region found beyond the observable Universe."*[399] And Linde too admits that *"we cannot see them"*, referring to the other universes.[400] A theory, however, that is not refutable by evidence is not a scientific theory, but science fiction, as I explain in Chapter 7.

Penrose rightly asserts that the inflationists' belief that inflation predicts the fine tuning at the start of the Universe provides a powerful driving force behind their position:

> *"[T]here is an extraordinary degree of precision in the way that the universe started, in the Big Bang, and this presents what is undoubtedly a profound puzzle... The view of the inflationists is different, namely that this puzzle is essentially 'solved' by their theory, and this belief provides a powerful driving force behind the inflationary position. However, I have never seen the profound puzzle raised by the Second Law [of increasing entropy] seriously raised by inflationists!"*[401]

Why is this belief that inflation 'solves' the problem of fine tuning such a *"powerful driving force"* – as Penrose says – behind the inflationary position? Guth himself admitted that *"the idea of a truly eternal universe – one that has always existed and will always exist – is very appealing, since it frees us from all questions about how the universe was created"*.[402] Guth welcomes the idea that, once the mystery of fine tuning is resolved, we can dispose of the dominant idea present in *"both Judeo-Christian tradition and scientific contexts"* that portrays *"the origin of the universe as a unique event"*.[403] At a workshop of 1982, Linde too admitted that his

---

[399] Alan Guth, *The Inflationary Universe* (1998): 245.
[400] Andrei Linde, *Inflation and Quantum Cosmology,* (1990): 26.
[401] Roger Penrose, *The Road to Reality* (2004): 754.
[402] Alan Guth, *The Inflationary Universe* (1998): 248–249.
[403] *Ibidem*: 251.

idea of eternal inflation, implying a multiverse with no beginning in time disposes of the worrying idea of creation. It no longer appeared to be necessary to assume that a first mini-universe really appeared out of nothing at some moment t=0, so that *"there might be no initial creation to worry about"*.[404] He repeated this in 2002, affirming that the multiverse resulting from chaotic inflation gets rid of the 'miracle' of fine tuning:

> *"Since we have all universes with all possible laws of physics described by our extended action, we will certainly find the universe where we live in... It is sufficient to consider an extended action represented by a sum of all possible actions of all possible theories in all possible universes. One may call this structure a 'multiverse'... Given the choice among different universes in this multiverse structure, we can proceed by eliminating the universes where our life would be impossible. This simple step is sufficient for understanding of many features of our universe, that otherwise would seem miraculous."*[405]

Linde's idea is that all *possible* universes *really* exist. It is obvious, though, that all that is real must be possible, but not all that is possible is necessarily real: though ten-kilometre high mountains are possible, they do not really exist on planet Earth. We shall see that Max Tegmark uses exactly the same sophism, stating that all that is mathematically possible, must exist physically.[406] Linde admits his theory is just speculation: *"We need to move carefully, constantly keeping in touch with solid... facts, but from time to time allowing ourselves to satisfy our urge to speculate."*[407]

---

[404] Cited in Helge Kragh, *Conceptions of Cosmos* (2007): 235.

[405] Andrei Linde, "Inflation, Quantum Cosmology and the Anthropic Principle", arXiv.org/pdf/hep-th/0211048 (2002): 16–17.

[406] The sophism that all that is possible is necessarily real is refuted in Section 8.1 of Chapter 8 and Linde's and Tegmark's use of this sophism is analysed in Section 8.3.

[407] Andrei Linde, "Inflation, Quantum Cosmology and the Anthropic Principle", arXiv.org/pdf/hep-th/0211048 (2002): 30.

Much of the speculation about inflation and the multiverse seems to be motivated by the desire to get rid of the idea of creation (see Section 8.3 of Chapter 8). This may be why so many other cosmologists too, in spite of the speculative status of the Guth-Linde theories of eternal inflation and the multiverse, accept it as if it were a matter of fact, as Penrose points out: "*In my own opinion, this picture [of inflation at the beginning of the Universe] must be regarded as very speculative..., although it is often presented as virtually established fact.*"[408]

Section 4.5 Superstrings and Susskind's theory of the multiverse

Susskind's theory of the multiverse departs from string theory in physics. In its turn, string theory started as a search for unifying the fundamental forces of nature. I will first treat the search for symmetry in physics (Section 4.5.1), and then strings, superstrings and the multiverse (Section 4.5.2).

Section 4.5.1 The search for symmetry

What is symmetry? To understand this, we must take into account that in physics there are laws that, in concrete situations, determine and predict certain results. If the relationship between cause and effect, determined by a physical law, doesn't change in an operation in which I substitute some elements of the natural or experimental situation for others, then we say that the law is symmetric under that operation. For example, there is a physical axiom that says that all physical laws must be symmetric with respect to the system of coordinates. The physical law does not change, even though the coordinate system changes. When we change the location of a physical system on just one of the axes of the coordinate system, it is called *translation*. When we change its location in three-dimensional space, it is called *rotation*. *Dilation* is a geometric transformation that is

---

[408] Roger Penrose, *The Road to Reality* (2004): 752.

implied by symmetry under rotations in *space*, for example, the law of conservation of momentum (= the velocity of an object multiplied by its mass); an example of symmetry under translation in *time* is the law of conservation of energy (the total energy of an isolated system remains constant). This symmetry with respect to the system of coordinates, proven by general relativity, means that physical laws are background-independent and is considered to be fundamental in physics.

There is another type of symmetry, in which we exchange the physical objects subject to that law and not the coordinate system. In an operation where we substitute some objects for others, without affecting the result of the law, the physical law is symmetric for that operation. For example, if we drop balls of the same size but a different weight (iron or wood), from the same height, they will all arrive at the Earth's surface at the same time. We say that Newton's law of gravity is symmetric in Galileo's experiment, that is, in the operation of substituting balls of one weight for balls of another weight, but identical size: the acceleration of both bodies is the same, even if they have a different mass. In all of these cases the cause, i.e. gravity, produces the same effect, that is, the same acceleration and the same travel time.

If we were to carry out the same experiment in a perfect vacuum, a third symmetry would be added, along with the two already pointed out, that is, the law of gravity produces the same acceleration independent of the size or shape of the object that is falling. When the astronauts dropped a feather and a ball from the same height on the Moon, both objects reached the surface of the Moon at exactly the same time. In this example, we increased the number of symmetries from two to three, *by changing the context of the experiment* from the Earth to the Moon, and, vice-versa, the third symmetry would be broken if we moved the experiment from the Moon to the Earth. By understanding more symmetries, we increase the information about the properties of the physical laws in question.

In the example of the symmetries of the law of gravity and the resulting acceleration of different types of objects, the symmetries give us partial, but not comprehensive, information about the gravitational force. When the characteristics of a physical force are comprehensively described and determined by its symmetries, we talk about *gauge* forces and the *gauge* principle applies. 'Gauge' means 'standard measure'.

The example given above shows us that symmetry and the breaking of symmetry depends on the physical context in which a physical law operates. It is not the same thing to drop a feather and a tennis ball in a vacuum as within an atmosphere. Generally, situations of symmetry are unstable, and the situation after symmetry breaking is stable. This explains why many symmetries are broken spontaneously (spontaneous symmetry breaking). For example, if I place a pencil in a perfectly vertical position on its tip on the table, when I let go, the law of gravity determines that the pencil will fall, but it doesn't tell us in which direction it will fall. The law of gravity is symmetric with respect to the direction of fall. But that symmetry is inherently unstable. Once the pencil has fallen, the symmetry has been broken by the final position of the pencil. The symmetry breaking has led to a stable situation.

The set of possible physical states after symmetry breaking implies a probability distribution that can have different shapes, whose sum is, by definition, one. In the example of the probability distribution of possible positions of the pencil after falling, all of them have the same probability and there is an equal distribution (whose graph is a horizontal straight line). In other cases, other probability distributions can exist, for example, a normal curve or a sinusoidal curve. It is necessary for symmetry to break, but the way in which it breaks is incidental: "*The laws [of physics] describe only the space of what possibly may happen; the actual world governed by those laws involves a choice of one realisation from many possibilities.*"[409] As such, the transition from a symmetric and unstable situation to an asymmetric and stable situation, and the way in which one of the many possibilities contained in the probability distribution is realised, depends not only on the physical law but also on the existing external, physical context that the law is operating in:

> "*[T]he symmetry transformations considered so far have been ones that leave the laws of physics invariant. Here the 'laws of physics' means the dynamical laws that govern how the state of the world evolves in time, expressed in classical physics by Newton's laws and Maxwell's equations, and in quantum physics by the Schrödinger equation. A subtle point about this is that, while the form of the equations may not change under symmetry transformations, in general the solutions to the equations will change [in different 'world states']. While the laws governing the evolution of the state of the world may be symmetric, the actual state of the world generally is not.*"[410]

As well as the universal symmetry of physical laws with respect to the system of coordinates, and partial symmetries with respect to the substitution of some elements for others in different world states, physicists extend the concept of symmetry even further, e.g. the symmetry of fundamental physical forces, also known as the unification of forces in very high energy situations, such as existed during the Big Bang, or can be generated in particle collision experiments. The four forces in physics are different, both in their range of action and their magnitudes. However, it is conjectured that at certain very high energy levels, some of these forces become interchangeable, and therefore, symmetric. So, in that case, it is no longer a matter of different forces but instead a single *unified force.*

---

[409] Lee Smolin, *The Trouble with Physics* (2006): 60.
[410] Peter Woit, *Not Even Wrong* (2006): 71.

The search for the unification of the electromagnetic and weak nuclear forces has been a big success.[411] At very high energy levels, the electromagnetic force and the weak nuclear force act as a single unified force. Above this energy level, the electroweak super force is symmetric. Below this level, the symmetry is spontaneously broken and the forces, with their respective carrier particles, are differentiated, both in magnitude and range as well as mass. The two forces referred to are gauge forces. When the electroweak force differentiates into two forces, the carrier particle of this force also differentiates, generating the so-called gauge bosons, that is, the γ photons for the electromagnetic force, and the Z, W^+ and W^- bosons for the weak nuclear force.

Weinberg asserts that, *"to complete the [electroweak] theory, we must now make some assumption about the mechanism of symmetry breaking. We want this mechanism to give masses not only to the $W^{\pm}$ and Z, but to the electron as well".*[412] He refers to the Higgs mechanism, and the particles corresponding to the Higgs field are the Higgs bosons. Higgs particles are the quanta of the Higgs field. Making the relevant corrections, in particular, the value of the fine-structure constant ($\alpha = 1/129$ instead of $\alpha = 1/137$), Weinberg obtained the following values for the masses of $W^{\pm}$ and Z^0:

$$m_W = \frac{38.4 GeV}{|\sin\theta|} \quad \text{and} \quad m_Z = \frac{76.9 GeV}{|\sin 2\theta|}$$

According to Weinberg, *"whatever the value of θ, these masses are too large for there to have been any hope of detecting the W or Z in the 1960s or early 1970s".*[413] Weinberg is referring to the fact that, in those days, when he first published his book, the particle accelerators were not powerful enough to achieve the high energies necessary for producing bosons with

---

[411] Steven Weinberg, "The Electroweak Theory", in: *The Quantum Theory of Fields, Volume II* (2005): 305–318.

[412] Steven Weinberg, *The Quantum Theory of Fields, Volume II, Modern Applications* (1996): 308.

[413] *Ibidem*: 310–311.

these masses. By 1983, all measurements of the angle θ had yielded a value of $sin^2 \theta = 0.23$, which allowed the masses of the two particles to be predicted, i.e. $m_W = 80.1 GeV$ and $m_Z = 91.4 GeV$.[414]

In 1983, the $W^\pm$ and Z^0 particles were discovered and Weinberg reported with justified pride that their respective masses had values "*in satisfactory agreement with the predictions of the electroweak theory*",[415] i.e. $m_W = 80.410 \pm 0.18\,GeV$ and $m_Z = 91.1887 \pm 0.0022\,GeV$. So, the theory of weak decay by Sheldon Glashow (a physicist from the United States born in 1932), the Pakistani physicist Abdus Salam (1926–1996), and Steven Weinberg (born in 1933), for which they received the 1979 Nobel Prize, was corroborated.

The Feynman diagrams[416] of this decay, as presented in Figure 4.1, summarise what is now possible to observe in particle accelerators:

Figure 4.1 The weak interaction mediated by bosons W^+, W^- and Z^0

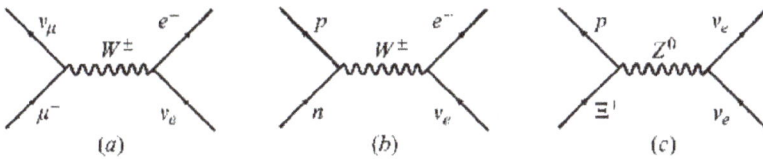

(a) $\mu^- + v_e \rightarrow v_\mu + e^-$; (b) $n + v_e \rightarrow p + e^-$; (c) $\Xi^+ + v_e \rightarrow p + v_e$.

These observations (where the proton mass is $m_p = 938.3 MeV/c^2$) allowed the mass of the Higgs boson to be estimated at:[417]

$$120 m_{proton} \approx 112.5 GeV < m_{Higgs} < 240 m_{proton} \approx 225 GeV.$$

---

[414] Steven Weinberg, *The Quantum Theory of Fields, Volume II, Modern Applications* (1996): 316.
[415] *Ibidem*: 316.
[416] See Richard Feynman, *Quantum Electrodynamics. The strange theory of light and matter* (2006).
[417] Fot upper and lower limits see Steven Weinberg, *The Quantum Theory of Fields, Volume II, Modern Applications* (1996): 316 and Lee Smolin, *The Trouble with Physics* (2006): 260, respectively.

The mass of the Higgs boson is the only mass in the Standard Model that does not have well defined limits: it is not 'protected' as quantum physicists say, as is the case for other bosons and fermions. But even so, in order to prevent the mass of the Higgs boson becoming too large, and to keep it within certain limits, the values of the free constants in the Standard Model (some 20) must be fine tuned with a precision of $1/10^{32}$. The slightest variation in these constants would make the Higgs boson much more massive than predicted by the electroweak theory. In July 2012, the Large Hadron Collider at CERN, that makes protons collide at nearly the speed of light, corroborated the mass of the Higgs boson as being within the limits posed by the predictions of the electroweak theory.

The Higgs particle is important, not only in the electroweak theory, but also in the Standard Model in general, into which it was integrated by Gerard 't Hooft and Maarten Veltman, Dutch physicists who won the Physics Nobel Prize in 1999, yielding the following results according to 't Hooft:[418] firstly, *"the need for one spin 0 particle to give the electro-weak force the symmetries it has"*; secondly, that *"this one Higgs particle now couples to the quarks and leptons to give them their masses"*; and thirdly, *"the same Higgs particle can also produce transitions between the various quark types"*.

From the point of view of the logic of scientific discovery,[419] these discoveries made by different physicists are valid:

1) The predictions of the electroweak theory can be confronted with the facts. So, the theory about the unified, electroweak force is science.

2) The electroweak theory has, as a matter of fact, been corroborated by observations, in particular, the characteristics of the carrier particles, $W^{\pm}$ and $Z^{0,}$ of the weak force, between 1962 and 1983, and the Higgs boson, at the CERN accelerator, in July 2012.

---

[418] Gerard 't Hooft, *In search of the ultimate building blocks* (1997): 115.
[419] See Karl Popper, *The logic of scientific discovery* (1980), and summary in Section 21.

Table 4.1 gives us the bosons, which carry the four fundamental forces. Note that the graviton is a hypothetical particle, as yet not observed in reality.

Table 4.1 The standard model: the fundamental force carrier bosons

force	carrier	symbol	mass*	charge	spin	range
electromagnetic	photon	γ	0	0	1	∞
weak nuclear	weak force carriers	W^-	80.4	-1	1	$10^{-19}m$
		W^+	80.4	+1		
		Z	91.2	0		
unified electroweak	Higgs boson	H^0	125		0	
strong nuclear	gluon	g	0	0	1	$10^{-15}m$
gravitational	graviton?					∞
*mass in GeV/c^2						

The fermions are the twelve elementary particles which make up all the matter in the Universe.[420] There are two types of fermions: leptons (among them electrons and neutrinos), and quarks (the up, down, charm, strange, top and bottom quarks). Isolated quarks have not been observed as such, but rather, it is assumed they exist inside hadrons and mesons. Part of the theory concerning quarks is called quantum chromo dynamics (QCD), where the word 'chromo' (=colour) means that it is conjectured that each of the three types of quarks, in each of its two versions (up & down, charm & strange, top & bottom), could have one of three 'colours' (referring to new quantum numbers, not to a colour), giving a total of six quarks, each one in three colours, without counting their respective anti-particles (see Table 4.2).

---

[420] All baryons (some 120) consist of three quarks and all mesons (some 140) of one quark and one anti-quark.

Table 4.2. The standard model: fermions (leptons and quarks)

	leptons						quarks					
type	electron	electron neutrino	muon	muon neutrino	tau	tau neutrino	up	down	charm	strange	top	bottom
symbol	e	v_e	μ	v_μ	τ	v_τ	u	D	c	S	t	b
mass GeV/c	$5.1^* \, 10^{-4}$	$<1^* \, 10^{-8}$	0.106	$<1.7^* \, 10^{-4}$	1.78	$<1.8^* \, 10^{-2}$	0.005	0.01	1.5	0.2	175	4.7
electric charge	-1	0	-1	0	-1	0	$+2/3$	$-1/3$	$+2/3$	$-1/3$	$+2/3$	$-1/3$
spin	½	½	½	½	½	½	½	½	½	½	½	½

Then there are the hadrons, divided between the baryons, which consist of three quarks, and the mesons, which consist of two quarks. I will only mention the two hadrons that have a longer life span (the proton and the neutron), omitting the very short-lived hadrons (the lambdas, sigmas, xis, omega, and deltas) and mesons (the pions, caons, psi, and D's), which decay soon after being produced in particle collisions. Protons are the very solid building stones of the Universe. Of a given quantity of protons, after about 10^{32} years, 50% will have decayed spontaneously. In the case of a neutron, the equivalent half-life is 11 minutes. The Universe, right now, has an age of $\pm 15^*10^9$ years, so very few protons have spontaneously decayed.

Table 4.3 The standard model: two baryons

name	symbol	mass GeV/c^2	charge	half-life time	quarks	quark sign	spin
proton	p	0.938	+1	$\approx 10^{32} \, yr$	uud	+1	0
neutron	n	0.940	0	$\approx 660 \, s$	udd	-1	0

From the 1970s on, the search began for unifying the strong nuclear force with the electroweak force. The theory of unification in this case seeks to unify *five* particles related to these three forces, that is, three types of quarks, each with two versions, and two types of leptons, i.e. electrons and neutrinos. For this reason, it is also known as the *SU(5) theory*.

The theory made a number of predictions specific to the Standard Model, which have now been corroborated. It also made other predictions which can be tested against reality. The new prediction has to do with proton decay. There are several ways in which protons may decay,[421] but the half-life of a proton is a multiple of 10^{32} years, so the decay event is very rare:

"One of these new predictions was that there had to be processes by which quarks can change into electrons and neutrinos, because in SU(5), quarks, electrons, and neutrinos are just different manifestations of the same underlying kind of particle. SU(5) indeed predicts such a process, which is similar to radioactive decay. This... prediction is characteristic of grand unification. It is required by the theory and is unique to it. The decay of a quark into electrons and neutrinos would have a visible consequence. A proton containing that quark is no longer a proton; it falls apart into simpler things. Thus, protons are no longer stable particles; they undergo a kind of radioactive decay. Of course, if this happened very often, our world would fall apart, as everything stable in it is made of protons. So, if protons do decay, the rate must be very small. And that is exactly what the theory predicted: a rate of less than one such decay every 10^{33} years."[422]

---

[421] Several ways a proton can decay into other particles: $p \rightarrow \pi^0 + e^+$; $p \rightarrow \pi^0 + \mu^+$; $p \rightarrow K^+ + \bar{\nu}$; $p \rightarrow K^0 + e^+$; and $p \rightarrow K^0 + \mu^+$, among others, with exposure ranking from 70 to 80 kiloton during one year (one kiloton is one million kilos, about one million liters of water). The half-life time of these decays is between 5 and 50 times 10^{32} years.

[422] Lee Smolin, *The Trouble with Physics* (2006): 63–64.

Given the rarity of proton decay events, an experiment was designed: filling subterranean tanks at great depths in abandoned mines with ultra-pure water (i.e. in Lake Erie, Ohio; the Kamiokande detector in Japan; and the Kolar Gold Fields mine, in southern India), where, given the extremely low proton-decay rate, a few decays per year might be observed. After initially disappointing results, Anthony Mann and Indian and Japanese teams published papers with apparent evidence in favor of proton-decay.[423]

But the results were later disproved, when the Super-Kamiokande team showed that the ν_μ/ν_e anomaly was caused by neutrino oscillations, not proton decay, and ruled out the proton lifetime proposed in these papers. 'Neutrino oscillation' refers to the phenomenon that neutrinos can spontaneously change their 'flavor', for example, from a muon neutrino ν_μ to an electron neutrino ν_e (see Table 4.2). In 2017, Mann referred me to a Kamiokande paper that refutes his prior interpretation: the paper states that *"trilepton proton decay modes were offered [by Mann et al.] as an explanation of the atmospheric neutrino flavor 'anomaly', before neutrino oscillations were established"*, and setting *"partial lifetime limits of* 1.7 ∗ 10^{32} *and* 2.2 ∗ 10^{32} *years for* $p \to e^+\nu\nu$ *and* $p \to \mu^+\nu\nu$, *respectively"* which *"provide strong constraints to the permitted parameter space [proposed by previous papers] which predict lifetimes of around* $10^{30} - 10^{33}$ *years"*.[424]

---

[423] Anthony Mann, T. Kafka & W. Leeson, "The Atmospheric Flux ν_μ/ν_e Anomaly as Manifestation of Proton Decay $p \to e^+\nu\nu$", in: *Physics Letters B*, vol. 291, 1992, pp. 200–205. See also H. Adarkar *et al.* (Tata Institute of Fundamental Research, Bombay) and Y. Hayashi *et al.* (Osaka City University), "Experimental Evidence for G.U.T. Proton Decay" (2000), arXiv:hep-ex/0008074; and M. Miura, "Search for Proton Decay via $p \to e+\pi^0$ and $p \to \mu+\pi^0$ in 0.31 Megaton-years Exposure of the Super-Kamiokande Water Cherenkov Detector", arXiv:1610.03597:hep-ex.

[424] Anthony Mann, private email to the author, October 12, 2017; Super-Kamiokande Collaboration, "Search for Trilepton Nucleon Decay via $p \to e^+\nu\nu$ and $p \to \mu^+\nu\nu$ in the Super-Kamiokande Experiment", in: *Physical Review Letters*, vol. 113 (September 2014): 101801-2. The 2015 Physics Nobel Prize was awarded to Takaaki Kajita of the Super-Kamiokande team and Arthur McDonald, a Canadian physicist, *"for the discovery of neutrino oscillations, which shows that neutrinos have mass"*.

Even if this part of the *SU(5)* theory would eventually be conclusively corroborated by the facts, it would leave two different entities without unification, i.e. bosons and fermions. Bosons are the carriers of the forces, such as photons, the $W^{\pm}$ and Z^0 particles, the gluons and the hypothetical gravitons, and fermions constitute matter, such as electrons, protons, neutrons, and neutrinos. Supersymmetry theories aim to unify forces and matter: bosons and fermions. Through supersymmetry we can substitute bosons with super-fermions and fermions with super-bosons, without changing the results of the experiment. It is conjectured that each fermion has a super-partner, namely, an unknown boson with the same mass and charge. For example, the partner of the electron would be the super-electron, or 'selectron'. This supersymmetry is somewhat difficult to achieve, because fermions, including electrons, obey Pauli's exclusion principle that prevents them from having the same quantum state, whereas bosons seek to be in the same quantum state. The theory was first proposed in the Soviet Union, by four Russian physicists, in 1971and 1972. Nobody in the West read Soviet academic journals, but the idea was re-proposed in the West by Julius Wess and Bruno Zumino in 1973.

Another problem is that the energies necessary to achieve this supersymmetry can't be reached in today's particle accelerators. While some 100 *GeV* is required for the Higgs, at least 10^{15} *GeV* is required to obtain spontaneous supersymmetry breaking. This in itself does not make it impossible to confront the theory with the facts at some point in the future. The Higgs boson too had to wait 48 years, from 1964 to 2012, before a particle accelerator was at hand, powerful enough to detect it.

Section 4.5.2 Strings, superstrings and the multiverse

In this matter, I choose to be led by experts on the subject, both those that propose string and superstring theory, e.g. Joseph Conlon, Michael Green, Brian Greene, Pierre Ramond, Joël Scherk, John Schwarz, Leonard

Susskind, Gabriele Veneziano, Edward Witten and others, as well as their critics, among them, Richard Feynman, Daniel Friedan, Sheldon Glashow, Gerard 't Hooft, Roger Penrose,[425] Lee Smolin[426] and Peter Woit.[427] Steven Weinberg, who dedicates an entire volume of his three volume work *The Quantum Theory of Fields* to supersymmetry, has only a few passing remarks on superstring theory, which is *"beyond the scope of this book"*.[428]

The work on the unification of the electroweak and strong nuclear forces, paved the way for string theory. As a matter of fact, string theory started as a theory of strong interactions.[429]

In 1968, Gabriele Veneziano introduced the mathematical concept behind 'strings', although he did not use the word 'string'. He used Euler's beta function, which was the first known scattering amplitude in string theory, presenting a set of equations that describe the probabilities of two particles colliding and scattering at different angles.[430] In the early 1970s, Yoichiro Nambu, Holger Nielsen and Leonard Susskind, each independently of the others, interpreted these equations using the old S-matrix theory. The S-matrix, where 'S' is for scattering, was conceived by John Wheeler in 1937 and developed by Werner Heisenberg in 1943. It is a mathematical object which tells us what happens when two particles on different courses collide: do they scatter, following their respective paths intact but with a different momentum, or do they annihilate, producing other particles?[431]

---

[425] Roger Penrose, *Fashion, Faith and Fantasy: on the New Physics of the Universe* (2016): Chapter 1, Fashion.

[426] Lee Smolin, *The Trouble with Physics. The Rise of String Theory, the Fall of Science, and What Comes Next* (2006).

[427] Peter Woit, *Not Even Wrong. The Failure of String Theory and the Search for Unity in Physical Law* (2006).

[428] Steven Weinberg, *Quantum Theory of Fields, Volume III Supersymmetry* (2005): 397.

[429] See Jospeh Conlon, "What was string theory?", in: *Why String Theory* (2016): Ch. 5.

[430] Gabriele Veneziano, "Construction of a Crossing Symmetric Regge-Behaved Amplitude for Linearly Rising Regge Trajectories", in: *Nuovo Cimento* (1968): 190–197.

[431] See Peter Woit, *Not Even Wrong* (2006): 139–149.

Instead of particles that can be conceived as points in space, string theory conceives *strings* that can stretch and contract like rubber bands. Strings can be closed, connecting the two ends, like loops, or open, with two endpoints, like shoelaces. When they stretch, their energy increases, and when they lose energy, they contract. These strings travel through space-time and can collide and exchange energy. Like rubber-bands, the strings can also oscillate: "*A string living in two spatial dimensions has one direction it can oscillate in; a string in three spatial dimensions has two directions it can oscillate in; a string in twentyfive spatial dimensions has twentyfour directions it can oscillate in.*"[432] The vibrating strings substitute for the point particles of the Standard Model that are produced in proton collisions in particle accelerators. It is either point particles or strings, the two theories being mutually incompatible.

The mathematics of string theory are complicated. It is to the merit of string theorists that they overcame these mathematical problems. But there remained some serious problems. On the theoretical level, there was the problem that string theory was only consistent with Einstein's special relativity and with quantum mechanics, if four conditions were satisfied, each of them being unacceptable in orthodox physicis:[433]

a. To be a consistent quantum theory, string theory, according to Claud Lovelace, required 25 spatial dimensions.

b. A particle called the tachyon should exist, appearing to move faster than the speed of light. Conlon somehow justifies the tachyon as a signal of instability.[434]

c. Besides the photon, other particles without mass must exist, i.e. particles that are never at rest, because mass is a measurement of the energy of a particle at rest.

---

[432] Joseph Conlon, *Why String Theory* (2016): 70.
[433] Lee Smolin, *The Trouble with Physics* (2006): 105; and Peter Woit, *Not Even Wrong* (2006): 146–147.
[434] Joseph Conlon, *Why String Theory* (2016): 75.

d. Fermions do not exist, implying that quarks do not exist.

The third condition (c) was a problem, because there were no known particles without mass subject to the strong interaction. The fourth condition (d) presented a huge problem, since quarks and fermions did appear to exist. But above all, the first two conditions (a and b) seemed unacceptable, since they violated two axioms of physics, firstly, that nothing travels faster than light, and secondly, that observable space has three dimensions, not 25.

By this time, *Quantum Chromo Dynamics* (QCD), which postulated particles, not strings, was shown to quite fit the Standard Model. The extravagant character of string theory and the success of QCD convinced people that "*string theory was the failed theory of strong interactions*".[435]

In the early seventies, Pierre Ramond tried to rescue the theory, by resolving three of the four problems of the four above mentioned. The tachyon and its apparent faster than light velocity were eliminated from the string theory, making the theory compatible with special relativity, and he reduced the number of spatial dimensions from 25 to nine, which was, of course, still six too many. Ramond also reintegrated fermions into the model, given that, in his model, "*operators appearing in the description of point particles in conventional theories must be thought of as averages over some internal motion when applied to a hadronic system*".[436] Both bosons and fermions appeared as specific states in the same complex system and were interchangeable at high energies. From that moment, string theory was revived as supersymmetry string theory, or theory of superstrings.

Another means of achieving the same result was presented by Andrei Neveu and John Schwarz, who added to the theory the possibility of the superstrings interacting. Later, Neveu and Joel Scherk postulated that the strings have vibrational states corresponding to bosons and, in 1974, Scherk and Schwartz made the necessary mathematical adjustments for some of

---

[435] Joseph Conlon, *Why String Theory* (2016): 73, 75.
[436] Pierre Ramond, "Dual Theory for free Fermions", in: *Physical Review D* (1971): 2415.

these bosons without mass to be gravitons, assuming that, in that way, the superstring theory was the long sought-after quantum theory of gravity.[437]

In this superstring theory, photons are produced by the vibrations of open or closed strings, and gravitons only by the vibrations of closed strings. The ends of open strings are equivalent to the pairs of particles with opposite charges, for example, an electron and a positron. The vibration of the string between its two ends describes the photon that transports the charge between the particle and its antiparticle. As such, the string yields both the particles and the forces of the Standard Model, in addition to gravity and the graviton. Closed strings originate in the collisions between the two ends of an open string, producing a photon in the consequent annihilation of the particle and antiparticle. The force field lines of classical physics are no more than vibrating strings. This was the principle of the duality of strings and force fields.

String theorists worried about the existence of anomalies[438] in the theory. This changed dramatically in 1984, when John Schwarz and Michael Green published an article which eliminated the anomalies present in the theory up until that time, and rectified previous calculations to make the superstring theory fit for both open and closed strings, also reducing the number of space-time dimensions from eleven to ten.[439] Before the article was published, Edward Witten asked them to send the manuscript to him. Such was Witten's prestige, that dozens of theoretical physicists also started studying it. String theory changed overnight from its marginal status to being the centre of attention of theoretical physics in the USA. In 1983, 16

---

[437] Joel Scherk & John Schwarz, "Dual Models for Non-Hadrons", in: *Nuclear Physics B*, vol. 51 (1974): 118–144. See also Joseph Conlon, *Why String Theory* (2016): 76.

[438] 'Anomalies' appear in quantum physics when renormalisation techniques are applied to eliminate infinities in the solution to the equations. For the anomaly problem in string theory, see Peter Woit, *Not Even Wrong* (2006): 122–124.

[439] Michael Green & John Schwarz, "Anomaly Cancellations in Supersymmetric D=10 Gauge Theory and Superstring Theory", in: *Physics Letters B*, vol. 149 (1984): 117–122.

articles were published on strings; in 1984, there were 51; in 1985, 316; and in 1986, 639.[440] String theory had bounced back.

At Harvard, the string theory seminar was called 'Postmodern Physics'. An important characteristic of postmodern physics was that the problem of how to experimentally test the theory was rarely discussed. Since observable space has only three spatial dimensions and the theory had nine of them, superstring theorists speculated that the additional six dimensions were so small they were not observable: *"There appeared to be no choice but to curl them up so that they were too small to be perceived."* [441] Hiding the additional dimensions generated a new problem, namely, that string theory was background-dependent, i.e. not symmetric under a change of coordinate systems and, in consequence, each coordinate system had its own string theory. In other words, each different way of curling up and hiding the extra dimensions generated a new theory:

> *"Because string theory is a background-dependent theory, what we understood about it at a technical level was that it gave us a description of strings moving in fixed-background geometries. By choosing different-background geometries, we got technically different theories... The physical predictions given by all these different theories were different, too.... when the strings are allowed to move in the complicated geometry of the six extra dimensions, there arise lots of different kinds of particles, associated with different ways to move and vibrate in each of the extra dimensions."*[442]

If there are an unlimited number of ways to envisage the geometry of this hidden six-dimensional space, then there are an unlimited number of string theories. Let me explain: the geometry of this six-dimensional hidden space

---

[440] Data from Peter Woit, *Not Even Wrong* (2006): 150.
[441] Lee Smolin, *The Trouble with Physics* (2006): 119.
[442] *Ibidem*: 119, 121.

fixed a list of free constants. These constants determined the particular characteristics of a specific geometry, for example, its volume. A typical string theory could have hundreds of constants that described how the strings travel through the hidden space, how they vibrate and how they interact. Each geometry and different mode of moving, vibrating and interacting of strings, implied different values of the constants and, as a consequence, produced different particle masses and different force magnitudes. Some of these particles form part of the Standard Model, but the majority are as unobservable as the geometry that fixes them: *"because there were a huge number of choices for the geometry of the extra dimensions, the number of free constants went up, not down"*.[443]

It is no surprise then that some constants were generated that were more or less compatible with the Standard Model, but other parts of the Standard Model were not reproduced. So, a key question emerged: was there a way to curl up the extra dimensions in such a way as to reproduce the Standard Model of particle physics completely? It was a matter of finding geometry for the world of the hidden spatial dimensions that, on a much larger, visible scale would be observed as our three-dimensional space. The answer seemed to be found in 1985, in an article by Witten and others,[444] who used the geometry of space developed by Eugenio Calabi and Shing-tung Yau, called Calabi-Yau space. Thus, it seemed to be possible to substitute the constants in the Standard Model, such as, for example, those determining the masses of different particles, for constants that described the geometry of a Calabi-Yau space.

---

[443] Lee Smolin, *The Trouble with Physics* (2006): 121.
[444] Philip Candelas, Gary Horowitz, Andrew Strominger & Edward Witten, "Vacuum Configurations for Superstrings", in: *Nuclear Physics B*, vol. 258 (1985): 46–74.

In the early 1990s, five types of string theories existed that were thought to be connected at a deeper level.[445] However, there was a problem. There are many Calabi-Yau spaces, at least several hundreds of thousands, according to Yau himself. Each of these possible spaces gives rise to a different variation of particle physics and a different list of free constants.[446] If there are that many ways to compact the ten-dimensional space, with its six hidden space dimensions, into the four dimensions of our world, then nothing seems to be really explained. There is no explanation why one should choose one Calabi-Yau space and not another, nor of how one gets from a particular Calabi-Yau space to our world: *"if our world was described by one of the Calabi-Yau geometries, there was no explanation for how it got that way"*.[447]

This was also the point made by Daniel Friedan, one of the founders of a renowned group of string theorists, at Rutgers University, who eventually got disillusioned with string theory:

"String theory, as it stands, has failed as a theory of physics because of the existence of a manifold of possible background space times. All potentially observable properties of string theory depend on the geometry and topology of the background space-time in which the strings scatter. In string theory, a specific background space-time has to be selected by hand, or by initial conditions, from among the manifold of possibilities.

---

[445] I.e. the following five: (1) Type I superstring theories, namely, the $SO(32)$, which Michael Green & John Schwarz had discovered in 1984 as eliminating anomalies, in "Anomaly Cancellations in Supersymmetric D=10 Gauge Theory and Superstring Theory", in: *Physics Letters B*, vol. 149 (1984): 117–122; (2) and (3) two variations of type II superstring theories; (4) heterotic string theory with two copies of E_8 symmetry; and (5) a variation of heterotic string theory, with $SO(32)$ symmetry. See Peter Woit, *Not Even Wrong* (2006): 154 and Lee Smolin, *The Trouble with Physics* (2006): 129 and Joseph Conlon, *Why String Theory* (2016): 83–91.
[446] Lee Smolin, *The Trouble with Physics* (2006): 122.
[447] *Ibidem*: 123.

Many continuously adjustable parameters must be dialled arbitrarily to specify the background space-time. The existence of a manifold of possible background space times renders string theory, as it stands, powerless to say anything definite that can be checked."[448]

Conlon too makes this point, but he asserts that this does not make superstring unscientific (I'll come back to this assertion shortly):

"How many correct ways are there to go from ten to four dimensions in string theory? The apparent answer is infinity. There are a large number of exact supersymmetric solutions which have continuous parameters, in particular, type II strings on Calabi-Yau geometries. As these parameters are continuous, they can take an infinite set of values. With an infinite set of values there are an infinite number of solutions... Whatever the status of string theory in ten dimensions, there are roughly 10^{500} consistent ways of curling up six of the ten dimensions to turn the theory into a four-dimensional one."[449]

Originally, superstring theory appeared to be the grand unified theory of physics, but the multiplication of Calabi-Yau hidden space-geometries meant that *"string theory itself was in need of unification"* [450] and consequently, *"by the early 1990s, interest in superstring theory was beginning to slow down"*.[451]

In March 1995, at a string theory conference in Los Angeles, Witten came to the rescue of superstring theory by suggesting that a theory should exist that unified string theory. He didn't present that theory, but did mention some characteristics that it should have, based on recent advances relating string theory to general relativity, amongst other things. To achieve this

---

[448] Daniel Friedan, "A Tentative Theory of Large Distance Physics", in: *Journal of High Energy Physics,* vol. 2003 (2003): 6.
[449] Joseph Conlon, *Why String Theory?* (2016): 97–98.
[450] Lee Smolin, *The Trouble with Physics* (2006): 129.
[451] Peter Woit, *Not Even Wrong* (2006): 154.

unification, Witten conjectured that a tenth spatial dimension must exist, giving a total of eleven space-time dimensions and potentially unifying string theory with a supersymmetric theory of gravity that also had eleven dimensions. This extra spatial dimension allowed the assertion that a string actually contained a hidden dimension, transforming it from a loop into something like a doughnut.

Witten's suggestion implied it was possible to integrate the string theories using a theory of membranes from the 1980s. Witten christened his theory to be the *M* theory and declared that the theory would be discovered in the near future.[452] The promise of a future *M* theory gave a new incentive to superstring theorists. A series of annual conferences on string theory began in Amsterdam in 1997; at Cambridge University in 2002, there were 445 participants; in Kyoto in 2003, 293; in Paris in 2004, 477; and in Toronto in 2005, 440. This way of promising a theory that does not yet fully exist, provoked criticism from other physicists:

"I would not be prepared to call string theory a 'theory' rather... just a hunch. After all, a theory should come together with instructions on how to deal with it to identify the things one wishes to describe, in our case the elementary particles, and one should, at least in principle, be able to formulate the rules for calculating the properties of these particles, and how to make new predictions for them. Imagine that I give you a chair, while explaining that the legs are still missing, and that the seat, back and armrest will perhaps be delivered soon; whatever I did give you, can I still call it a chair?"[453]

Other renowned physicists also criticised this game of hidden spatial dimensions, for example, Sheldon Glashow, a renowned particle physicist and, with Steven Weinberg, a Nobel Prize winner:

---

[452] Peter Woit, *Not Even Wrong* (2006): 175–176.
[453] Gerard ′t Hooft, *In search of the ultimate building blocks* (1997): 163.

"[P]hysicists have not yet shown that superstring theory really works. They cannot even show that the standard theory, our successful description of the 'low energy' world, is a necessary and logical consequence of string theory. They can't even be sure that their formalism includes a description of such things as protons and electrons. There is not yet even one teeny-tiny experimental prediction. Why, you may ask, do the string theorists insist that space is nine-dimensional? It is not a consequence of elegant arguments… It is simply that string theory doesn't make sense in any other kind of space…

The historial connection between experimental physics and theory has been lost as far as superstring theory is concerned. Until the string people can explain and interpret perceived properties of the real world, they are simply not doing physics. Should they be paid by physics departments and be permitted to pervert impressionable students? Will young Ph.D.s, whose expertise is limited to superstring theory be employable if and when the string snaps? String thoughts may be more appropriate to departments of mathematics or even to schools of divinity. How many angels can dance on the head of a pin? How many dimensions are there in a compactified manifold, thirty powers of ten smaller than a pinhead?"[454]

Joseph Conlon, however, sees a future for string theory.[455] He does not deny that there are some 10^{500} ways of getting from the 10 space-time dimensions of superstring theory to the four dimensions of our observable world, but, accepting Popper's point that falsifiability is a necessary trait of science, he foresees more sophisticated instruments of observation in future, capable of testing this set of theories, which all imply that strings exist. For

---

[454] Sheldon Glashow, *Interactions. A Journey through the Mind of a Particle Physicist and the Matter of this World* (1988): 334–335.

[455] Joseph Conlon thinks the Ads/CFT correspondence is promising, see "What is String Theory", in: *Why String Theory* (2016): 93–96; Roger Penrose, however, is very sceptical about it, see his "Ads/CFT", in: *Fashion, Faith and Fantasy* (2016): 104–116.

example, in the future there might be particle colliders capable of reaching the energy level necessary for supersymmetry, which is 10^{15} GeV. While some 100 GeV was required for the Higgs, at least 10^{15} GeV is required to obtain spontaneous supersymmetry breaking, by means of "*a hypothetical accelerator*" at a cost "*of 10^{20} times more money than we could possibly raise*".[456] In the same vein, Conlon expects super microscopes to be invented in the not very near future, capable of seeing strings. Today's most sophisticated microscope is the electron microscope, through which we can observe very small objects, like atoms, and viruses. What kind of microscope do we need to see strings? Conlon gives us a clue:

> "*String theory is most famous as a theory of quantum gravity and a candidate theory of fundamental interactions at the smallest possible scales, scales that are as small compared to an atomic nucleus as an atomic nucleus is to a person.... Once you have a microscope that is capable of resolving sufficiently small lengths, there is no mystery about how to test the relative claims that the electron is a particle or the electron is a string. You use the microscope, and you go and look. Indeed, no agonising about falsifiability occurred when string theory in its original incarnation was proposed as a theory of the strong force, and the characteristic length of strings was thought to be a femtometre. The reason string theory was originally ruled out as an account of the strong force was precisely because as more experimental data arrived, its predictions totally and spectacularly failed to accord with this data..*"[457]

The point Conlon makes is that even though string theory cannot be tested at the present moment, because of our limited testing technologies, it may be possible to test it by future experiments, once more sophisticated technology is available. This has happened several times in the history of

---

[456] David Gross, in: Paul Davies & Julian Brown, eds, *Superstrings* (2000): 141
[457] Joseph Conlon, *Why String Theory?* (2006): 5–6, 217–218.

science. We saw examples of it in this chapter, when treating the corroboration of the predictions of the electroweak force, and of the mass of the Higgs boson. So, if string theory can make predictions that can be tested in a faraway future, then it is science.

I agree with Conlon that, once such super microscopes or super particle colliders are available, we can put to test the superstring set of 10^{500} theoretical-mathematical ways of getting to our four-dimensional Universe. Once put to test, in a not very near future, there are two and only two possible outcomes: it will then appear that one of those 10^{500} theories is true, without us knowing which one of them (since we put to test the whole set, not one of them), or it will appear that all of them are false.

Why is string theory so popular? The reason appears to be some pysicists' fascination with mathematical miracles. Critics within the physics community maintain that string theory is *"little more than science fiction in mathematical form"*.[458] This is also Smolin's criticism:

> *"[T]here could only be one consistent theory that unified all of physics, and since string theory appeared to do that, it had to be right. No more reliance on experiment to check our theories. That was the stuff of Galileo. Mathematics now sufficed to explore the laws of nature. We had entered the period of postmodern physics."*[459]

Most string theorists, Conlon being an exception, do not deny that, but for them, mathematical consistency, without the possibility to test the theory empirically, is enough to accept their theory as science. For example, four string theorists, in an article about *M* theory, affirm exactly that:

> *"Our strongest evidence for the conjecture is a demonstration that our model contains the excitations which are widely believed to exist in M*

---

[458] John Horgan, *Rational Mysticism* (2003): 175. Cited in Peter Woit, *Not Even Wrong* (2006): 257.
[459] Lee Smolin, *The Trouble with Physics* (2006): 116–117.

theory, super gravitons and large metastable classical membranes... The way in which these excitations arise is somewhat miraculous, and we consider this to be the core evidence for our conjecture."[460]

At the end of the article, they mention other possible tests of the theory that can lead to a corroboration of their conjecture but they are not referring to empirical evidence, but rather to mathematical implications. Susskind, one of the four authors, conveyed his fascination with the mathematical miracles realised in string theory:

"Excitingly, all of String Theory's consequences have unfolded in a mathematically consistent way. String Theory is a very complex mathematical theory with very many possibilities for failure. By failure I mean internal inconsistency. It is like a huge high-precision machine, with thousands of parts. Unless they all fit perfectly together in exactly the right way, the whole thing will come to a screeching halt. But they do fit together, sometimes as a consequence of mathematical miracles."[461]

In regard to these mathematical miracles, Penrose advises caution. He states that *"the irresistible allure of what are frequently termed [mathematical] 'miracles'... has strongly influenced the direction of theoretical research"*.[462] In analysing this fascination with mathematical miracles, Penrose distinguishes between *physical-mathematical* miracles, where the mathematically consistent theory produces physical statements that can be corroborated or refuted by evidence, aiding the progress of science, and *purely mathematical* miracles, where the theory does not produce physically falsifiable statements. Penrose counts string theory and Witten's *M* theory amongst these latter miracles:

[460] T. Banks, W. Fischler, S. Shenker & L. Susskind, "*M* theory as a Matrix Model: A Conjecture", in: *Physical Review D* (1997): 5112.

[461] Leonard Susskind, *The Cosmic Landscape* (2005): 124.

[462] Roger Penrose, *The Road to Reality* (2004): 1038.

"I am sure that string theory and M theory have themselves been guided by a great many such miracles... Are such apparent miracles really good guides to the correctness of an approach to a physical theory?... One must be exceedingly cautious about such things. It may well be that Dirac's discovery that his relativistic wave equation automatically incorporated the electron's spin seemed like such a miracle... and likewise Einstein's realization that his approach to gravity through the curved space of general relativity actually gave the correct answer for the perihelion motion of Mercury – which had puzzled astronomers for over 70 years previously. But these were clearly appropriate physical consequences of the theories that were being put forward, and the miracles supplied impressive confirmation of the respective theories. It is less clear what the force of the purely mathematical miracles is."[463]

He repeats this warning in his latest book:

"[String theorists feel] that the physical theory that gave rise to such powerful and subtle mathematics might also be likely to have some deeper validity as physics. Yet, we should be very cautious about coming to such conclusions. There are many instances of powerful and impressive mathematical theories where there has been no serious suggestion of any links with the workings of the physical world."[464]

Penrose is right, because while it is true that logical and mathematical consistency is necessary for a theory to be considered scientific, this consistency in itself is not enough and by itself it does not corroborate the theory. It must also make predictions that are falsifiable by empirical evidence. Susskind quite understands that his theory of superstrings makes the confrontation with the observations of physical reality impossible, and solves that problem by eliminating the falsifiability criterion from the

---

[463] Roger Penrose, *The Road to Reality* (2004): 1040–1041.
[464] Roger Penrose, *Fashion, Faith amd Fantasy* (2016): 94.

philosophy of science, ridiculing the "*Popperazzi*" and "*Popperism*"[465] and proposing his own philosophy of science as a substitute for Popper's: "*Good scientific methodology is not an abstract set of rules dictated by philosophers. It is conditioned by, and determined by, the science itself and the scientists who create the science.*"[466] This rather arrogant assertion puts him outside the realm of science and inside the realm of science fiction.

I will now turn to Susskind's multiverse theory. Before doing so, it is important to notice that superstring theory, even though capable of being attuned and adjusted so as to be compatible with aspects of the Standard Model, does not make predictions about the Universe at large, though it pretends to be a theory of everything. Daniel Friedan, in his criticism of string theory as it stands, argues that since string theory can't predict any large distance physics, and it is for that reason, among other, a failure. Since he is very knowledgeable in string theory, being one of its founders, his criticism is especially valuable:

"*In units of the Planck length, $l_p = (1 * 10^{19} GeV)^{-1} 1$, the smallest distance probed by feasible experiments is a very large dimensionless number, on the order of $1 * 10^{16} = (1 * 10^3 GeV\ l_p)^{-1}$, or perhaps $1 * 10^{14} = (1 * 10^5 GeV\ l_p)^{-1}$. In any theory of physics in which space-time distances are dimensionless numbers and in which the unit of distance lies within a few orders of magnitude of the Planck length, the only theoretical explanations and predictions that can be checked against experiment are those made in the large distance limit of the theory. The long-standing crisis of string theory is its complete failure to explain or predict any large distance physics. String theory, as it stands, cannot say anything definite about large distance physics. String theory, as it stands, is incapable of*

---

[465] Leonard Susskind, *The Cosmic Landscape. String Theory and the Illusion of Intelligent Design* (2006): 192, 195.
[466] *Ibidem*: 194.

determining the dimension, geometry, particle spectrum and coupling constants of macroscopic space-time. String theory, as it stands, cannot give any definite explanations of existing knowledge of the real world and cannot make any definite predictions... String theory, as it stands, has no credibility as a candidate theory of physics.

Recognizing failure is a useful part of the scientific strategy. Only when failure is recognized can dead ends be abandoned and useable pieces of failed programs be recycled. Aside from possible utility, there is a responsibility to recognize failure. Recognizing failure is an essential part of the scientific ethos. Complete scientific failure must be recognized eventually. String theory as it stands, fails to explain even the existence of a macroscopic space-time, much less its dimension, geometry and particle physics. The size of the generic possible background space-time is of order 1 in dimensionless units. Large distances occur only in macroscopic space times, which are found near the boundary of the manifold of background space times. String theory, as it stands, cannot explain the existence of a macroscopic space-time, being incapable of selecting from among the manifold of possible background space times."[467]

Though string theory as it stands is incapable of making predictions about macroscopic space-time, Susskind did present speculations about the universe, loosely connected with superstring theory. It is, however, not a scientific prediction, since it is impossible to confront it with evidence obtained by observation, as he himself admits. I am referring to Susskind's theory of the *multiverse*.[468] Multiverse theory speculates that each of the 10^{500} ways in which we can get from the ten dimensions, postulated by

---

[467] Daniel Friedan, "A Tentative Theory of Large Distance Physics", in: *Journal of High Energy Physics,* vol. 2003 (2003): 7–8.
[468] Leonard Susskind, *The Cosmic Landscape. String Theory and the Illusion of Intelligent Design* (2006).

superstring theory, to a four-dimensional universe, is realised in 10^{500} different universes, which by definition are unobservable:

"In the last several years, however, there has been a complete turnaround in how many string theorists think. The long-held hopes for a unique theory have receded, and many of them now believe that string theory should be understood as a vast landscape of possible theories, each of which governs a different region of a multiple universe."[469]

I am not going to analyse this science-fiction in detail. If anyone is interested in the details, he can read Susskind or Kaku.[470] No wonder that in Kaku's book, not only do references to publications on string theory abound, but also references to many works of science fiction, recognised as such. It would have been more fitting if he had given his book the subtitle of *The Science-fiction of Alternative Universes and our Future in the Cosmos* instead of *The Science of Alternative Universes and our Future in the Cosmos*.

More critical physicists are aware that Susskind's multiverse theory is not science. For example, George Ellis, criticising Susskind's book, affirms the following:

"A phalanx of heavyweight physicists and cosmologists are claiming to prove the existence of other expanding universe domains even though there is no chance of observing them, nor any possibility of testing their supposed nature except in the most tenuous, indirect way. How can this be a scientific proposal, when the core of science is testing theories against the evidence?"[471]

---

[469] Lee Smolin, *The Trouble with Physics* (2006): 149.

[470] Michio Kaku, *Parallel Worlds. The Science of Alternative Universes and our Future in the Cosmos* (2005).

[471] George Ellis, "Physics ain't what it used to be", in: *Nature*, vol. 438 (2005): 739–740.

Some string theorists, who accept Popper's philosophy of science, reject Susskind's speculation about the multiverse. For example, the Nobel Prize winner David Gross, who himself is a leading string theorist, pointed out in an interview by Geoff Brumfiel that Susskind's multiverse speculation is impossible to falsify and, therefore, not science:

"It is impossible to disprove. Because our Universe is almost by definition, everything we can observe, there are no apparent measurements that would confirm whether we exist within a cosmic landscape of multiple universes, or if ours is the only one. And because we can't falsify the idea, Gross says, it isn't science."[472]

In general, as Barnes correctly asserts, *"we cannot observe any of the properties of a multiverse, as they have no causal effects on our universe"*[473], and, recently, Joseph Conlon too, who does see a future for string theory, rejects the multiverse for being beyond any observational test:

"It is the sheer utter extravagance of the speculation, uncoupled from either rigorous calculation or experimental test. The argument requires the physical existence of 10^{500} additional universes, none of which we can probe experimentally... every time with different laws of physics and histories. None of these other universes are observable."[474]

The driving force behind Susskind's science fiction is his desire to eliminate any allusion to an intelligent creator from the possible explanations of the fine tuning of physical constants in the Big Bang, as is revealed by the subtitle of his book: *The Illusion of Intelligent Design.* I shall come back to this point of what motivates some cosmologists to embrace the multiverse speculation in Section 8.3 of Chapter 8.

---

[472] Geoff Brumfiel, "Our Universe: Outrageous Fortune", in: *Nature*, vol. 439 (2006): 11.
[473] Luke Barnes, "The Fine Tuning of the Universe for Intelligent Life", in: *Publications of the Astronomical Society of Australia*, vol. 29 (2012): 559.
[474] Joseph Colon, *Why string theory?* (2016): 101.

Section 4.6 Conclusions on multiverse theories

We have analysed six different multiverse theories:

1.   Hoyle's creation fields in the observable Universe.
2.   Wheeler's theory of the Big Crunch.
3.   Smolin's theory of the multiverse in black holes.
4.   Barrow 's theory of variation of the constants.
5.   The Guth-Linde eternal inflation multiverse.
6.   The Susskind multiverse.

The first four theories are scientific, because they have observable implications. These are Hoyle's theory of creation fields; Wheeler's Big Crunch theory; the theory of variation of the constants by Barrow; and Smolin's theory of the multiverse in black holes. As a matter of fact, these four scientific theories have been refuted by empirical evidence, as we saw in Sections 4.1, 4.2, 4.3 and 4.4. In Sections 4.5 and 4.6, we analysed two theories which were shown to be non-testable by observable, real evidence, for which reason they should be classified as science fiction, according to the criteria of the philosophy of science that I explain in Chapter 7. These are the Guth-Linde science-fiction of eternal inflation and the multiverse; and the science-fiction of superstrings and the multiverse by Susskind, Kaku and others.

So, we have four scientific theories refuted by empirical evidence; and two speculations belonging to the realm of science fiction. These theories attempt, in one way or another, to solve the problem of the fine tuning of the Universe at the moment of the Big Bang.

The conclusion of this review is that theories about the multiplication of universes in space and/or time are either scientific theories falsified by empirical evidence, or are unfalsifiable science fiction, so that in either case they do not constitute a scientific explanation of the fact of fine tuning.

CHAPTER 5

FACT: ENTROPY AND EVOLUTION OF THE UNIVERSE

In this chapter,[475] we will first analyse the first law of thermodynamics, then the second law, and then see some implications for a thermal machine, the Universe at large and planet Earth.

Section 5.1 The first law of thermodynamics

In thermodynamics, we affirm that a system (also called a 'control volume') realises work (W), if it exports energy that is not heat (Q), for example, by generating potential electrical energy or, at the frontiers of the system, generating mechanical movement. The heat is a form of energy transferred at the frontiers of the system through a differential of temperature (T).

A system never *contains* work or heat, though it may contain energy. In general, for non-adiabatic systems, commonly referred to as non-conservative systems (I will explain this term shortly), we know that:

1) Work and heat are phenomena and forms of energy that we encounter *at the frontiers of a system*, expressed in Joules $(N * m)$.
2) Work and heat are transient phenomena, thus work and heat are differential processes represented by ∂W and ∂Q, respectively.

---

[475] This chapter owes much to Richard Sontag & Gordon Van Wylen, *Introduction to Thermodynamics. Classical and Statistical* (1991) and my colleagues at the Universidad Iberoamericana, Santa Fe, i.e. Dr Erich Starke and Dr Alejandro Mendoza.

3) Work and heat are not point functions, which means that both work ∂W, performed or absorbed by a given system, and heat ∂Q transferred, are not exact differentials.

4) Work and heat are not state variables, so that they cannot be used to describe the equilibrium states of a thermodynamic system.

That is why we can say that for adiabatic systems, work and heat are not path-dependent. To demonstrate this, we suppose there are two possible *states* of a thermodynamic system, namely state 1 and state 2 and three state changes or *trajectories*, namely A, B and C. A trajectory is being traversed by means of a process. This results in two possible thermodynamic *cycles*, as can be seen in the Figure 5.1, namely, cycle $1 \rightarrow A \rightarrow 2 \rightarrow B \rightarrow 1$ and cycle $1 \rightarrow A \rightarrow 2 \rightarrow C \rightarrow 1$:

Figure 5.1. The thermodynamic cycles $1 \rightarrow A \rightarrow 2 \rightarrow B \rightarrow 1$ and $1 \rightarrow A \rightarrow 2 \rightarrow C \rightarrow 1$ in a system with two states, state 1 and state 2

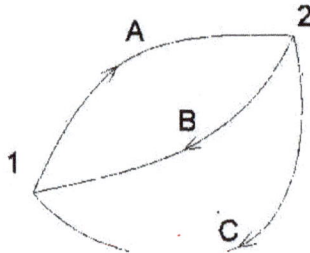

The net flow of heat in the frontiers of states 1 and 2 of the system is equal to the net export (export minus import) of work in these frontiers. So for the thermodynamic cycle AB, we have equation $1 \rightarrow A \rightarrow 2 \rightarrow B \rightarrow 1$ and for the cycle AC, $1 \rightarrow A \rightarrow 2 \rightarrow C \rightarrow 1$. For thermodynamic cycle $1 \rightarrow A \rightarrow 2 \rightarrow B \rightarrow 1$ we have:

(1) $\int_{1A}^{2A} \partial Q + \int_{2B}^{1B} \partial Q = \int_{1A}^{2A} \partial W + \int_{2B}^{1B} \partial W.$[476]

---

[476] We use $\int_1^2 \partial Q$ and not $\int_1^2 dQ$, to indicate that the total differential ∂Q is inexact.

We use $\int_1^2 \partial Q$ and not $\int_1^2 dQ$, to indicate that the total differential ∂Q is inexact. For thermodynamic cycle $1 \to A \to 2 \to C \to 1$ we have:

(2) $\int_{1A}^{2A} \partial Q + \int_{2C}^{1C} \partial Q = \int_{1A}^{2A} \partial W + \int_{2C}^{1C} \partial W.$

Subtracting (5.2) from (5.1), we obtain:

(3) $\int_{2B}^{1B} \partial Q - \int_{2C}^{1C} \partial Q = \int_{2B}^{1B} \partial W - \int_{2C}^{1C} \partial W.$

From equation (3) we obtain:

(4) $\int_{2B}^{1B} \partial Q - \int_{2B}^{1B} \partial W = \int_{2C}^{1C} \partial Q - \int_{2C}^{1C} \partial W.$

Reversing the direction of the thermodynamic cycle we obtain:

(5) $\int_{1B}^{2B} \partial Q - \int_{1B}^{2B} \partial W = \int_{1C}^{2C} \partial Q - \int_{1C}^{2C} \partial W.$

Rewriting equation (5) we obtain:

(6) $\int_{1B}^{2B} (\partial Q - \partial W) = \int_{1C}^{2C} (\partial Q - \partial W).$

To understand what follows, we must know three basic distinctions:

A)   In the first place, we distinguish between:

a) An *open system*, through which a quantity of mass is passing, per time unit $(dm/dt \neq 0)$;

b) a *closed system* with a constant quantity of mass $(dm/dt = 0)$.

B)   In the second place, we distinguish between:

a) An *adiabatic system* in which the quantity of heat energy in the system is constant, which means that it does not receive, nor dissipate heat Q in relation to its surroundings ($\partial Q = 0$);

b) a *non-adiabatic system*, which receives and dissipates heat Q in interaction with other systems or with its surroundings ($\partial Q \neq 0$).

C)   In the third place, we distinguish between:

a) *Reversible thermodynamic cycles,* where the final state of the cycle is the initial state of the next cycle, with identical initial conditions, in every cycle that repeats itself:

$(1 \rightarrow A \rightarrow 2 \rightarrow B \rightarrow 1)_{cycle\ n+1} = (1 \rightarrow A \rightarrow 2 \rightarrow B \rightarrow 1)_{cycle\ n};$

b) *irreversible thermodynamic cycles,* which repeat themselves, but with different initial conditions, so that each new cycle starts from different initial conditions:

$(1 \rightarrow A \rightarrow 2 \rightarrow B \rightarrow 1)_{cycle\ n+1} \neq (1 \rightarrow A \rightarrow 2 \rightarrow B \rightarrow 1)_{cycle\ n}.$

We can observe that the equation (6) is valid if, and only if, the system considered is adiabatic. We have proven that the trajectories B and C are quantitatively equal and for that reason it follows that *the quantitative relations between heat and work* ($\partial Q - \partial W$) *are independent of the trajectory* and depend exclusively on the initial and final states of a system and, therefore, constitute a function of state called *energy*:

(7) $dE = (\partial Q - \partial W).$

The energy E has three components, namely kinetic energy K, potential gravitational energy U_g, and internal energy U. In thermodynamics, the internal energy of a system is the energy contained within the system, excluding the kinetic energy K, and the potential gravitational energy U_g of the system:

(8) $dE = dK + dU_g + dU.$

Equation (8) supposes the existence of the internal energy state variable U, whose variation in adiabatic systems is equal to the adiabatic work performed. Here dU is an exact variable.

From equations (7) and (8) we obtain:

(9) $\partial Q - \partial W = dK + dU_g + dU.$

We will now define kinetic energy:

(10) $K = \frac{1}{2}mv^2 \Rightarrow \frac{dK}{dR} = mv \Rightarrow dK = mvdR$,

where $v = \frac{dR}{dt} = \dot{R}$ is velocity, and R, distance.

And we also define potential gravitational energy:

(11) $U_g = -\frac{GMm}{R} \Rightarrow \frac{dU_g}{dR} = \frac{GMm}{R^2} \Rightarrow dU_g = \frac{GMm}{R^2}dR,$

where G is the gravitational constant, which has a value of $G=6.673*10^{-11}$ m^{3}s^{-2} kg^{-1}.

We substitute (10) and (11) in (9), obtaining:

(12) $\partial Q - \partial W = mvdR + \frac{GMm}{R^2}dR + dU.$

Integrating (12) we obtain:

(13) $Q_{1\rightarrow2} - W_{1\rightarrow2} = \int_1^2 \left(mvdR + \frac{GMm}{R^2}dR + dU\right) = \int_1^2(mvdR) +$ $\int_1^2 \frac{GMm}{R^2}dR + \int_1^2 dU.$

We resolve (13) to obtain *the first law of thermodynamics* for a thermodynamically closed system, also known as *the law of the conservation of energy*:

(14) $Q_{1\rightarrow2} - W_{1\rightarrow2} = \frac{1}{2}m(v_2{}^2 - v_1{}^2) + \left(\frac{GMm}{R_2} - \frac{GMm}{R_1}\right) + U_2 - U_1.$

In the case of a closed system, the subscript 1 refers to the initial state and the subscript 2 refers to the final state of the thermodynamic cycle.

Section 5.2 The second law of thermodynamics

In what follows, I will analyse the second law of thermodynamics, in which the fundamental concepts are entropy and reversibility. Let us consider the relations between heat and work in a thermal machine. Let us suppose a thermal machine where the temperature of the source is

T_H and the temperature of the surroundings where the residual heat Q_l is dissipated is T_L. In a thermal machine, the work that is realised is:

(15) $W = Q_H - |Q_L|.$

Figure 5.2 The Carnot machine: thermal machine operating in cycles between two thermal containers

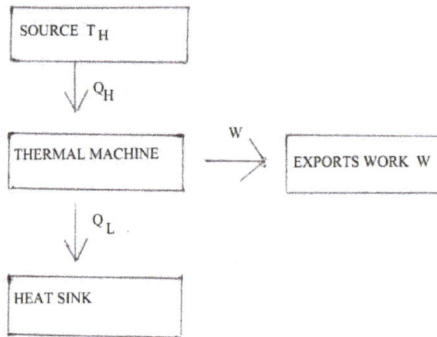

By way of example, let us consider $Q_H = 200\ kJ$ and $Q_L = -180\ kJ$. This means that the work that is exported is: $W = 20\ kJ$. We define the thermal efficiency η_t of a thermal machine as the proportion of the heat that is received (Q_H) being transformed into useful work W. The rest of the heat is dissipated (Q_L). In this case the thermal efficiency η_t of the thermal machine is:

(16) $\eta_t = \dfrac{W}{Q_H} = \dfrac{Q_H - |Q_L|}{Q_H} = 1 - \dfrac{|Q_L|}{Q_H}.$

We see that the thermal efficiency is adimensional. In the equation $\eta_t = W/Q_H$, the work W is that required of the thermal machine. The heat provided by the source is Q_H and represents the cost, in terms of fuel, of operating the thermal machine. In the example work $W = 200 - |-180| = 20$ and thermal efficiency, $\eta_t = \dfrac{20}{200} = 1 - \dfrac{|-180|}{200} = 10\%$.

Not every reversible cycle is a Carnot cycle, but every Carnot cycle is a reversible cycle. In the case of a Carnot cycle, which represents an ideal case that does not exist in reality, like a limit to which an open system may tend, the efficiency of a thermal machine is at its maximum, though never a 100%, meaning that a complete conversion of heat into another type of energy is not possible, according to the second law as defined by Lord Thomson, Baron of Kelvin (1824-1907) a British mathematical physicist. In a cycle of a thermal machine we have:

$$(17) \qquad \oint \partial Q = \int_1^2 \partial Q + \int_2^1 \partial Q = |Q_H| - |Q_L|.$$

We divide the terms by the temperature T:

$$(18) \qquad \frac{1}{T_H}\int_1^2 \partial Q + \frac{1}{T_L}\int_2^1 \partial Q = \frac{|Q_H|}{T_H} - \frac{|Q_L|}{T_L}.$$

In 1865, Rudolf Clausius (1822-1888), a German mathematical physicist, used for the first time the concept of 'entropy', defining the variation of entropy in a system as the ratio of the flow of heat and absolute temperature.[477] A closed system has reached thermodynamic equilibrium when all material objects have reached the same temperature, through the automatic transfer of heat from warmer to colder objects. We define the entropy variation of a *process* as the entropy of state 2 (S_2) minus the entropy of state 1 (S_1), so that $S_2 - S_1 \geq \int_1^2 \frac{\partial Q}{T}$, where "*the equality holds for a reversible process and the inequality for an irreversible process*".[478]

Since T_H and T_L are constants, we obtain the entropy variation ΔS in a *cycle* of a thermal machine, which is the second law of thermodynamics for a thermal machine:

---

[477] See Helge Kragh, *Entropic Creation* (2008): 29–30.
[478] Richard Sontag & Gordon Van Wylen, *Introduction to Thermodynamics* (1991): 199.

$$(19) \quad \Delta S_{cycle\ thermal\ machine} = \frac{1}{T_H}\int_1^2 \partial Q + \frac{1}{T_L}\int_2^1 \partial Q = \frac{|Q_H|}{T_H} - \frac{|Q_L|}{T_L}.$$

In what follows, I will analyse simple cycles which consist of two processes, namely, a process that goes from state 1 to state 2, and another process that returns from state 2 to state 1. The cycle $1 \rightarrow 2 \rightarrow 1$ can be *reversible* or *irreversible*. Given the three above mentioned distinctions, there exist, in theory, various combinations of these different properties, of which I shall analyse the following:

a) The thermal machine as a thermodynamically closed and non-adiabatic system. We will see both reversible and non-reversible cycles.

b) The Universe as a thermodynamically closed and adiabatic system, which, in theory, can be in a reversible cycle, first expanding, then collapsing, with **closed geometry**; or in an irreversible process with adiabatic expansion, with a flat or open **geometry**.

Section 5.2.1 Entropy of a thermal machine with a reversible cycle

We have the following equations for work W and entropy variations $S_2 - S_1$ and $S_1 - S_2$ of a thermal machine with a reversible cycle:

$$(20) \quad W_{rev} = \oint \partial Q = |Q_H| - |Q_L| > 0.$$

$$(21) \quad \Delta S_{process\ A} = S_2 - S_1 = \int_1^2 \frac{\partial Q}{T} = \frac{|Q_H|}{T_H}.$$

$$(22) \quad \Delta S_{process\ B} = S_1 - S_2 = \int_2^1 \frac{\partial Q}{T} = \frac{|Q_L|}{T_L}.$$

According to Kelvin's proposal, which has been corroborated by experimental facts, the ratios of the heat flows $\frac{|Q_H|}{|Q_L|}$ and of high and low temperatures $\frac{T_H}{T_L}$, in a Carnot cycle, are equivalent.

(23) $\dfrac{|Q_H|}{|Q_L|} = \dfrac{T_H}{T_L}$.

So, by definition, the Carnot cycle is reversible. From (23), we obtain:

(24) $\dfrac{|Q_H|}{T_H} = \dfrac{|Q_L|}{T_L}$.

From (21), (22) and (24), we obtain the entropy variation in a reversible cycle:

(25) $\Delta S_{reversible\ cycle} = \oint \dfrac{\partial Q}{T} = \dfrac{|Q_H|}{T_H} - \dfrac{|Q_L|}{T_L} = 0$.

Given the fact that the entropy variation in the surroundings has to do with the entropy variation in a non-adiabatic system, and since that entropy variation is zero, it follows from (25) that (*surr* = *surroundings*):

(26) $\Delta S_{surr} = 0$.

Given the fact that the total entropy variation is the sum of the system's entropy variation and the surroundings' entropy variation, it follows from (25) and (26) that the total entropy variation in a cycle of Carnot is zero:

(27) $\Delta S_{total} = \Delta S_{rev\ cycle} + \Delta S_{surr} = 0$.

Section 5.2.2 Entropy of a thermal machine with an irreversible cycle

The term W_{irre} refers to the work done by a thermal machine with an irreversible cycle and Q_{Lirre} refers to the heat that is dissipated by and irreversible thermal machine:

(28) $W_{irre} = \oint \partial Q = |Q_H| - |Q_{L\ irre}| > 0$.

Let us remember (20):

(20) $W_{rev} = \oint \partial Q = |Q_H| - |Q_{L\ rev}| > 0$.

Now we know that a reversible thermal machine delivers more work than an irreversible thermal machine:

(29) $W_{rev} > W_{irrev}.$

From (20), (28) and (29), comparing reversible and irreversible processes, we obtain:

(30) $Q_{L\,irre} > Q_{L\,rev}.$

Let us remember and rewrite equation (25) as equation (31):

(31) $\Delta S_{rev\,cycle} = \oint \frac{\partial Q}{T} = \frac{|Q_H|}{T_H} - \frac{|Q_L|_{rev}}{T_L} = 0.$

From equations (30) and (31), it follows that for an irreversible cycle the entropy variation is negative:

(32) $\Delta S_{irre\,cycle} = \oint \frac{\partial Q}{T} = \frac{|Q_H|}{T_H} - \frac{|Q_L|_{irre}}{T_L} < 0.$

Equations (31) and (32) constitute Clausius' inequality for any cycle: $\oint \frac{\partial Q}{T} \leq 0$, where "*the equality holds for a reversible process and the inequality for an irreversible process*".[479] Since the entropy variation of the surroundings is always positive, and the positive entropy variation of the surroundings overtakes the negative entropy variation of the irreversible cycle:

(33) $\Delta S_{surr} > 0$ and

(34) $|\Delta S_{surr}| > |\Delta S_{irre\,cycle}|,$

and since, by definition:

(35) $\Delta S_{total} = \Delta S_{irre\,cycle} + \Delta S_{surr},$

it follows, from (32), (33), (34) and (35), that:

(36) $\Delta S_{total} = \Delta S_{irre\,cycle} + \Delta S_{surr} > 0.$

---

[479] Richard Sontag & Gordon Van Wylen, *Introduction to Thermodynamics* (1991): 188.

Now let us remember the entropy variation in a reversible cycle (equation 27):

(37) $\Delta S_{total} = \Delta S_{rev\ cycle} + \Delta S_{surr} = 0.$

From (36) and (37), we derive *the second law of thermodynamics, which states that, in a cycle, the total entropy of a system plus its surroundings either increases or remains the same, but never decreases*:

(38) $\Delta S_{total} \geq 0 |^{=0\,reversible\ cycle}_{>0\ irreversible\ cycle}.$

Section 5.3 The Universe as a closed and adiabatic system

Both the Universe (Section 5.3) and the solar system (Section 5.4) are involved in irreversible processes, the first one adiabatic, and the second one, non-adiabatic, as we shall now see.

Section 5.3.1 The first law of thermodynamics for the Universe

In this section, I shall analyse the implications of the first and second laws of thermodynamics for our Universe.

With respect to the first law, we have to take into account, in the first place, that the Universe is a thermodynamically *closed* system ($\frac{dm}{dt} = 0$) and *adiabatic*, which means that it does not receive, nor dissipate heat Q in relation with its surroundings or other systems:

(39) $q_{1 \rightarrow 2} = w_{1 \rightarrow 2} = 0.$

In the second place, we suppose that the Universe's internal energy has a very low participation in the total energy and varies little with time ($U_2 \approx U_1$), so that, in leaving out the internal energy from the equation, we obtain a result that is a good approximation to reality:

(40) $U_{internal\ energy} \ll E_{universe}.$

We remember equation (14):

$$(14) \qquad Q_{1 \to 2} - W_{1 \to 2} = \tfrac{1}{2} m(v_2{}^2 - v_1{}^2) + \left(\frac{GMm}{R_2} - \frac{GMm}{R_1}\right) + U_2 - U_1.$$

From equations (14), (39) and (40), we obtain *the first law of thermodynamics for the Universe*:

$$(41\ A) \quad \tfrac{1}{2} m(v_2{}^2 - v_1{}^2) + \left(\frac{GMm}{R_2} - \frac{GMm}{R_1}\right) = 0.$$

From (41 A), we obtain (41 B):

$$(41\ B) \quad \tfrac{1}{2}(v_2{}^2 - v_1{}^2) = -GM \left(\frac{1}{R_2} - \frac{1}{R_1}\right).$$

From (41 A & B) we see that in our Universe, involved in an adiabatic expansion process, the variation in the total amount of kinetic and gravitational energy is zero, though one can increase at the other one's expense.

Section 5.3.2 The second law of thermodynamics for the Universe

We will now analyse the second law of thermodynamics for the Universe, in the case of two cosmological alternatives, as analysed in section 1 of Chapter 3, where I explained the meaning of k and Ω:

1)   A cosmologically closed Universe ($k = +1$, $\Omega > 1$), conceived as a thermodynamically closed system ($\frac{dm}{dt} = 0$), adiabatic ($\partial Q = 0$, $\partial W = 0$) and *reversible*, with successive cycles of expansion and collapse.

2)   A cosmologically flat ($k = 0$, $\Omega = 1$) or open ($k = -1$, $\Omega < 1$) Universe, conceived as a thermodynamically closed system ($\frac{dm}{dt} = 0$), adiabatic ($\partial Q = 0$, $\partial W = 0$), but, in this case, involved in an *irreversible process*, with eternal expansion.

In both cases, the Universe is a thermodynamically closed and adiabatic system, so that the entropy of the system is equal to the total entropy:

(42) $\Delta S_{system} = \Delta S_{total}$.

We will first see the case of reversible expansion (a succession of big bangs and big crunches), and then the case of irreversible expansion (one Big Bang followed, after a long time, by the entropy death of the Universe).

Section 5.3.2.1 The Universe in adiabatic and reversible expansion

As we saw, the entropy variation of a reversible cycle is zero:

(25) $\Delta S_{reversible\ cycle} = \oint \frac{\partial Q}{T} = \frac{|Q_H|}{T_H} - \frac{|Q_L|}{T_L} = 0$.

So, if the Universe were involved in a reversible cycle of big bangs and big crunches, the entropy variation of the Universe as a system would be zero:

(43) $\Delta S_{universe,system} = \oint \frac{\partial Q}{T} = \frac{|Q_H|}{T_H} - \frac{|Q_L|}{T_L} = 0$.

From equations (42) and (43), we obtain the total entropy variation:

(44) $\Delta S_{total} = \Delta S_{universe,system}$.

From (43) and (44) we obtain:

(45) $\Delta S_{universe,total} = \oint \frac{\partial Q}{T} = \frac{|Q_H|}{T_H} - \frac{|Q_L|}{T_L} = 0$.

So, we have proved that the entropy variation in a Universe with successive cycles of expansion and collapse and, for that reason, with a closed geometry, the total entropy variation is zero (see also Chapter 6).

Section 5.3.2.2 The Universe in adiabatic and irreversible expansion

In a reversible cycle, the entropy variation is:

(46) $\oint \frac{\partial Q}{T} = \int_1^2 \frac{\partial Q}{T} \big|_{1\to2}^{rev} + \int_2^1 \frac{\partial Q}{T} \big|_{2\to1}^{rev} = 0.$

In an irreversible cycle the entropy variation is:

(47) $\oint \frac{\partial Q}{T} = \int_1^2 \frac{\partial Q}{T} \big|_{1\to2}^{irre} + \int_2^1 \frac{\partial Q}{T} \big|_{2\to1}^{irre} < 0.$

A reversible and an irreversible cycle are identical in the trajectory that they follow the first time, from state 1 to state 2, so that $\int_1^2 \frac{\partial Q}{T} \big|_{1\to2}^{irre} = \int_1^2 \frac{\partial Q}{T} \big|_{1\to2}^{rev}$). The difference occurs in the trajectory in the opposite direction, from state 2 to state 1, so that $\int_2^1 \frac{\partial Q}{T} \big|_{2\to1}^{irre} \neq \int_2^1 \frac{\partial Q}{T} \big|_{2\to1}^{rev}$. This means that we can subtract (5.46) from (5.47), to obtain:

(48) $\int_2^1 \frac{\partial Q}{T} \big|_{2\to1}^{irre} - \int_2^1 \frac{\partial Q}{T} \big|_{2\to1}^{rev} < 0 \Rightarrow$

(49) $\int_2^1 \frac{\partial Q}{T} \big|_{2\to1}^{irre} < \int_2^1 \frac{\partial Q}{T} \big|_{2\to1}^{rev} \Rightarrow$

(50) $\int_2^1 \frac{\partial Q}{T} \big|_{2\to1}^{rev} > \int_2^1 \frac{\partial Q}{T} \big|_{2\to1}^{irre}.$

In a reversible cycle:

(51) $\int_2^1 \frac{\partial Q}{T} \big|_{2\to1}^{rev} = \int_1^2 \frac{\partial Q}{T} \big|_{1\to2}^{rev}.$

From (50) and (51) it follows that:

(52) $\int_1^2 \frac{\partial Q}{T} \big|_{1\to2}^{rev} > \int_2^1 \frac{\partial Q}{T} \big|_{2\to1}^{irre}.$

By definition, the variation of entropy in a reversible process in the trajectory from state 1 to state 2 (see equation 21):

(53) $(S_2 - S_1)|_{rev} = \int_1^2 \frac{\partial Q}{T} \big|_{1\to2}^{rev}.$

From (52) and (53) we obtain:

(54) $(S_2 - S_1)|_{rev} > \int_2^1 \frac{\partial Q}{T} |_{2\to1}^{irrev}$.

Since the process that moves the first time from state 1 to state 2 is identical in a reversible and an irreversible cycle, so that $\int_1^2 \frac{\partial Q}{T} |_{1\to2}^{irre} = \int_1^2 \frac{\partial Q}{T} |_{1\to2}^{rev}$), it follows that:

(55) $(S_2 - S_1)|_{rev} = (S_2 - S_1)|_{irrev}$.

From equations (54) and (55), we obtain:

(56) $(S_2 - S_1)|_{irrev} > \int_2^1 \frac{\partial Q}{T} |_{2\to1}^{irrev}$.

Since we are speaking of an adiabatic process, it follows that:

(57) $\partial Q = 0$.

Since the definite integral of zero, is zero[480], it follows from (57) that:

(58) $\int_2^1 \frac{\partial Q}{T} |_{2\to1}^{irrev} = 0$.

Combining equations (5.56) and (5.58), we obtain:

(59) $(S_2 - S_1)|_{irrev} > 0$.

So, *the variation of entropy of a closed system involved in an adiabatic and irreversible process, in the trajectory from state 1 to state 2, is positive: its entropy increases with time.* Any process occurring in a closed system either increases the entropy of the system or leaves it constant. For irreversible processes, the entropy increases; for the reversible ones, the entropy remains constant.

---

[480] If $Q = 0$, the integral of $\int_2^1 0 . dQ = C - C = 0$.

We may conclude the analysis of the implications of the second law for the Universe as follows. The Universe is a *thermodynamically* closed system. Since, as a matter of fact, the Universe is *cosmologically* speaking open (according to Wiltshire's *time-scape* model: see Section 2 of Chapter 2 of this book) or flat (according to the standard ΛCDM model), in any case, it will expand forever, so that, *thermodynamically* speaking, the Universe is a closed $\left(\frac{dm}{dt} = 0\right)$ and adiabatic ($\Delta Q = 0$) system, involved in a process that is the first part of an *irreversible* cycle. So, it follows that the variation of entropy in our Universe, from the Big Bang to its final heat death, is positive and increases with time:

(60) $(S_2 - S_1)|_{universe} > 0.$

Since the Universe is either open or flat, it will continue to expand forever and heat death will occur, when the cooling of the Universe approaches equilibrium at a very low temperature after a very long time. Heat death means that temperature differences can no longer be exploited to perform work, as in our solar system at the present moment. All stars will have died, all black holes will have evaporated, all protons will have decayed, and very low energy photons will have been spread over an infinite space. This is the point where the Universe reaches thermodynamic equilibrium, which means that it has reached its state of maximum entropy.

Section 5.4 The entropy of the solar system and the Earth

The Earth is a subsystem of the Universe that is thermodynamically open, meaning it receives and expels energy-matter ($dm/dt \neq 0$), and is not adiabatic, meaning that it receives and dissipates heat ($\Delta Q = 0$). It is special case, because of the evolution of complex life and its associated eco-system. The Earth is comparable to an irreversible thermal machine. Such a machine receives heat from a source, in this case the Sun; it transforms part of the heat it receives into work, in this case the work of Earth's eco-system, which includes living organisms and industrial machines; and the rest of the heat is dissipated in a heat sink, in this case the Universe. The Earth receives a minimal part of the heat the Sun radiates; the rest of it being dissipated in the cosmos. Therefore:

$$(61) \quad W_{earth} = \oint \partial Q = |Q_H| - |Q_L| > 0.$$

The entropy variation of the Earth, with the increase of the number of living organisms, and industrial machines and by trapping heat through the emission of CO_2, has a negative sign, meaning that we import more energy in the form of heat than we dissipate into the cosmos. This way, the Earth is an irreversible, non-adiabatic thermal machine (figure 5.3).

$$(62) \quad \Delta S_{earth} = \oint \frac{\partial Q}{T} = \frac{|Q_H|}{T_H} - \frac{|Q_L|}{T_L} < 0.$$

Figure 5.3. The Earth conceived as a thermal machine

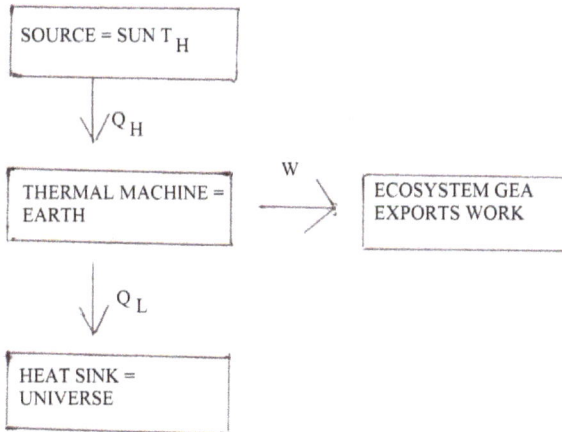

This phenomenon of a negative variation of entropy (equation 62) is what Prigogine has called "*exporting entropy*".[481]

On the other hand, Earth's surroundings, especially the Sun, dissipate enormous amounts of heat into the cosmos, so the Sun's entropy variation is positive:

$$(63) \quad \Delta S_{solar\ system} > 0.$$

Obviously, in absolute terms, the increase of the entropy of the solar system is much bigger than the decrease of entropy on planet Earth:

$$(64) \quad |\Delta S_{solar\ system}| > |\Delta S_{earth}|.$$

For that reason, the total entropy of the solar system, including the Earth, is increasing:

$$(65) \quad \Delta S_{total} = \Delta S_{earth} + \Delta S_{solar\ system} > 0.$$

This way, I have demonstrated that the entropy variation of some subsystems of the Universe, like the Earth's ecosystem, comparable to the entropy variation of an irreversible, non-adiabatic thermal machine, can be negative for a long time, until it succumbs to the increase of the entropy of the total solar system and the entire Universe.

---

[481]   Ilya Prigogine, *From Being to Becoming. Time and Complexity in the Physical Sciences* (1980); and *Order out of Chaos. Man's New Dialogue with Nature* (1984).

CHAPTER 6

MYTH: CONFORMAL CYCLICAL COSMOLOGY

Section 6.1 Summary of the theory of conformal cyclical cosmology

At the end of his most recent book, Roger Penrose presents a cosmology of his own, trying to explain *"the extraordinary suppression of gravitational degrees of freedom in the Big Bang"*,[482] which he labels *"conformal cyclic cosmology"* (henceforward CCC).[483] He makes it clear that his model is heavily dependent on the cosmological constant: *"CCC does require a positive cosmological constant Λ"*.[484]

Conformal field theory (CFT) is at the heart of quantum field theory (QFT). CFT is a QFT that is *conformally invariant*. When zooming in on very small, local regions of space-time, conformal transformations of space-time constitute the set of transformations that locally preserve angles between any two lines, though not necessarily distances, which may be rescaled. Conformal transformations include the group of Poincaré transformations, which constitute the symmetry group of relativistic field theory in flat space.[485] In three spatial plus one time dimensions, conformal symmetry has **15 degrees of freedom**: ten for the Poincaré group, four for conformal transformations, and one for a dilation.

---

[482] Roger Penrose, *Fashion, Faith and Fantasy* (2016): 371.
[483] *Ibidem*: 371–390.
[484] *Ibidem*: 381.
[485] See Joshua Qualls, *Lectures on Conformal Field Theory*, arXiv:1511.04074v2, 2016

Penrose uses CFT, because at the very high temperatures present in the Big Bang, the rest masses of the particles concerned become insignificant in relation to the very high kinetic energy of these particles in motion, so that *"the relevant physics at the Big Bang, being in effect the physics of massless particles, will be conformally invariant physics"*.[486] Helmut Friedrich evaluates Penrose's use of the conformal structure of the gravitational field[487] and thinks it is useful in a global analysis of solutions to the Einstein field equations: *"Since important open problems of general relativity are questions about the global conformal structure of the gravitational field, the understanding of the conformal structure of Einstein's field equations should prove profitable"*.[488]

Penrose's CCC proposal is about the possibility of *a pre-Big Bang physical reality*, that would explain the fact of the low gravitational entropy in the Big Bang, which is not explained in the eternal universe speculations of Wheeler, Smolin and Guth-Linde, which Penrose, for that very reason, rejects. The eternal universe of Penrose is an eternal succession of aeons. Each universe, after a very long time, say 10^{100} years, is almost massless, because most baryons are swallowed up by black holes, which eventually swallow each other up and then evaporate; neutrinos survive, but are stretched out over a spatially infinite space; 50% of protons decay after 10^{32} years; then, after another 10^{32} years, 50% of the remaining protons also decay; obviously, after 10^{100} years, few protons remain; after decaying, protons leave a remnant of positrons, which together with the remaining electrons allow for the law of conservation of charge being obeyed, but, when colliding with them, annihilate. For all these reasons a universe ends up consisting mostly of very low frequency, low energy photons.

[486] Roger Penrose, *Fashion, Faith and Fantasy* (2016): 377.
[487] Helmut Friedrich, "Einstein's Equation and Conformal Structure", in: S. Huggett *et al.*, eds., *The Geometric Universe: Science, Geometry, and the Work of Roger Penrose* (1998): 81–98.
[488] *Ibidem*: 95.

The transition of one aeon into the next one is "*a smooth conformal continuation*", since the beginning and end of each aeon is a very smooth distribution of "*entirely massless particles*",[489] so that each universe begins and ends with low gravitational entropy. The beginning is as follows: "*Over the passage of time density irregularities became gravitationally enhanced... to produce stars, these being gathered in galaxies, with massive black holes in galactic centres, this clumping being ultimately driven by relentless gravitational influences*" and "*this indeed would have presented a vast entropy increase*".[490]

But then the contrary process takes place, and degrees of freedom are lost, returning to a state of low gravitational entropy: "*by the time all the black holes have completely evaporated away in an aeon (after some 10^{100} years since its big bang), the entropy definition that would initially be employed as appropriate would have become inappropriate after that period of time, and a new definition, providing a far smaller entropy value, would have become relevant some while before the crossover into the next aeon*".[491] The gravitational version of the Second Law employed by Penrose is a far echo from its thermodynamic cousin and is more like a third law, with entropy first increasing, then diminishing. This is how, in CCC, each aeon starts and finishes with low gravitational entropy. A Higgs mechanism allows for the reappearance of mass following the crossover, starting with "*the dark matter that is required for consistency with astrophysical observation*".[492]

Penrose proposes two observational tests for his CCC.[493] Firstly, the encounters between super-massive black holes in an aeon previous to the next one result in bursts of gravitational wave energy that become visible as circular irregularities in the CMBR of the next aeon. Secondly, the magnetic

---

[489] Roger Penrose, *Fashion, Faith and Fantasy* (2016): 378.
[490] *Ibidem*: 255.
[491] *Ibidem*: 386.
[492] *Ibidem*: 387, 389.
[493] Roger Penrose, *Fashion, Faith and Fantasy* (2016): 389–390.

fields sometimes present in the large voids of intergalactic space are taken by Pentose to be primordial, i.e. already present in the early Big Bang, representing remnants from the previous aeon.

Section 6.2 Criticism of the conformal cyclical cosmology

I will now express a few critical commentaries on Penrose's CCC.

1) Penrose's conjecture requires a positive cosmological constant. The positive cosmological constant, and the dark energy behind it, appear, however, to be a myth, as I explained in Chapter 2.[494] It is worth clarifying that Friedrich's analysis of the global conformal structure of the gravitational field is compatible with $\Lambda > 0$ [495], with $\Lambda < 0$ [496], but also with $\Lambda = 0$ [497], so, from that point of view, we can do without the cosmological constant if we wish to.

2) Dark matter, though it is not an essential requirement of Penrose's proposal, is considered by him to be part of the beginnings of every new aeon. Dark matter, however, is a myth too, as I showed in Chapter 2.[498]

3) Even though the smooth gravitational transition from one aeon to the next one is explained, no explanation is given of how a big bang occurs at the beginning of each new aeon. An aeon ends with a

---

[494] See in this book Section 2.2 of Chapter 2.

[495] Helmut Friedrich, "Existence and structure of past asymptotically simple solutions of Einstein's field equations with positive cosmological constant", *Journal of Geometry and Physics,* vol. 3 (1986): 101–117; and *idem,* "On the Existence of n-Geodesically Complete or Future Complete Solutions of Einstein's Field Equations with Smooth Asymptotic Structure", in: *Communications in Mathematical Physics* (1986).

[496] Helmut Friedrich, "Einstein Equations and Conformal Structure: Existence of Anti-de Sitter-type Space-times", in: *Journal of Geometry and Physics*, vol. 17 (1995): 125–184; *idem,* "Einstein's Equation and Geometric Asymptotics", in: *Proceedings of the 15th International Conference on General Relativity and Gravitation* (Dec. 1997): 16–21.

[497] Helmut Friedrich & Gabriel Nagy, "The Initial Boundary Value Problem for Einstein's Vacuum Field Equation", in: *Communications in Mathematical Physics* (1999) 619–655.

[498] See in this book Section 2.1 of Chapter 2.

smooth distribution of particles, mainly very low energy photons, in an infinite space, which is something quite different from the smooth distribution of very high energy photons in the very reduced space of a big bang at the beginning of a new aeon. Penrose does not address the question of how this comes about.

4) I would like to stress the previous point, in a more stringent way, in terms of the Second Law of Thermodynamics. Penrose conjectures that gravitational entropy is low at the beginning of each aeon, then increases with time when matter lumps together and black holes are produced at the centre of each galaxy, and finally diminishes because of black holes evaporating at the end of each aeon, making possible a smooth transition from one aeon to the next one. Strictly speaking, however, 'gravitational entropy' is not part of the Second Law. We should not lose sight of the Second Law *in the strict sense of thermodynamics*. In Section 5.2 of Chapter 5, I gave physical-mathematical proof of the fact that the variation of entropy of a closed system – 'closed', not in the cosmological-geometric, but the thermodynamic meaning of the word – involved in an adiabatic and irreversible process, is positive; and in Section 2.2 of Chapter 2 and Section 5.3 of Chapter 5, I gave proof that our Universe is a case of adiabatic, irreversible expansion. The consequence of these two facts is that from the point of view of the Second Law, *in the strict thermodynamic meaning of the word*, the entropy variation of our Universe is positive and, therefore, increases with time. Even if we accept that gravitational entropy is equally low at the beginning and the end of each aeon, it is not clear how a smooth transition is made from a state of strictly *thermodynamic* higher entropy at the end of each aeon, to a state of strictly *thermodynamical* lower entropy in the big bang at the beginning of each new aeon.

5) I will now turn my attention to the two ways, proposed by Penrose, to check his CCC with empirical observations. I start with the circular irregularities in the CMBR. Though it is theoretically possible that circular irregularities in the CMBR of our aeon have their origin in the spheroidal gravitational waves produced by encounters between massive black holes in a previous aeon, there are explanations of these spheroidal waves within our own aeon. Koranda and Allen explain primordial spheroidal gravitational waves in the CMBR, that are generated during the mixed phase occurring after inflation when the universe smoothly transforms from being radiation to dust dominated.[499] They do so in the context of inflation theory, but the same can hold in the transition from a radiation-dominated to a dust-dominated universe, without previous inflation. Ockham's razor suggests we may prefer the present-aeon explanation over the previous-aeon conjecture.

6) Penrose takes the presence of magnetic fields sometimes present in the large voids of intergalactic space as remnants from a previous aeon, thereby corroborating his conjecture. Beck and others, however, have offered an explanation within the time of our own aeon. They explain the possible magnetization of cosmic voids by isolated galaxies in voids, which result in a major fraction of the void's volume being filled with magnetic fields of a minimum strength.[500] For the same reason as in the previous point − Ockham's razor − we may prefer the present-aeon explanation over the previous-aeon conjecture.

---

[499] Scott Koranda & Bruce Allen, "CBR Anisotropy from Primordial Gravitational Waves in Two-component Inflationary Cosmology", in: *Physical Review D52* (1995): 1902–1919.

[500] A. Beck, M. Hanasz, H. Lesch, R. Remus and F. Stasyszyn, "On the magnetic fields in voids", in: *Monthly Notices of the Royal Astronomical Society*, vol. 429 (2013): L60–L64.

7) There is a problem with Penrose's reduction of the fine tuning of the physical constants of our Universe to the very low number of gravitational degrees of freedom in the Big Bang. There are other instances of fine tuning that Penrose knows of, but are simply glossed over in a very cavalier manner and not explained in his conjecture. He knows, for example, Hoyle's analysis of the extraordinary fine tuning of some physical constants required for the nuclear fusion processes which transform hydrogen into deuterium and deuterium into helium, and then helium into carbon and oxygen,[501] but attributes this fine tuning to our *"favorable location within our given space-time universe"*.[502] As a matter of fact, however, the triple-alpha nuclear fusion process is not due to some favorable location in our Universe, but is the same all over our observable Universe and so are the finely tuned physical constants necessary for it, and this fine tuning is not explained by any physical theory, as Hoyle correctly argued.[503] Hoyle's own explanation of creation fields in the observable Universe has been refuted by the facts, as I demonstrated in Section 4.1 of Chapter 4. In an earlier book, however, Penrose rejected the fact of fine tuning, arguing that it is not inexplicable *per se*, but rather due to our ignorance:

> *"Now we can suggest answers to questions as to why the physical constants or the laws of physics generally, are specially designed in order that intelligent life can exist at all. The argument would be that if the constants or the laws were any different, then we should not be in this particular universe, but we should be in some other one! In my opinion, the strong anthropic principle has a somewhat dubious character, and it tends to be invoked by*

---

[501] Roger Penrose, *Fashion, Faith and Fantasy* (2016): 317–319.
[502] *Ibidem*: 319.
[503] See Section 1.3.3 of Chapter 1 and Section 3.4.1 of Chapter 3.

theorists whenever they do not have a good enough theory to explain the observed facts."[504]

Here, Penrose is clearly on the defensive, and his defense has a somewhat dubious character, for two reasons. In the first place, the two instances of fine tuning he admits to, i.e. the masses of fundamental particles and the triple-alpha process, have not been explained by any physical theory, and his idea that a future theory might explain all this, superseding our present ignorance, is not falsifiable at the present moment, for the simple reason that this theory does not (yet) exist. In the second place, the fact of the fine tuning of physical constants in our Universe has only two rational explanations, i.e. the multiverse, which is science fiction, as I have argued in Chapter 4, or an intelligent cause, as I shall argue in Chapter 8. The third instance of fine tuning Penrose admits to is the suppression of gravitational degrees of freedom in the Big Bang, which he tries to explain by his CCC, without succeeding in doing so, as I have argued in this Chapter.

8) Penrose himself does not take his own CCC conjecture too seriously, which is why, with characteristic modesty, in a display of self-irony, he labels it *"conformal crazy cosmology"*.[505]

---

[504] Roger Penrose, *The Emperor's New Mind* (1991): 434.
[505] Roger Penrose, *Fashion, Faith and Fantasy* (2016): 371, my underlining.

PART II

METAPHYSICS

CHAPTER 7

HOW TO DISTINGUISH SCIENTIFIC FACT
FROM MYTH

This chapter contains six parts:

1. The demarcation line between science and non-science.
2. The philosophy of the three worlds.
3. The orderly, hidden structure of the physical world.
4. Scientific and metaphysical determinism and indeterminism.
5. The frontier between science and science fiction.
6. The fascination with mathematical miracles.

Section 7.1 The demarcation line between science and non-science

The most important statements that emerged in the history of cosmology – which are the object of this study – are evaluated from the point of view of Popper's philosophy of science. Sir Karl Popper (1902–1994) was born in Vienna, Austria, where he joined the Vienna Circle, whose logical positivism he strongly criticised. In 1935, he published *Logik der Forschung*,[506] in which he proposed the principle of falsifiability as the demarcation line between empirical science and non-science (metaphysics, theology, science fiction, etc.).[507]

---

[506] In this chapter, I use Karl Popper, *The Logic of Scientific Discovery*, e-book (2005).
[507] Karl Popper, *The Logic of Scientific Discovery* (2005): 17-20 and 57–73.

He was a strong critic of Plato's idealism, Aristotle's essentialism, Marx's historicism and Freud's verificationism in books such as *The Open Society and Its Enemies* (1945) and *The Poverty of Historicism* (1957). After Hitler came to power, Popper left Vienna and taught Philosophy and Logic in New Zealand (1937–1945) and at the London School of Economics (1946–1969). A big part of Popper's work is philosophy of science, but an important part, for instance, where he discusses indeterministic causality and realism, is metaphysics.

In this chapter, I shall explain what this philosophy of science of Popper is about and why other philosophies of science, which in their time were very popular, like, for example, idealism and positivism, today are no longer considered to be valid points of view. Largely, these philosophies have been marginalised, because Popper himself criticised them with arguments that convinced the majority of today's scientists. Among other things, Popper expressed justified pride for having 'killed' positivism, though recognising, at the same time, that some less informed academics might erroneously think he supports positivism.[508]

According to Popper, true science develops theories or sets of universal statements which are – in varying degrees – logically related among themselves, together with basic statements, which are logically derived from these universal statements and are capable of refuting them. The universal statements are intended to be valid in all space-time regions of the Universe; in contrast, the basic statements, only in one or various well defined and, therefore, limited space-time regions. By definition, the universe contains an infinite number of space-time regions. Hence, there are three types of statements:

---

[508] Malachi Hacohen, *Karl Popper: The Formative Years, 1902–1945: Politics and Philosophy in Interwar Vienna* (2000): 212–213.

1) *Universal statements* about the orderly and hidden structure of the Universe, which are falsifiable but not verifiable. Their refutation is actively sought via the verification of basic statements that contradict them logically.

2) *Basic statements* refute, logically, universal statements and are, themselves, true or false by virtue of empirical observation of the facts in a space-time region.

3) *Existential statements* contradict, logically, universal statements and can be verified by basic statements, but cannot be falsified.

The following example serves to explain these distinctions:

a) Universal statement: *"all swans are white"*.

b) Basic statement: *"there is a black swan right here, right now"*.

c) Existential statement: *"black swans exist, somewhere in this or another universe"*.

Logically, the verification of the basic statement (b) refutes universal statement (a) and verifies existential statement (c). On the other hand, the refutation of basic statement (b) does not falsify existential statement (c), nor does it verify universal statement (a), because nothing prevents the existence of other space-time regions with black swans. We therefore say that universal statements can be falsified but cannot be verified; and their logical counterpart, the existential statements, can be verified, but not falsified.

Given that the basic and universal statements are falsifiable, they are scientific. On the other hand, it is characteristic of non-scientific statements (for instance, existential statements, and those of science fiction, metaphysics or theology) that they are not falsifiable.

Hence, there are different scenarios of falsifiability and non-falsifiability for the three different types of statements:

Table 7.1 Falsifiability and non-falsifiability of different statements

	Universal statement	Basic statement	Existential statement
Falsifiable by facts	Yes	Yes	No
Verifiable by facts	No	Yes	Yes

The falsifiability criterion, according to Popper:

"[Marks] the line of demarcation between those statements and systems of statements which could be properly described as belonging to empirical science, and others",[509] i.e. *"a line (as well as this can be done) between the statements, or systems of statements, of the empirical sciences, and all other statements – whether they are of a religious or of a metaphysical character, or simply pseudo-scientific"*.[510]

There is a type of scientific statement which is not universal, but historical, i.e. statements about the evolution of the Universe, about the evolution of life on Earth, or about human history. The statements *"the French Revolution took place at the end of the 18th century"*, or *"the evolution of the Universe began with the Big Bang at the time t = 0"*, or *"our solar system began some five billion years ago with a supernova"* are not universal, but historical. However, they are falsifiable, and, therefore, belong to the field of science.

The growth process of a scientific theory by progressive corroboration and refutation of its statements should not be confused with the process of cultural assimilation of scientific theories. In the latter case, it is a historical process of replacing old '*paradigms*'[511] by others. The work of Thomas Kuhn (1922–1996), a sociologist of science from the United

---

[509] Karl Popper, *Conjectures and Refutations* (1989): 255.
[510] *Ibidem*: 63–64.
[511] Thomas Kuhn, *The Structure of Scientific Revolutions* (1996), 'Paradigm' means 'fundamental theory'.

States, titled *The Structure of Scientific Revolutions*, published in 1962, explored the cultural and institutional factors that contribute to the adherence to false paradigms, though these may contain some manifest 'anomaly', or to the partial or complete substitution of old paradigms by new ones, on some occasions by the refutation of a part of the old paradigm, or by the explanation of an 'anomaly' of the old paradigm.

We see, for instance, that pre-Socratic theory, which was heliocentric, competed for some time with the Aristotelian-Ptolemaic theory, which was geocentric, and though the latter was false, it replaced the earlier one, which was true, for nearly two thousand years. Kuhn's work is a contribution to the *sociology of science* on the interaction of scientific theories and culture, but not to *the philosophy of science*. His relativism, i.e. the suggestion that science is more a matter of fashion than facts, presented a gift to those who wish to deny the importance of science.

Kuhn does not appear to distinguish between these two approaches.[512] His commentary that scientific theories have been historically incomplete, and are mixes of true and false statements, is correct. The history of statements about the origin and the evolution of the universe illustrate this historical-sociological truth. For instance, Copernicus refuted the Ptolemaic theory of the Earth being the centre of the universe, but maintained its epicycles; Kepler's theory refuted part of Copernicus' theory (the epicycles), but rescued the true part (the Earth orbiting the Sun); Newton's theory corroborated part of Kepler's theory and reformulated Kepler's third law deriving it from more general principles; Newton's universal law of gravity refuted another part of his own theory, namely, the statement about the static universe, although no one realised it for two hundred years; yet another part of the Newtonian theory, about absolute space and time, ended up being refuted by

---

[512] Thomas Kuhn, *The Structure of Scientific Revolutions* (1996): 146–147.

Einstein's theory of general relativity; another part of Einstein's theory, about a static universe and the cosmological constant, was refuted by Friedmann-Lemaître's theory of the expanding universe, etc.

However, Kuhn's criticism of Popper, as if the latter only allows for the refutation of an *entire* theory is incorrect. Popper's philosophy of science does admit that a theory may be partly false and partly true. It all depends on the degree of *logical coherence* of a theory. If the entire theory is logically integrated, so that all its statements are derived from axioms, it follows that, indeed, the refutation of one of the statements of the system refutes the entire system. This is an 'axiomatised system'. Historically, however, a 'theory' usually has various parts that may not have a mutual, logical inter-dependency. In such a case, from the strictly logical point of view, this involves two or more theories which are historically presented as one theory, by a single author in a single book. In this case, the refutation of a part of the theory does not imply that its other part or other parts are false:

"The axioms are chosen in such a way that all the other statements belonging to the theoretical system can be derived from the axioms by purely logical or mathematical transformations... In a theory, thus axiomatized... we may investigate whether a certain part of the theory is derivable from some part of the axioms. Investigations of this kind... have an important bearing on the problem of falsifiability. They make it clear why the falsification of a logically deduced statement may sometimes not affect the whole system but only some part of it, which may then be regarded as falsified. This is possible because... the connections between its various parts may yet be sufficiently clear to enable us to decide which of its sub-systems are affected by some particular falsifying observation."[513]

---

[513] Karl Popper, *The Logic of Scientific Discovery* (1980): 50–51.

An example of an axiomatised system is the entire set of equations of electromagnetic theory, including Maxwell's four electrodynamical ones, which can be deduced from just two axioms: the Lorentz force law and the charge conservation law, as I have proven elsewhere.[514]

If a theory has survived numerous attempts to refute it, it becomes increasingly robust or 'probably true'. The word 'probably' that I just used here has a figurative meaning. Suppose a hypothesis H about a causal relationship between X and Y $(X \longrightarrow Y)$, in the sense that X is the only cause of Y $(H(X \longrightarrow Y) \equiv [p(\exists Y|\exists X) = 1 \wedge p(\exists Y|\nexists X) = 0])$. The fact that it has been corroborated many times, with accumulated evidence E from many space-time regions, does not mean that we know it is true, because the number of attempts to refute it is limited, and the number of space-time regions in which we can put it to test is necessarily infinite $(if\ p(X \longrightarrow Y|E_{lim}) = 1 \Rightarrow 0 < p(X \longrightarrow Y|E_{inf}) < 1)$. This is the reason why we say that universal statements can be falsified, but cannot be verified. That is, in brief, Popper's argument against inductivism.[515]

In his *Theory of Probability*, published in 1939, Harold Jeffreys, an English Mathematician, applied his ideas of Bayesian probability to the methodology of science, rejecting the path of deduction in science and opting for the path of induction. According to Jeffreys, deductive logic only permits three possible scenarios for any proposition: definite confirmation; refutation; or ignorance regarding its truth or falsity. However, no number of previous instances of confirmation of a scientific law offers deductive proof that the scientific law will be upheld in a new instance. There is no guarantee that a law, which has been upheld in all previous instances, will not be refuted in some future instance.[516]

---

[514] See John Auping, "Las ecuaciones de Maxwell un un solo sistema axiomático", in: *El Origen y la Evolución del Universo*", e-book (2016): 567–602.
[515] Karl Popper, *Realism and the Aim of Science* (1994): 217–261.
[516] Harold Jeffreys, *Theory of Probability, Third Edition* (2003): 1–3.

Paradoxically, the fact that the repeated corroboration of a basic statement that is deduced from a universal statement does not verify the universal statement, is the reason why Popper says that universal statements can never be definitively verified, and is also the very reason why Jeffreys rejects the deductive method in science and, therefore, Popper´s philosophy of science.

For this reason, Jeffreys, and others, like Adolf Grünbaum, opt for the inductive method, which allows us to calculate, on the basis of observations in a random sample, the probability that a statement about phenomena in a concrete space-time region, from which the sample is drawn, is true or false. It so happens, that both supporters of the deductive method and supporters of the inductive method agree that it is impossible to definitively corroborate universal statements, but this circumstance encourages the first ones to abandon the hope of definitively verifying universal statements, conceiving them, though, as the cornerstone of science, whereas it leads the latter ones to banish universal statements altogether from scientific theory.

I think that Jeffreys, Grünbaum and inductivism in general, demand too much of science. Even though we can never be sure whether a universal statement is true, we can corroborate it again and again, thus making the theory more robust, and we can know whether it is false. I believe, with Popper, that we should not abandon the search for universal statements in scientific theory, but rather consider it to be the supreme goal of science.

I also believe, however, that there is a place for induction in scientific research, given the fact that basic statements are usually corroborated or refuted by inductive, probabilistic methods, that is to say, by drawing a sample from a population in a limited space-time region, testing the hypothesis in that sample, and then calculating the probability that the same holds true in the population at large. Induction can never verify universal statements, but can verify or refute basic statements.

Section 7.2 The two worlds and three worlds philosophies

Before analysing the frontier between science and metaphysics, it is necessary to first explain Popper's philosophy of the three worlds, which is best understood in contrast with the philosophies of the two worlds, which preceded it. According to Plato (428–348), the knowledge of ideas in the soul is innate, whereas opinions are dependent on the shifting world that is captured by the senses. These ideas, and indeed the soul itself, are thought to be eternal and immutable, whereas opinions are ephemeral, as is the world of the senses. René Descartes (1596–1650), like Plato, distinguishes between 'convictions' (called opinions by Plato) and 'knowledge': knowledge consists of convictions based on reasons so strong that it is impossible for us ever to have any reason for doubting what we are convinced of. He admitted though, that a person having such certainty can be mistaken! It seems he was more concerned with subjective certainty, as opposed to doubt, than with objective truth, as opposed to error.

I share Popper's criticism of Descartes, in the sense that strong convictions can never be a substitute for testing universal statements, though reason is an important source of *creating* these statements.

In his *Treatise of Human Nature* (1739), David Hume (1711–1776) argued against idealism and the existence of innate ideas. For him, all human knowledge was based solely on experience. He even was sceptical about the concept of causal relations in nature, arguing that the human mind invents them and imposes them on nature. This way, he paved the way for positivism, as conceived by Ludwig Wittgenstein (1889–1951) and Rudolf Carnap (1891–1970), who deemed universal statements to be meaningless, because they cannot be verified. Positivism reduces science to a series of basic statements about phenomena of the physical world, which must be verified and certified by a sort of notarial protocol. In adhering to positivism, science gives up really understanding these phenomena.

I share Popper's criticism of positivism: instead of abandoning universal theories, because they cannot be verified, Popper abandons the requirement of verifiability, proposing the criterion of falsifiability for universal statements to be accepted as science.

In his *Critique of Pure Reason* (1781) and *Prolegomena to Any Future Metaphysics* (1783), Immanuel Kant (1724–1804) rejected Descartes' idealism, where true knowledge is embedded in the eternal human soul, but also criticized Hume's empiricist, sceptical view of causal relations, arguing that the mind is not a blank page written upon by the physical world, but, on the contrary, 'our intellect does not draw its laws from nature but imposes its laws upon nature', as I shall explain shortly. Says Kant:

> "*[T]hen a light dawned upon all natural philosophers. They learnt that our reason can understand only what it creates according to its own design: that we must compel Nature to answer our questions, rather than cling to Nature's apron strings and allow her to guide us. For purely accidental observations, made without any plan having been thought out in advance, cannot be connected by a law which is what reason is searching for.*"[517]

Impressed by the success of Newton's cosmology, Kant not only asserted that the knowledge of the laws of nature is a priori, prior to our observations (which is correct), but that it is also a priori true (which is mistaken).

Popper agrees with Kant's idealism, in as far as it states that human knowledge is structured by human thought, and with his realism, in as far as it states that objects exist independently of our mental representations of them, but rejects Kant's view that our ideas about natural laws automatically deliver truth. Popper argues that their truth is not self-evident, and that nature itself must tell us whether these universal laws are true or false, which leads him to the criterion of falsifiability in his philosophy of science:

---

[517] Immanuel Kant, preface to the 2nd edition of the *Critique of Pure Reason*, quoted in Karl Popper, *Conjectures and Refutations* (1989): 189.

"When Kant said, 'Our intellect does not draw its laws from nature but imposes its laws upon nature', he was right. But in thinking that these laws are necessarily true, or that we necessarily succeed in imposing them upon nature, he was wrong. Nature very often resists quite successfully, forcing us to discard our laws as refuted; but if we live we may try again...

Kant assumed, correctly I think, that the world as we know it is our interpretation of the observable facts in the light of theories that we ourselves invent. As Kant puts it: 'Our intellect does not draw its laws from nature, but imposes them upon nature'... I feel that it is a little too radical, and I should therefore like to put it in the following modified form: 'Our intellect does not draw its laws from nature, but tries – with varying degrees of success – to impose upon nature laws which it freely invents'. The difference is this. Kant's formulation not only implies that our reason attempts to impose laws upon nature, but also that... they must be true a priori... Yet we know since Einstein that very different theories and very different interpretations are also possible, and that they may even be superior to Newton's... Since Kant believed that it was our task to explain the truth of Newton's theory, he was led to the belief that this theory followed inescapably and with logical necessity from the laws of our understanding.

The modification of Kant's solution which I propose... frees us from this compulsion. In this way, theories are seen to be the free creations of our own minds, of an attempt to understand intuitively the laws of nature. But we no longer try to force our creations upon nature. On the contrary, we question nature, as Kant taught us to do; and we try to elicit from her negative answers concerning the truth of our theories: we do not try to prove or to verify them, but we test them by trying to disprove or to falsify them, to refute them."[518]

---

[518] Karl Popper, *Conjectures and Refutations* (1989): 47–48, 93, 190–191.

In short, our theories impose their laws on nature, but it is not tyranny: if nature does not agree, it kicks back, falsifying our a priori statements.

Both idealism and positivism share a two-world philosophy. In Kant's idealism, the mind imposes universal laws upon physical reality, accepting them as true without testing them; and in Wittgenstein's and Carnap's positivism, physical reality suggests to our mind basic statements that reflect things we observe. In the first case, our universal theories are thought to be true a priori, in the second case, they are rejected as being meaningless, because they cannot be verified.

Figure 7.1 The philosophy of two worlds

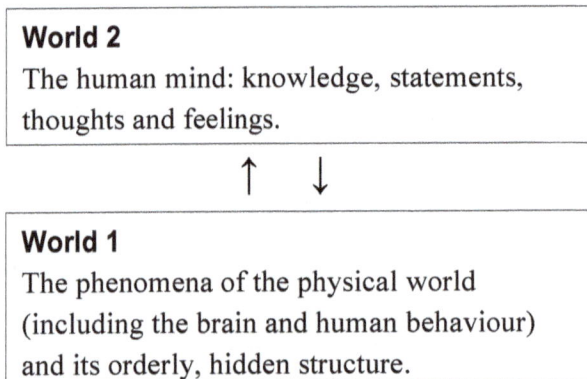

World 2
The human mind: knowledge, statements, thoughts and feelings.

↑ ↓

World 1
The phenomena of the physical world (including the brain and human behaviour) and its orderly, hidden structure.

On the other hand, in the philosophy called *realism* (with no adjectives),[519] there are three and not just two worlds. It is the philosophy of the three worlds of Karl Popper[520] and Roger Penrose.[521] In their view, theories constitute a world on its own (world 3) that does not belong, neither to physical reality (world 1), nor to the subjective ideas and thoughts of the mind (world 2). In this philosophy of the three worlds is it possible to contrast universal laws (world 3), conceived by

[519] Karl Popper, *Realism and the Aim of Science* (2000). Not Kant's *ontological* realism.
[520] Karl Popper, *The Open Universe* (2000): 113–123.
[521] Roger Penrose, *The Road to Reality* (2004): 17–21, 1027–1033.

the human mind (world 2), with the reality of the physical world (world 1), which will then decide whether they are corroborated or falsified.

In world 3, we find not only scientific theories, but also metaphysics, ethics, theology, etc. All these theories are objective, in as far as they speak of reality, both the reality of things going on in world 1, and the reality of the interactions between the three worlds.

Figure 7.2 The philosophy of three worlds

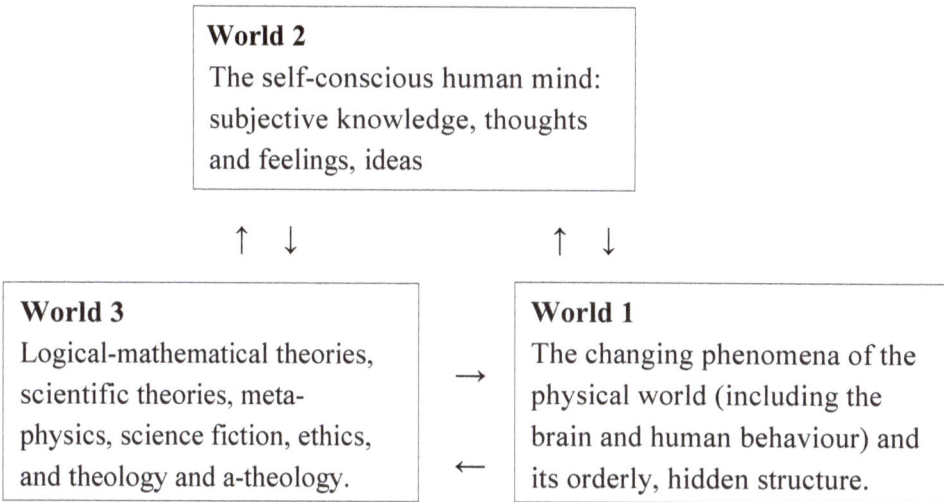

World 2
The self-conscious human mind: subjective knowledge, thoughts and feelings, ideas

↑ ↓   ↑ ↓

World 3		World 1
Logical-mathematical theories, scientific theories, meta-physics, science fiction, ethics, and theology and a-theology.	→ ←	The changing phenomena of the physical world (including the brain and human behaviour) and its orderly, hidden structure.

This is the moment to define the term 'real'. Something 'really' exists when it exists in one of the three worlds or as a relation between the three worlds. Therefore, something 'really' exists if it is part of the cause-and-effect network that extends in and between the three worlds. We must not confuse 'real' with 'true'. Theories in world 3 may be *false*, but being false does not prevent them from being *real* and having very *real* effects in world 1. The destructive power of false theories, such as communism which holds that persons are determined by the social entities to which they belong, or fascism which claims that persons are determined by their race, confirms that false theories can have real and destructive influence in human history.

Metaphysics treats the dynamic interaction of the three worlds. For this reason, metaphysical theories are not falsifiable with data of world 1, which doesn't mean that they cannot be criticised and are not worthy of criticism, because they deal with real problems that arise in the interaction of the three worlds.

$1 \leftrightarrow 3$. The interaction of worlds 1 and 3 goes in both directions and is the realm of the philosophy of science. In the logic of scientific discovery, the influence of world 1 in world 3 $(1 \rightarrow 3)$ is how scientific theories are put to test. World 1 facts determine whether world 3 theories are corroborated or falsified. When confronted with theories, nature hits back if they are not true.

Scientific theories are corroborated or falsified with data from world 1, whereas metaphysical theories, which also belong to world 3, cannot be corroborated or falsified with such data. So how can we distinguish between true and false metaphysical theories? According to Popper, although metaphysical theories cannot be falsified with empirical data, they can be criticised by considering their rationality, i.e. their capacity to analytically or ethically *solve problems* that develop in the interaction of the three worlds:

"[I]s it possible to examine irrefutable philosophical theories critically? If so, what can a critical discussion of a theory consist of, if not of attempts to refute the theory? In other words, is it possible to assess an irrefutable theory rationally – which is to say, critically?... Every rational theory, no matter whether scientific or philosophical, is rational in so far as it tries to solve certain problems. A theory is comprehensible and reasonable only in its relation to a given problem-situation, and it can be rationally discussed only by discussing this relation. Now if we look upon a theory as a proposed solution to a set of problems, then the theory immediately lends itself to critical discussion – even if it is non-empirical and irrefutable. For we can now ask questions such as: Does it solve the problem? Does it

solve it better than other theories? Has it perhaps merely shifted the problem? Is the solution simple? Is it fruitful? Does it perhaps contradict other philosophical theories needed for solving other problems?"[522]

There is also the influence in the opposite direction, from world 3 to world 1 (3 → 1), firstly, because we impose theories on nature, asking it to tell us whether they are true or false, and secondly, because the scientific, metaphysical and ethical theories from world 3 transform world 1 in very powerful ways, independently of their truth or falsehood. They transform physical and human reality through the application of science in innovating technologies, productivity, society's infrastructure and institutions, but also through the destruction of our ecosystems. And depending on whether social theories are true or false, they transform society, for better or for worse.

(2 ↔ 3). The influence of world 3 on world 2 (3 → 2) has been explored by philosophers such as Bernard Lonergan (1904–1984) who wants to help us *"thoroughly understand what it is to understand"*,[523] and Javier Zubiri (1898–1993), who looks for a way to avoid both modern subjectivism and empiricism. The latter understands 'reality', not as reality in itself or by itself, but as being perceived and apprehended by man's sentient intelligence.[524] Here, we are at the frontier of philosophy and psychology.

Where worlds 2 and 3 interact, another problem emerges, in the opposite direction (2 → 3): how does the human mind create and conceive the objective theories of science and metaphysics? Steven Pinker has argued in favour of a computational theory of the human mind.[525] On the other hand, Penrose has convincingly argued that the human mind's creation of scientific and metaphysical ideas does not follow the path of algorithms, as

---

[522] Karl Popper, *Conjectures and Refutations* (1989): 198–199.
[523] Bernard Lonergan, *Insight: A Study of Human Understanding*, in: *Collected Works*, vol. 3 (1992): 22, 769.
[524] Javier Zubiri, *Inteligencia Sentiente: Inteligencia y Realidad* (1980).
[525] Steven Pinker, *How the Mind Works* (1997).

would be the case if the mind were a computer in the flesh, but rather the opposite: first a universal idea is intuited, and then algorithmic procedures are put in place to create a logically consistent theory out of these intuitions and put them to test.[526]

(1 ↔ 2). The metaphysical problem, first expressed in Descartes' dualism, of the interaction of the self-conscious mind, which belongs to world 2, and the brain, which belongs to world 1, in the processes of perception, decision making and initiating behaviour, lies in the frontier of neuro-psychology and metaphysics. Epiphenomenalism only accepts the existence of a one-way traffic, from the brain to the self-conscious human mind (1 → 2), but categorically denies any influence in the opposite direction, from the self-conscious human mind on cerebral operations and, through them, on human behaviour (2 ↛ 1). This is, for example, Susan Greenfield's theory.[527] Humans' being conscious of what they experience and decide, is considered to be an epiphenomenon. This way, epiphenomenalism denies the existence of conscious human liberty in decision-making. Of course, we can agree with Greenfield that many times humans act impulsively, without making a conscious decision, but rather, being conditioned by previous experience.

Another school of thought, with powerful empirical evidence to support its ideas, is that of interactionist dualism, that argues in favour of a two-way traffic and influence, from the brain to the self-conscious mind (1 → 2) and from the self-conscious mind to the brain (2 → 1). This theory has been proposed by authors like John Eccles and Karl Popper,[528] among others, where Eccles makes the neuro-physiological argument, and Popper, the philosophical one. I have treated this problem at some length in another

---

[526] Roger Penrose, *The Emperor's New Mind. Concerning Computers, Minds, and the Laws of Physics* (1991).
[527] Susan Greenfield, *Journey to the Centers of the Mind. Toward a Science of Consciousness* (1995).
[528] Karl Popper & John Eccles, *The Self and Its Brain*, Springer Verlag, New York, London, 1981. See also works by authors like Gazzaniga, Sperry and Zeier.

book of mine.[529] Though neuro-psychology goes a long way in exploring the interaction of the self-conscious mind and its brain, ultimately this problem cannot be resolved on that level, and is a metaphysical problem.

Section 7.3 The orderly, hidden structure of the physical world

It is time to define the term 'orderly, hidden structure' of the physical world, that I have used before. The physical reality of world 1 has three levels. The *first level* is the one of its observable, changing phenomena, which we can represent in *images*. This is the superficial level within the reach of our senses, where we find observable and variable phenomena, as Heraclitus (535–475) said: "*Everything changes and nothing remains*".[530]

The *second level* is the one of reconstructing reality in models and constructs which represent real structures of physical objects in world 1. This intermediate level shares with the first level the fact that things can be observed, sometimes with the help of technological innovations like the electron microscope, which can see down to individual atoms, or the Hubble telescope which can look at the beginnings of the Universe, and comes close to the third level, which imposes scientific laws on nature.

The *third level* is the orderly, hidden structure of the physical reality of world 1, which Heraclitus conceived as the Reason (λογος), i.e. the fundamental, underlying and persisting order of the cosmos, which nowadays, thanks to the progress of science, can be represented by a set of physical-mathematical laws. We impose them on nature, and if found to be true, are thought to guide it by means of their immutable and constant action, determining the observable changes in the world of transient and variable phenomena. This set of scientific laws belongs to world 3. By way of example, let us have a look at *water* and *glucose*.

---

[529] See John Auping, "La interacción de mente y cerebro", in: *Una revisión de la teoría psicoanalítica a la luz de la ciencia moderna*, e-book (2000): 162–175.
[530] παντα χωρει, και ουδεν μενει.

Image 7.1 Picture of two observable phenomena: sugar and water[531]

a) The superficial level within the reach of our senses is that of the sugar and water we see and use, which we can represent by pictures or images, as the one shown above.

b) On the intermediate level, we encounter models of the molecular structure of the glucose and water molecules, which can be partly observed with modern high resolution microscopes:

Figure 7.3 Model that reconstructs the molecular structure of water[532]

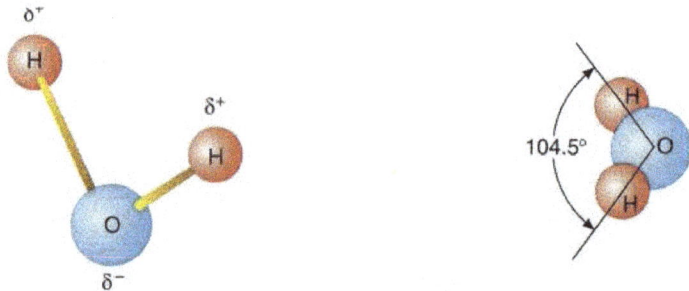

---

[531] Picture taken by Joaquín Cabeza.
[532] Figure created by Trudy & James McKee, *Biochemistry* (2003): 66.

Figure 7.4 Model of the molecular structure of glucose

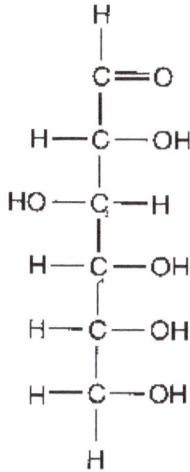

$$
\begin{array}{c}
\text{H} \\
| \\
\text{C}=\text{O} \\
| \\
\text{H}-\text{C}-\text{OH} \\
| \\
\text{HO}-\text{C}-\text{H} \\
| \\
\text{H}-\text{C}-\text{OH} \\
| \\
\text{H}-\text{C}-\text{OH} \\
| \\
\text{H}-\text{C}-\text{OH} \\
| \\
\text{H}
\end{array}
$$

We see that each carbon atom C has four bonds with other atoms; each oxygen atom O, two; and each hydrogen atom H, one; these numbers (one, two, four) prepare the terrain for the mathematical representations of the molecule at level three.

c) The third level is the mathematical expression of the orderly, hidden structure of reality. In the case of water and glucose, it is a set of physical laws, partly belonging to nuclear physics, which determine how hydrogen ($^1H^1$) and helium ($^4He^2$) were produced in the Big Bang, and how stars produce helium ($^4He^2$) from hydrogen ($^1H^1$), and oxygen ($^{16}O^8$) and carbon ($^{12}C^6$) from helium ($^4He^2$);[533] and partly belonging to quantum physics and chemistry, which tell us how electrons in an atom behave, and how they come together in molecules. In Math box 7.1, I give a few examples of these mathematical laws of physics and chemistry in the case of water and glucose, revealing their orderly, underlying structure.

---

[533] I explained this in more detail in Section 3.3 of Chapter 1.

Math box 7.1 The orderly, hidden structure of water and glucose

Ernest Rutherford (1871–1934), and Hans Geiger (1882–1945) discovered that the atom is not a dense little object, but more like a micro planetary sysem, mostly a void, with a positively charged nucleus in the centre and negatively charged electrons orbiting the nucleus. Rutherford identified the positively charged proton as being part of the nucleus, and James Chadwick (1891–1974), the neutron[534]. Niels Bohr (1885–1962) integrated Rutherford's model with the quantum theory that had been proposed by Max Planck (1858-1947).

There are two versions of the Planck constant – which is the quantum of action – h and $\hbar$, that are related as follows (n is a quantum number):

(1) $n\hbar = \dfrac{nh}{2\pi}$,

where Planck's constant has the following values for $n = 1$:

(2) $h = 6.6260755 * 10^{-34} \, Js$ and $\hbar = \dfrac{h}{2\pi} = 1.0546 * 10^{-34} \, Js$.

Bohr discovered that if a hydrogen atom (one proton with one electron) absorbs a photon, the electron jumps from its inner orbit to a more exterior orbit, and vice versa, when the electron returns to its original orbit, energy is released in the form of a photon. A photon (the word is due to Albert Einstein) is a quantum of electromagnetic energy. The lowest energy orbit of the electron is called the ground state. After absorbing energy, the electron jumps to a higher energy orbit, which is called an excited state. The change in the energy state of the electron, when changing orbits, is:

---

[534] See Ernest Rutherford, *The Scattering of Alpha and Beta Particles by Matter and the Structure of the Atom* (1911); James Chadwick, "Bakerian Lecture. The Neutron", *Proceedings of the Royal Society A: Mathematical, Physical and Engineering Sciences*, vol. 142 (1933).

$$(3)\ \Delta E = h\nu = E_{n_2} - E_{n_1},$$

where E is the electron's energy state and ν the frequency of the photon, which can be more or less energetic: the higher its frequency, the more energetic it is. Bohr's discovery led to the fundamental notion that in atoms, electrons go around the nucleus in different orbits. The different possible orbits of an electron correspond to the quantum number n, which can vary from $n = 1$ to $n = 7$, which represent the seven rows of the periodic table of elements, disovered by Dmitri Mendeleev (1837–1904). Each orbit allows for one to four orbitals, s, p, d and f, respectively, depending on the value of l, another quantum number: s for $l = 0$; p for $l = 1$; d for $l = 2$; and f for $l = 3$.

Wolfgang Pauli (1900–1958), a student of Bohr's, discovered that each orbit and each orbital can only have a limited, maximum number of electrons. This is known as the 'Pauli exclusion principle'.[535] In the case of electrons in atoms, it means that it is impossible for two electrons in the same orbital to have the same values for more than three of the four quantum numbers, which are n, the principal quantum number; l, the angular momentum quantum number; m_l, the magnetic quantum number; and m_s, the spin quantum number. For example, if two electrons go around in the same orbital, and if their n, l and m_l values are identical, then their spin m_s numbers must be different, and since electrons can only have half-integer spins, the two electrons must have opposite spin values of +1/2 and −1/2, respectively.

According to Friedrich Hund (1896–1997), within the boundaries of the Pauli exclusion principle, two electrons 'prefer' occupying two different orbits with identical spins, over occupying one orbit with

---

[535] Wolfgang Pauli, "Über den Zusammenhang des Abschlusses der Elektronengruppen im Atom mit der Komplexstruktur der Spektren", *Zeitschrift für Physik*, vol. 31 (1925): 765–783. See for a popular explanation Sheldon Glashow, *Interactions* (1988): 54–57.

paired spins, i.e. spins with contrary values, one positive, one negative. Atoms can share electrons with other atoms, in molecules, so as to obtain fully occupied outer orbits. The last column of the periodic table (see Table 7.2) registers elements that have outer orbits with the maximum number of electrons and are, therefore, incapable of binding to other elements. These are the so called inert gases. We notice that the maximum number of electrons per atom, in the fourth column, is equal to the atomic number at the right hand side of the inert gas, in the fifth colum, which registers the number of protons, which is always equal to the number of electrons in the nucleus of this particular element.

Table 7.2 The periodic table of the elements

periodic table	quantum number	orbitals s, p, d, f	maximum number of electrons per atom	last column period. table
I	$n=1$	2	2	Helium $^4He^2$
II	$n=2$	2, 6	2+8=10	Neon $^{20}Ne^{10}$
III	$n=3$	2, 6	2+8+8=18	Argon $^{40}Ar^{18}$
IV	$n=4$	2, 6, 10	2+8+18+8=36	Kripton $^{84}Kr^{36}$
V	$n=5$	2, 6, 10	2+8+18+18+8=54	Xenon $^{131}Xe^{54}$
VI	$n=6$	2, 6, 10, 14	2+8+18+32+18+8=86	Radon $^{222}Rn^{86}$
VII	$n=7$	2, 6, 10, 14	unstable	unstable

From all these rules of quantum physics and chemistry, among others, it follows that, if any atom's outer orbit/oribital has 'too many' electrons, or 'too few' electrons, it can bond together with other atoms, with 'too few electrons' or 'too many' electrons in their outer orbit, respectively, thereby sharing electrons with each other in overlapping orbits, so as to produce molecules. When different atoms share electrons to produce a molecule, the atoms' nuclei cannot be too close (because of the the repulsion between the positively charged nuclei), nor at too great a distance (which would prevent the atoms from having overlapping orbits). For example, in the periodic table, the elements of the first column have only one electron in the outer orbit, which makes them susceptible to chemically bonding with the elements of the penultimate column, which have one electron 'missing' in their outer orbit. So, lithium ($^7L^3$, 2+1 electrons), sodium ($^{23}Na^{11}$, 2+8+1), potassium ($^{39}K^{19}$, 2+8+8+1), rubidium ($^{85}Rb^{37}$, 2+8+18+8+1), cesium ($^{133}Cs^{55}$,2+8+18+18+8+1) and francium ($^{222}Fr^{87}$, 2+8+18+32+18+8+1) can easily bond with fluorine ($^{19}F^9$, 2+7), chlorine ($^{35}Cl^{17}$, 2+8+7), bromine ($^{81}Br^{35}$ or $^{79}Br^{35}$, 2+8+18+7), iodine ($^{127}I^{53}$, 2+8+18+18+7) and astatine ($^{210}At^{85}$, 2+8+18+32+18+7).

Let us now come back to the two molecules of water and glucose. It so happens that hydrogen ($^1H^1$) has one proton and one electron in orbit one, making it possible for two hydrogen atoms to bond with one oxygen atom ($^{16}O^8$) which has eight neutrons, eight protons and eight electrons, two of them in orbit 1, and six in orbit 2, making it possible to bond with two hydrogen atoms, thereby reaching the maximum of eight electrons in orbit 2. This is how we get water.

Carbon ($^{12}C^6$) has six neutrons, six protons and six electrons, two of them in orbit 1, and four in orbit 2, making it possible to bond with other atoms, sharing with them a total of four electrons, until carbon reaches the maximum of eight electrons in orbit 2. This characteristic makes carbon especially apt for bonding with other atoms to produce organic material. In glucose, one carbon atom bonds with another carbon atom plus one oxygen atom plus one hydrogen atom; or with two other carbon atoms plus one oxygen atom plus one hydrogen atom; or with one other carbon atom plus one oxygen atom plus two hydrogen atoms. In all cases, one hydrogen atom shares *one* electron with one other atom; one oxygen atom shares *two* electrons with one or two other atoms; and one carbon atom shares *four* of them with two, three or four other atoms, as we see in the case of glucose, reproduced in Figure 7.4.

Mathematical physics permits us also to express in mathematical equations the energy of one electron, its velocity inside the atom, and the distance of its orbit from the nucleus. To get to these equations, we need four physical constants (4 to 7) and five equations (8 to 12):

(4) electron mass $m_e = 9.1094 * 10^{-31}\ kg$.

(5) electron charge $e = 1.6022 * 10^{-19}\ C$ ($=3 * 10^9\ g^{1/2}cm^{3/2}s^{-1}$).

(6) vacuum permisivity $\epsilon_0 = 8.854 * 10^{-12}\ N^{-1}m^{-2}C^2$.

(7) Planck's   constant $h = 6.6261 * 10^{-34}\ Js$ ($=6.6261 * 10^{-27}\ gcm^{-2}s^{-1}$).

(8) Coulomb's law (about the force interacting between two equally charged particles): $F_{p \leftrightarrow e} = \dfrac{1}{4\pi\epsilon_0} \dfrac{e^2}{r_e^2}$.

(9) Newton's second law: $F_{p \leftrightarrow e} = m_e a_e$, where a_e is the accelaration (see equation 11).

(10) the kinetic energy of a point mass: $K = \dfrac{1}{2}mv^2$.

(11) the centripetal acceleration of an electron's circular orbit: $a_e = v_e^2/r_e$.

(12) the wavelength of a quantum particle: $\lambda = \dfrac{h}{p} = \dfrac{h}{mv}$, where $p = mv$ is the particle's momentum and h, the Planck constant.

(13)   the electron's angular momentum is the product of its momentum and distance from the nucleus: $L_e = m_e v_e r_e$.

According to Bohr, an electron orbiting the atom's nucleus can only 'choose' an orbit at such distance r_e from the nucleus that the circumference $2\pi r_e$ is equal to an integer multiple of its wavelength λ:

(14) $2\pi r_e = n\lambda = n\dfrac{h}{p} = n\dfrac{h}{m_e v_e} = n\dfrac{\hbar 2\pi}{m_e v_e} \implies$

(15) $r_e = n\dfrac{\hbar}{2\pi m_e v_e} = n\dfrac{h}{2\pi m_e v_e}$.

From (13) and (15), we obtain:

(16) $L_e = m_e v_e n\dfrac{\hbar}{2\pi m_e v_e} = n\dfrac{\hbar}{2\pi}$.

From (8), (9) and (10), we obtain:

(17) $\dfrac{1}{4\pi\epsilon_0}\dfrac{e^2}{r_e^2} = \dfrac{m_e v_e^2}{r_e} \implies v_e^2 = \dfrac{1}{4\pi\epsilon_0}\dfrac{e^2}{m_e r_e}$.

From (15) and (17), we obtain the velocity of an electron in an atom:

(18) $v_e = \dfrac{e^2}{2\epsilon_0 nh} \xRightarrow{for\ n=1} v_e = \dfrac{e^2}{2\epsilon_0 h}$.

From (15) and (18), we obtain the electron's distance from its nucleus:

(19) $r_e = n\dfrac{nh}{2\pi m_e}\dfrac{2\epsilon_0 nh}{e^2} = \dfrac{n^2\epsilon_0 h^2}{\pi m_e e^2}$.

This means that the distance from the nucleus in the first orbit is:

(20) $r_0 = \dfrac{\epsilon_0 h^2}{\pi m_e e^2} = 52.9 * 10^{-12}\ m$ (with $n = 1$).

We will now proceed to calculate the energy of the electron, which is the sum of its kinetic energy and the energy resulting from the electromagnetic force exerted on the electron by the proton. Since energy is the product of force and distance, and the force exerted on the electron by the proton has a negative sign, we obtain from (9):

$$(21) \quad U_n = F_{p \leftrightarrow e}(-r_e) = -\frac{1}{4\pi\epsilon_0}\frac{e^2}{r_e}.$$

From (10) and (21), we obtain the total energy of the electron:

$$(22) \quad E_n = K_n + U_n = \frac{1}{2}mv^2 - \frac{1}{4\pi\epsilon_0}\frac{e^2}{r_e}.$$

If we substitute (18) and (19) in (22), we obtain:

$$(23) \quad E_n = \frac{1}{2}m_e\left(\frac{e^2}{2\epsilon_0 nh}\right)^2 - \frac{e^2}{4\pi\epsilon_0}\frac{\pi m_e e^2}{n^2\epsilon_0 h^2} = m_e\left(\frac{e^4}{8\epsilon_0^2 n^2 h^2} - \frac{2e^4}{8\epsilon_0^2 n^2 h^2}\right) = -\frac{m_e e^4}{8\epsilon_0^2 h^2}\frac{1}{n^2}.$$

If we substitute in (23) the values of (4), (5), (6) and (7), we obtain, for example, the electron's energy in orbit 1 (with $n = 1$):

$$(24) \quad E_1 = -\frac{m_e e^4}{8\epsilon_0 h^2} = -2.167 * 10^{-11} \, gcm^2 s^{-2} = -2.167 * 10^{-18} \, J \cong -13.598 \, eV.$$

The equation (24) permits resolving equation (3) about the energy variation of an electron when it changes its orbit.

$$(25) \quad \Delta E = h\nu = E_{n1} - E_{n2} = -\frac{m_e e^4}{8\epsilon_0^2 h^2}\left(\frac{1}{n_2^2} - \frac{1}{n_1^2}\right) = \frac{m_e e^4}{8\epsilon_0^2 h^2}\left(\frac{1}{n_1^2} - \frac{1}{n_2^2}\right).$$

The reader might have another look at the images of water and glucose, as reproduced above, in Figure 7.1, and marvel how transient and variable phenomena at the surface of reality, like water and glucose, are determined by immutable, mathematical-physical laws at a deeper level. One might ask, "*how can concrete reality become abstract and mathematical?*"[536] Penrose himself, following Popper, answers his question with the three worlds model, which we saw above. Both the orderly, hidden structure of world 1 and the abstract mathematical-physical laws of world 3 are real, and they are connected in such a way that the laws of world 3 can be put to test by science confronting them with the very real, hidden structure of world 1, with the result that they are either falsified or corroborated.

Section 7.4 Determinism and indeterminism

I just said that 'transient and variable phenomena at the surface of reality are *determined* by immutable, mathematical-physical laws'. The use of the word 'determined' does not imply that I share the interpretation of causality proposed by scientific and metaphysical 'determinism'. On the contrary, I share Feynman's and Popper's points of view on scientific and metaphysical 'indeterminism'.[537] We shall now address this matter, starting with the principle of 'universal causation'.[538] I define it, in the words of Pierre Simon Laplace (1749–1827), as follows: "*a thing cannot occur without a cause which produces it*".[539] I shall explain in Chapter 8, that four things are implied in this principle:[540]

1) An event or set of events *A*, is a cause of another event or set of events *B*, if the event or set of events *A* is a necessary initial condition, although not always a sufficient one, for the event or set of events *B*.

---

[536] Roger Penrose, *The Emperor's New Mind* (1991): 430.
[537] Karl Popper, *The Open Universe. An Argument for Indeterminism* (2000); Richard Feynman, *The Character of Physical Law* (1967).
[538] Karl Popper, *The Open Universe. An Argument for Indeterminism* (2000): 10.
[539] Pierre Simon Laplace, *A philosophical Essay on Probabilities* (2015): 11.
[540] See Section 1 of Chapter 8.

2) All events and sets of events have a cause, or various causes, i.e. for all B belonging to the set of real events E_r, there must exist an event or set of events A that also belongs to E_r such that A is cause of B.

3) If an event or set of events A is cause of another event or set of events B, it follows that the initial time T_A of the cause A is earlier than the initial time T_B of the effect B.

4) No event or set of events A is cause of itself. [541]

Section 7.4.1 Scientific determinism and scientific indeterminism

Scientific determinism conjectures that, in a closed physical system, given a scientific theory (that establishes the relation between cause and effect) and initial conditions (the cause), we can *exactly* predict what will happen (the effect), once we overcome our ignorance. In recent times, René Thom has defended scientific determinism, arguing that science would lose its usefulness if we give up our endeavor to predict events.[542] This is a rather dogmatic argument, since we can use science to make predictions, with certain error margins, without the pretence of being able to make long term predictions that are 100% exact. From the point of view of scientific determinism, which necessarily presupposes metaphysical determinism, it is due to our present lack of knowledge of some hidden variables involved, that we cannot (yet) make 100% exact, long term predictions.

It is important to clarify that even if we adhered to metaphysical determinism, scientific determinism would be unsustainable, as we shall now see. There are four arguments against scientific determinism, in the first place, special relativity; secondly, the theory of chaos; thirdly, the emergence of new things in the history of the Universe; and fourthly, the proper character of universal statements needed to make predictions.

---

[541] Except the Cause of the Universe. I treat that question in Section 1 of Chapter 8.
[542] René Thom, "Halte au hazard, silence au bruit", in: Stefan Amsterdamski, *La querelle du déterminisme* (1990): 61–78.

Special relativity. Special relativity teaches us that the signals of any event occurring anywhere in the Universe cannot travel faster than light. If the Sun had a solar explosion in the direction of Earth, the electromagnetic waves would disrupt our electronic systems, but we would not be aware of it until eight minutes after the solar explosion, which is the time the sunlight needs to reach us. This means we would be unable to predict the disruption of our electronic and electrical systems. At the moment that it occurs, the solar explosion is outside our light cone, making it impossible for us to observe it at that very moment and predict its consequences. This is what Stephen Hawking argues, representing his argument in Figure 7.5.

Figure 7.5 The light cones of observer $P1$[543]

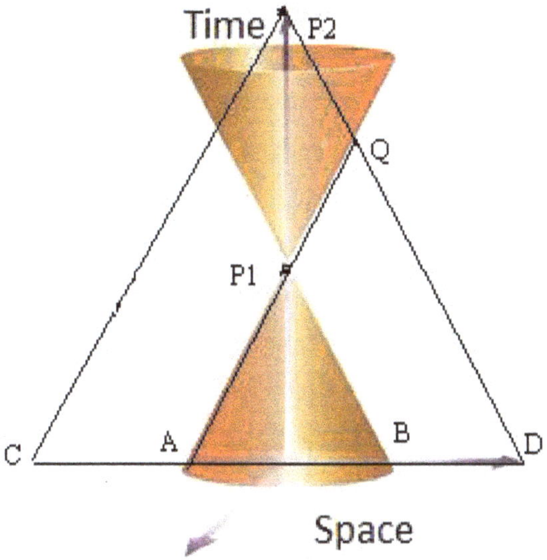

The observer P_1 cannot predict the event Q, because most of the initial conditions that cause Q to occur are in the light cone AQD, outside P_1's

[543] Image by Stephen Hawking, *The Illustrated A Brief History of Time* (1996): chapter 2.

light cone AP_1B. It is only at some future moment, that observer P_2 can fully explain the event Q, because all the initial conditions, that caused Q to happen are in the light cone AQD, which is inside P_2's light cone $CP_2 D$. This means that science can explain everything in its past light cone, but cannot predict anything in its future light cone, because it is impossible to know all the initial conditions of future events, until the future has arrived.

The theory of chaos. Another argument against scientific determinism is the theory of chaos. Mathematical chaos was discovered by Edward Lorenz (1917–2008) in 1961.[544] He had a computer capable of doing sixty runs in one second, and a meteorological model of twelve equations. At one moment, in order to save time, he only used three decimals, instead of the six he had used in a previous run. He realized after a few trials that the results of the two series of runs were completely divergent. Since there are important numbers that have an infinite number of decimals, for example π, or for that sake, $1/3$ or $1/7$, and since not even the most advanced modern computer can contain an infinite number of decimals, this means we have to make the cut somewhere, so that after a certain number of runs, predictions become wildly divergent, depending on the number of decimals we use.

Lorenz went one step further, showing that the impossibility of making exact predictions is not only an effect of rounding decimals. He showed that unpredictability is inherent to any non-linear model, even without rounding decimals. With that goal in mind, he constructed the famous Lorenz water wheel – which the reader can look up in the internet – proving that gradually increasing the force of the stream of water falling into the buckets, the behaviour of the wheel passes from steady state to predictable behaviors of periodicity two, then four, then eight, etc., finally reaching a chaotic and unpredictable behaviour, without any periodicity, with the intervals between phase changes in phase space becoming ever shorter.

---

[544] Edward Lorenz, *The Essence of Chaos* (1995).

A phase change occurs when the wheel's behaviour changes from steady state to periodicity; every time it duplicates its periodicity; and finally, when it changes to chaotic behaviour. Though the wheel's behaviour in the chaotic phase does not repeat itself ever, this chaotic behaviour occurs within certain limits. Willem Malkus (1924–2016) proved that the same occurs in convection flows of fluids when gradually more heat is put into the system.

David Ruelle, a Belgian-French mathematician born in 1935 coined the name 'attractor',[545] which is a set of numerical values that constitute the limits within which a system exhibits its periodic or chaotic behaviour, and can be represented geometrically in two- or three-dimensional figures, in case the variable develops in two- or three-dimensional phase space. If the system is chaotic, so that we cannot predict its exact future behaviour, it is at the same time stable in other aspects, in as far as its behaviour remains on the attractor when moving forward in time. David Ruelle and Floris Takens (1940–2010), a Dutch mathematician, coined the term 'strange attractor' for attractors in chaotic systems, with a fractal structure.[546]

Some systems, represented by quadratic functions with one variable X_n and one bifurcation parameter a_p, develop, given certain initial conditions, period-doubling bifurcations, as for example the Mandelbrot set of quadratic polynomials[547] and May's logistic difference equation,[548] which is used in evolutionary biology. The succession of phase changes in phase space converge to a numerical constant δ named after Mitchell Feigenbaum a mathematician from the USA, born in 1944, who discovered it in 1978.[549]

---

[545] David Ruelle, *Chance and Chaos* (1991).

[546] David Ruelle & Floris Takens, "On the Nature of Turbulence", in: *Communications of Mathematical Physics*, vol. 20 (1971): 167–192.

[547] See Benoît Mandelbrot, *Fractals and Chaos. The Mandelbrot Set and Beyond* (2004).

[548] Robert May (born 1936) is an Australian physical mathematician.

[549] Mitchell Feigenbaum, "Quantitative Universality for a Class of Nonlinear Transformations", in: *Journal of Statistical Physics*, vol. 19 (1978): 25–52; and "The

Figure 7.6 is a good representation of the phase changes of such a system in time (the four images in the upper row) and in phase space (the four images in the lower row).

Figure 7.6 Dynamic systems in time and in phase space[550]

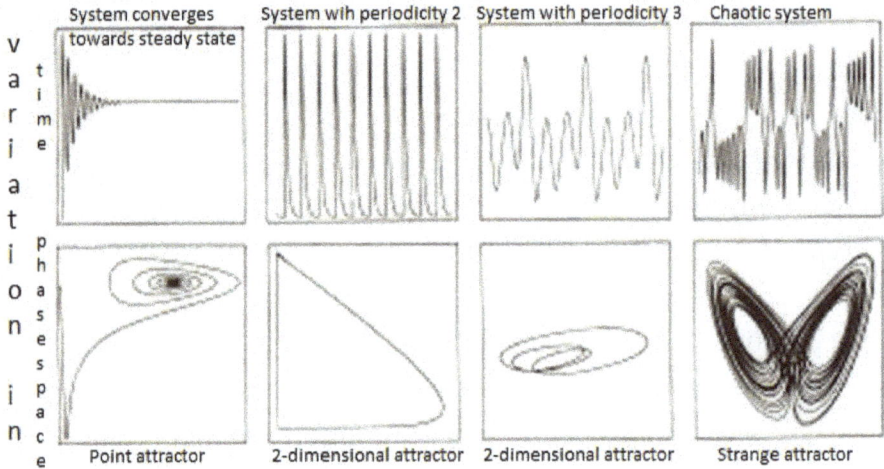

I ran May's logistic difference equation with *Mathematica*, with $X_n = 0.2$ and initial conditions $0 \leq a_p \leq 4.1$. The equation stops period-doubling at $a_p \geq 4$ entering unpredictable chaos, and then diverges towards negative infinity. The Mandelbrot set too, can diverge towards negative infinity, if initial conditions are changed, for example with $X_1 = 2$. In Math box 7.2, we can observe the values of a_p where a change in phase space occurs and can corroborate the approximate value of the Feigenbaum constant.

Math box 7.2 Logistic difference equation and Feigenbaum's constant

Robert May analysed the logistic difference equation:

Universal Metric Properties of Nonlinear Transformations", in: *Journal of Statistical Physics*, vol. 21 (1979): 669–706.
[550] Figure by Irving Epstein in: James Gleick, *Chaos. Making a New Science* (1988): 50.

1) $X_{n+1} = r_a[X_n(1 - X_n)] = rX_n - rX_n^2$.

We can see from Table 7.3 that a change in phase space occurs, the first time at $a_p = 1.02$ (from a zero steady state to a non-zero steady state); then at $a_p = 3.03$ (from steady state to periodicity two); at $a_p = 3.46$ (from periodicity two to periodicity four); at $a_p = 3.55$ (from periodicity four to periodicity eight); and finally at $a_p = 3.57$ (from periodicity to chaos).

Table 7.3 May's logistic difference equation $X_{n+1}=aX_n(1 - X_n)$

value of a_p	convergence to steady state	convergence to period of two	convergence to period of four / eight	convergence to chaos
$0 \le a < 1$	$X_{SS} = 0$			
from $a_1 \cong 1.02$	$X_{SS} \cong 0.091$			
to a = 2.00	$X_{SS} = 0.5$			
to a = 3.00	$X_{SS} \cong 0.663$			
from $a_2 \cong 3.03$		$X_{P1} \cong 0.608$ $X_{P2} \cong 0.722$		
to a $\cong$ 3.40		$X_{P1} \cong 0.452$ $X_{P2} \cong 0.842$		
from $a_3 \cong 3.46$			$X_{P1} \cong 0.839$ $X_{P2} \cong 0.468$ $X_{P3} \cong 0.861$ $X_{P4} \cong 0.413$	
to a $\cong$ 3.53			$X_{P1} \cong 0.822$ $X_{P2} \cong 0.517$ $X_{P3} \cong 0.882$ $X_{P4} \cong 0.369$	
from $a_4 \cong 3.55$			$X_{P1} \cong 0.8127$ $X_{P2} \cong 0.5405$ $X_{P3} \cong 0.8817$ $X_{P4} \cong 0.3703$	

			$X_{P5} \cong 0.8278$ $X_{P6} \cong 0.5060$ $X_{P7} \cong 0.8874$ $X_{P8} \cong 0.3548$	
from $a_5 \cong 3.57$				0.3427 $\leq X_{chaos}$ ≤ 0.8924
to $a \cong 4.0$				0.00001 $\leq X_{chaos}$ ≤ 0.9999
from $a \geq 4.01$				$X_{div} = -10^{\infty}$

Since these values are mostly numbers with an infinite number of decimals, they had to be rounded, leading us to approximate values ($\cong$). In the case of the logistic difference equation $X_{n+1} = a_p X_n (1 - X_n)$, I obtained the following approximation of the Feigenbaum constant:

2) $\delta = \left(\dfrac{F_{n+1}-F_n}{F_{n+2}-F_{n+1}}\right) \cong \left(\dfrac{a_{p+1}-a_p}{a_{p+2}-a_{p+1}}\right) \cong \left(\dfrac{3.03-1.02}{3.46-3.03}\right) \cong \left(\dfrac{3.46-3.03}{3.55-3.46}\right) \cong$ $\left(\dfrac{3.55-3.46}{3.57-3.55}\right) \approx 4.66.$

The exact Feigenbaum constant is derived from the following function:

3) $\delta = \lim\limits_{n \to \infty} \left(\dfrac{X_{n+1}-X_n}{X_{n+2}-X_{n+1}}\right) = 4.669201609102990671863 \ldots$

The emergence of new things. The fact that in the phase of chaos or disorder, things are not entirely chaotic, since chaos, at least for some time, occurs within the limits of an attractor, leads Polkinghorne to the important point that in this frontier area between chaos and order *new things* emerge:

"The interlacing of order and disorder is precisely what seems to be needed for the creative emergence of novelty. New things happen in regimes that we have learned to identify as being 'at the edge of chaos'.

Too far on the orderly side of that frontier and things are too rigid for there to be more than a shuffling rearrangement of already existing entities. Too far on the disorderly side, and things are too haphazard for any novelties to persist. An example of this principle is afforded by biological evolution. Without a degree of genetic mutation, life would be frozen into the existing range of forms. Too high a mutation rate, and there would be no quasi-stable species on which natural selection could operate."[551]

Popper takes the fact of new and unpredictable things emerging in the history of the Universe, which he refers to as 'emergent evolution', as an argument against scientific determinism, since many emerging events were not foreseeable, and therefore, unpredictable:

"The usual materialist and physicalist view is that all the possibilities which have realized themselves in the course of time and of evolution must have been, potentially, preformed, or pre-established, from the beginning … But if it is suggested that the future is and always was foreseeable, at least in principle, then this is a mistake, for all we know, and for all that we can learn from evolution. Evolution has produced much that was not foreseeable, at least not for human knowledge… Jacques Monod… speaks of the unpredictability of the emergence of life on earth, of the unpredictability of the various species, and especially of our own human species: 'we were unpredictable before we appeared', he says."[552]

Popper mentions five things *"that are altogether unpredictable or emergent"*:[553]

1) The production of heavier elements in the stars
2) The emergence of living organisms.

---

[551] John Polkinghorne, *Exploring reality* (2005): 27.
[552] Karl Popper & John Eccles, *The Self and Its Brain* (1981): 15–16.
[553] *Ibidem*: 16.

3) The emergence of sentient animal life.

4) The emergence of human consciousness and theories of self and death in humans.

5) The emergence of explanatory myths, religion, scientific theories, and works of art.

Popper's argument here against scientific determinism is quite simple. If we look at the expanding cloud of hydrogen and helium after the Big Bang, on the one hand, and then at planet Earth and its multiple life forms, on the other hand, no extraterrestrial scientist from another universe – with entirely different natural laws and elements – could have predicted the latter events on the basis of the first one, not even the emergence of water.

Universal statements in scientific predictions. There is a fourth argument against scientific determinism. Scientific prediction has three elements, i.e. a universal statement (establishing the relation between cause and effect), initial conditions (the cause), and a basic statement that constitutes the prediction (the effect). Now we saw in Section 7.2, that universal statements, because of the very fact that they are universal, can be corroborated, again and again, and can also be falsified, but can never be definitively verified. This means that, if there is any case of the prediction being wrong, the universal statement is falsified. This means that our scientific theory can never guarantee the truth of our predictions, but rather the other way around, the verification or falsification of our predictions decides whether the universal theory is being corroborated or falsified.

Each of the four arguments against scientific determinism – special relativity, the theory of chaos, the emergence of unpredictable, new things in the history of the Universe, and the character of universal statements in scientific predictions, – is in itself enough to refute it. But the fact that *scientific* determinism is wrong does not mean that *metaphysical* indeterminism is necessarily right. It could be that, even though *scientific* determinism is wrong, *metaphysical* determinism might be right.

We shall now turn our attention to the problem of metaphysical determinism versus metaphysical indeterminism. The latter school of thought sustains that physical reality is governed by laws which allow for "*pure chance without hidden variables*", whereas the first one holds that physical laws determine exact outcomes, which we would be able to predict if only we knew all "*fully determining hidden variables*".[554]

Section 7.4.2 Metaphysical determinism and indeterminism

To understand determinism, the example of the roulette can help. The wheel has 38 numbers, from 1 to 36 and two zeros. We all know that predictions in this game of chance are probabilistic. We say there is a chance of one in 38 that the ball falls say on number 14, and one in 19 that it falls on zero. The determinists argue that it is because of our lack of knowledge of hidden variables in the experimental set up, that we cannot make exact predictions. If we would know and could control the exact position of the wheel when the croupier starts pushing it, the exact amount of kinetic energy endowed by the croupier to the wheel and the ball, the exact location of the point where the croupier lets the ball loose, the ball's angular momentum, the little irregularities in the wheel, the temperature, the airflows around the wheel, etc., we would be able to exactly predict on what number the ball would fall.

In their book on the ontological interpretation of quantum theory, David Bohm and Basil Hiley present an outspoken defense of metaphysical determinism.[555] The following quote is revealing:

---

[554] Lawrence Sklar, "Probability", in: *Physics and Chance* (1993): 90–127.
[555] David Bohm & Basil Hiley, *The Undivided Universe. An Ontological Interpretation of Quantum Theory* (2003).

"We have thus far been explaining quantum probabilities in terms of chaotic motions that are implied by the quantum laws themselves, with pure ensembles representing chaotic motions of the particles and mixed ensembles bringing in also chaotic variations in the quantum field. Whenever we have statistical distributions of this kind, however, it is always possible that these chaotic motions do not originate in the level under investigation, but rather that they arise from some deeper level. For example, in Brownian motion, small bodies which may contain many molecules undergo chaotic velocity fluctuations as a result of impacts originating at a finer molecular level. If we abstract these chaotic motions and consider them apart from their possible causes, we have what is called a stochastic process which is treated in terms of a well-defined mathematical theory.

There are two attitudes to such a stochastic process. The first is that it is a result of deeper causes that do not appear at the level under discussion. The second is that there is some intrinsic randomness in the basic motions themselves. In so far as we apply the ordinary mathematical treatment, we need not commit ourselves to either attitude. But of course, if we are thinking of possible models for the process then our attitude may make a difference, because the assumption of deeper causes implies that the stochastic treatment will break down at the finer level at which these causes are operating."[556]

In essence, the argument of the authors can be summarized in two points:

1) On the level of physical-mathematical *equations*, especially in quantum theory, both determinists and indeterminists produce probabilistic equations. As far as the procedures of scientific investigation are concerned, there is no difference.

---

[556] David Bohm & Basil Hiley, *The Undivided Universe* (2003): 203.

2) On the level of *interpretation* of these equations, there is a significant difference. Determinists attribute the probabilistic quality of physical laws to our ignorance with respect to hidden variables or 'deeper causes'. Once we know and control them, probability will disappear from physics, chemistry, and, for that sake, human sciences.

The deterministic argument of 'hidden variables' and its assumption of 'deeper causes' is an existential statement, which, by its very nature, can never be refuted. Even if probabilistic physical laws, including newly discovered variables, are corroborated again and again, the ontological determinist maintains, again and again, that the element of chance is due to other hidden variables which we ignore, but do exist. There is of course no way to refute this kind of argument.

How then can we discuss irrefutable philosophical theory critically? Popper treats this problem in his *Conjectures and Refutations*, as we saw in Section 7.2 of this chapter. He argues that philosophical theories are in principle irrefutable, but we can discuss them critically, asking ourselves whether this theory helps us to solve some problem. Now it seems to me that metaphysical determinism does not solve any problems at all by making existential statements about the existence of hidden variables. It leads us nowhere. If this philosophy of science were the basis of science, we would have theories like this one: "*various unknown variables, once known and manipulated, will produce fixed effects that cannot as yet be observed*". This statement cannot be falsified, meaning that it is not a scientific theory.

On the contrary, metaphysical indeterminism leads to probabilistic explanations, like the following one: "*one or various known variables produce the following observed probabilistic effects*". Now this is certainly a scientific hypothesis, because it can be corroborated or falsified by confronting it with reality.

In direct opposition to metaphysical determinism, metaphysical indeterminism,[557] also known as intrinsic indeterminism,[558] sustains that exact predictions of certain events are impossible, for two reasons, which I will discuss further below:

1) New and unpredictable events occur: the fact that emergent evolution and downward causality exist, refutes the reductionist principle of upward causation.[559]

2) Physical causality itself is indeterministic and, therefore, probabilistic, allowing for a cause to produce a range of possible effects, each of them with its own probability,[560] the sum of which is one.

Emergent evolution and downward causality. Metaphysical determinism only accepts causation from lower to higher levels, which is the reductionist principle of 'upward causation'. For example, sociology and economics are reduced to psychology, psychology is reduced to biology, biology is reduced to chemistry, chemistry is reduced to physics, and physics is reduced to elementary particle physics. This reductionist principle of upward causation would imply that we could reduce *"the ups and down of the British trade deficit"* to biology and, ultimately, to particle physics.[561]

Against this rather outlandish claim, made by determinism, Popper argues that many events cannot exclusively be explained by 'upward causation', from the lower level to the higher level, since 'downward causation', from the higher level to the lower level, plays an important part in the generation of events. For example, in the process of diffraction, by a grating or crystal,

---

[557] Karl Popper, *The Open Universe. An Argument for Indeterminism* (2000); Richard Feynman, *The Character of Physical Law* (1967): 7–8.

[558] John Polkinghorne, *Exploring Reality* (2005): 13.

[559] Karl Popper and John Eccles, *The Self and Its Brain* (1981): 14–21.

[560] Richard Feynman, *The Character of Physical Law* (1967) and Karl Popper and John Eccles, *The Self and Its Brain* (1981): 11–35.

[561] Karl Popper and John Eccles, *The Self and Its Brain* (1981): 18.

"*the macrostructure may, qua whole, act upon a photon or an elementary particle*" and we also encounter downward causation in other tools and machines: "*When we use a wedge, for example, we do not arrange for the action of its elementary particles, but we use a structure, relying on it to guide the actions of its constituent elementary particles to act, in concert, so as to achieve the desired result.*"[562] Stars, by producing heavier elements, are a kind of machines and "*an excellent example of downward causation, of the action of the whole structure upon its constituent particles*", and finally, "*the most interesting examples of downward causation are to be found in organisms and in their ecological systems, and in societies of organisms*",[563] which, to some degree, act independently of the behaviour of their individual members and influence this behaviour in powerful ways.

All these examples suggested by Popper verify the following basic statement, corroborated in many space-time regions: 'in many processes in this Universe we observe downward causation'. This basic statement falsifies the reductionist, deterministic universal statement: 'all causality is always and everywhere upward causation'.

Physical causality itself is indeterministic. Indeterminism argues that there is chance and basic randomness all the way down to the deepest level of reality, and that probability is not just subjective, related to our ignorance, but rather objective, i.e. a characteristic of physical reality itself. Even if we had perfect knowledge of all physical laws implied in producing an event, and of all initial conditions (the cause), the effect would be a range of probable outcomes, the sum of which is one, and not one outcome with probability one. Feynman has put forward this principle of indeterminism very forcefully:

---

[562] Karl Popper and John Eccles, *The Self and Its Brain* (1981): 19
[563] *Ibidem*: 20

"[W]hat we are proposing is that there is probability all the way back: that in the fundamental laws of physics there are odds... [T]he hidden variable theory... cannot be true; it is not due to lack of detailed knowledge that we cannot make a prediction... It is not our ignorance of their internal gears, of the internal complications, that makes nature appear to have probability in it. It seems to be something intrinsic."[564]

What Feynman proposes, can be summarised in two points:

1) Mathematical-physical laws are probabilistic, in the sense that given a cause, the outcome will not be an effect of probability 1 (corroborating the law), or 0 (falsifying the law), but a range of more or less probable outcomes, the sum of which will be 1.

2) The probabilistic character of mathematical-physical laws is not due to our subjective ignorance, but there is intrinsic randomness and objective probability in nature itself (here his vision is diametrically opposed to ontological determinism).

Popper too asserts that physical laws are probabilistic, in the sense that given a physical law and initial conditions, the outcome will not be an effect of probability 1 (corroborating the law), or 0 (falsifying the law), but a range of more or less probable outcomes. This does not mean that we will never find outcomes with a probability of 1. Probabilistic physical laws can explain propensities of 1, but deterministic physical laws can never explain probabilities varying somewhere between 0 and 1:

"New atomic theory – quantum mechanics – has jettisoned strict determinism. It has enriched physics by introducing objective probability statements into the theory of elementary particles and atoms. As a consequence of this, we ought to abandon Laplacean determinism. Indeed, many of the former strictly causal statements of classical physics about macroscopic objects have been re-interpreted as probability statements

---

[564] Richard Feynman, *The Character of Physical Law* (1967): 147

that assert probabilities close to 1. Causal explanation has been at least partly replaced by probabilistic explanation.... It is important to realize that statements asserting probabilities or propensities other than 0 or 1 cannot be derived from causal laws of a deterministic type (together with initial conditions) ... A probabilistic conclusion can be derived only from probabilistic premises; for example, premises about equal propensities. But it is possible, on the other hand, to derive statements asserting propensities equal to, or approaching, 0 or 1 − and therefore of causal character − from typically probabilistic premises."[565]

Secondly, Popper states that there is intrinsic randomness and objective probability in nature itself. It is not so that the structure and behaviour of a particular cloud in the sky, if we only knew all the initial conditions, could be fully explained as if it were a clock, but, rather the contrary, clocks appear to behave like clouds, as becomes clear in the behaviour of atoms:

"The interpretation of the atomic nucleus as a system of particles in rapid motion and of the surrounding electrons as an electron cloud is sufficient to destroy the old atomistic intuition of a mechanical determinism. The interaction between atoms or molecules has a random aspect, a chance aspect; chance not only in the Aristotelian sense in which it is opposed to purpose, but chance in the sense in which it is subject to the objective probabilistic theory of random events, rather than to anything like exact mechanical laws. Thus, the thesis that all physical systems including clouds, are, in reality, clocks, [is] mistaken. According to quantum mechanics we have to replace it by the opposite thesis, as follows: All physical systems, including clocks, are, in reality, clouds."[566]

In Math box 7.3, I do exactly that: transform the traditional deterministic hypothesis about the atom's behaviour, that we saw in Math box 7.1, into a

---

[565] Karl Popper and John Eccles, *The Self and Its Brain* (1981): 24–25.
[566] *Ibidem*: 34.

probabilistic one, and then show that this probabilistic trajectory not only holds for electrons orbiting an atomic nucleus, but also for a tennis balls, meaning that clocks are clouds and, indeed, tennis balls behave like electrons…

Math box 7.3 The probabilistic equation for determining an electron's orbit

We saw in Math box 7.1 that the atom was conceived by Ernest Rutherford (1871–1934) and Niels Bohr (1885–1962) as a mini solar system, where the behaviour of the electrons orbiting the nucleus could be explained by the deterministic laws of classical, Newtonian, mechanics. It is true though that Bohr sensed there was some randomness going in all this, speaking of *preferred* orbits, but he did not create the probabilistic equations needed to express this randomness. It was Ludwig Boltzmann (1844–1906), who developed statistical mechanics, which explains how the properties of atoms determine the physical properties of matter, and Albert Einstein (1879–1955), who applied the Boltzmann concept of the *"probability of the state of a system"*[567] to electromagnetic radiation, conceiving it as a system of particles (photons), with which probability functions are associated, which graphically have the form of waves.

In his doctoral thesis, published in 1925, Louis de Broglie (1892–1987), a French physicist and 1929 Nobel Prize winner, applied Einstein's revolutionary concept of the photon to the electron.[568] Instead of Bohr's Newtonian model of the atom, he conceived an equation which expresses the probability of an electron being found at a certain distance from the

---

[567] Albert Einstein, "Concerning a Heuristic Point of View Toward the Emission and Transformation of Light", in: *Annalen der Physik*, vol. 17 (1905) and in: *American Journal of Physics,* vol. 33 (1965): 9.

[568] Louis de Broglie, "Recherches sur la théorie de quanta", in: *Annales de Physique*, Tome III (1925). Louis de Broglie, *Ondes et mouvements* (1926); "La mécanique ondulatoire et la structure atomique de la matiére et du rayonnement", in: *Le Journal de Physique*, Tome VIII (1927): 225–241.

nucleus. James Huheey and Ellen and Richard Keiter comment: "*According to the Bohr theory, this was an immutable radius, but in wave mechanics it is simply the 'most probable' radius for the electron to be located*".[569] These probabilistic wave functions (ψ) of an electron, in a system of spherical polar coordinates, have three variables: $\psi\,(r, \varphi, \theta)$. Let us see, for example, the $\psi\,(r)$ function, where Z is the number of protons in the nucleus:

Hydrogen in its normal state has orbital $1s$ ($n = 1, l = 0, m = 0$):

(1) $\psi\,(r) = 2 \left(\dfrac{Z}{a_0}\right)^{3/2} e^{-Z(r/a_0)}$.

Hydrogen in its excited state has orbital $2s$ ($n = 2, l = 0, m = 0$):

(2) $\psi\,(r) = \dfrac{1}{2\sqrt{2}} \left(\dfrac{Z}{a_0}\right)^{3/2} \left(2 - \dfrac{Z\,r}{a_0}\right) e^{-Z(r/2a_0)}$.

In the case of orbital $2p$ ($n = 2, l = 1, m = 0$), we obtain:

(3) $\psi\,(r) = \dfrac{1}{2\sqrt{6}} \left(\dfrac{Z}{a_0}\right)^{3/2} \left(\dfrac{Z\,r}{a_0}\right) e^{-Z(r/2a_0)}$.

The orbit of an electron in an excited hydrogen atom, has the following eigenfunction (with $Z = 1$, and quantum numbers $n = 2, l = 0, m = 0$):

(4) its radius is: $r_n = \dfrac{\varepsilon_0 n^2 h^2}{\pi m_e e^2}$.

(5) and its 'normal' orbit ($n = 1$) is: $a_0 = \dfrac{\varepsilon_0 h^2}{\pi m_e e^2} = 5.29 * 10^{-11}\ m$.

The electron's probability to be at a certain distance from the nucleus is:

(6) $P_{point} = [\psi(r)]^2$.

The probability of finding it in a spherical orbit is:

(7) $P_{sphere} = 4\pi r^2 [\psi(r)]^2$.

[569] James Huheey and Ellen and Richard Keiter, *Inorganic Chemistry* (1993): 11.

With the help of *Mathematica*, I graphed the wave functions of the hydrogen electron in a normal and an excited state (see Graphs 7.1 and 7.2).

Graph 7.1 The probability of finding the electron at a certain distance from the nucleus, when the hydrogen atom is in a normal state

probability

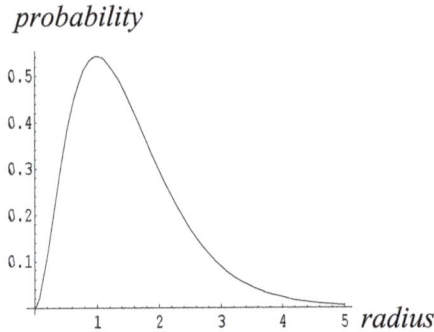

Explanation: vertical axis: probability (divided by its radius); horizontal axis: distance in multiples of Bohr radius

Graph 7.2 The probability of finding the electron at a certain distance from the nucleus when the hydrogen atom is in an excited state
probability

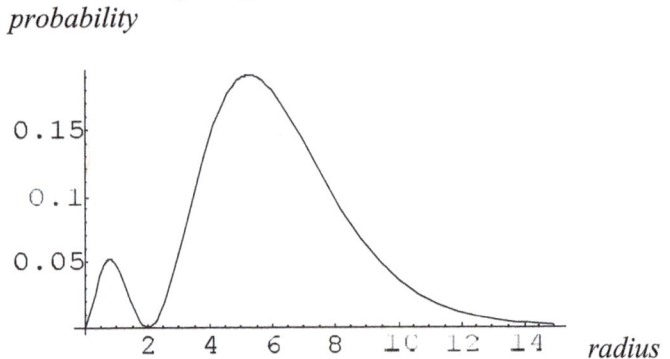

Explanation: vertical axis: probability (divided by its radius); horizontal axis: distance in multiples of Bohr radius

Both graphs show that there is no fixed, exactly predictable distance of the electron from the nucleus. The graphs also help to understand how people started conceiving photons and electrons as a wave. This is nonsense. The electron and the photon are point particles and not waves, but the behaviour of photons and electron can be represented by probability equations, which, when graphed, as in the two examples above, have the form of a wave.

Instead of using a matrix with two axes (where we see waves), we could also represent the same probability equations in the form of an onion (where we see layers), with different shades of grey between white (probability zero) and black (probability one), as in Graph 7.3. This graph does not mean, of course, that electrons are like onions, and have different layers. It is a representation of the different probabilities of an electron behaving in a certain way, not a representation of the structure of the particle.

Graph 7.3 The probability of finding an electron at a certain distance from the nucleus of hydrogen in an excited state ($n = 2, l = 0, m = 0$)

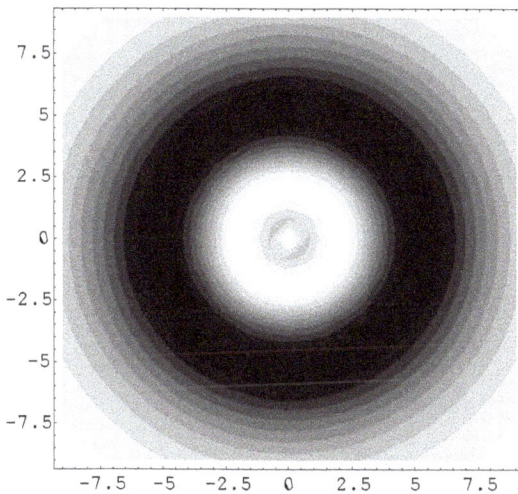

Explanation: white areas: probability=0; black areas: probability≈1; grey areas: o<probability<1

We can observe, comparing the second and the third graph, that the probability of the electron orbiting at a certain distance has two peaks in the second graph, which correspond to the two areas of darker grey in the third graph, one close to the centre, and one farther away and wider.

Another, frequently expressed idea, which is also not true, is the idea that only the behaviour of quanta is probabilistic. Probabilistic equations are not the exclusive domain of quantum physics, but extend also to the domain of classical physics. As a matter of fact, the whole of physics is probabilistic, as we shall now see in the case of the behaviour of a tennis ball.

I shall first reproduce the figure of the trajectories of a tennis ball, between A and B. Path number 1 is the classical, Newtonian path. Paths number 2, 3, 4, 5 and 6 represent other possible trajectories, each with their own amplitude.

Figure 7.7 Possible trajectories of a tennis ball, thrown from A to B [570]

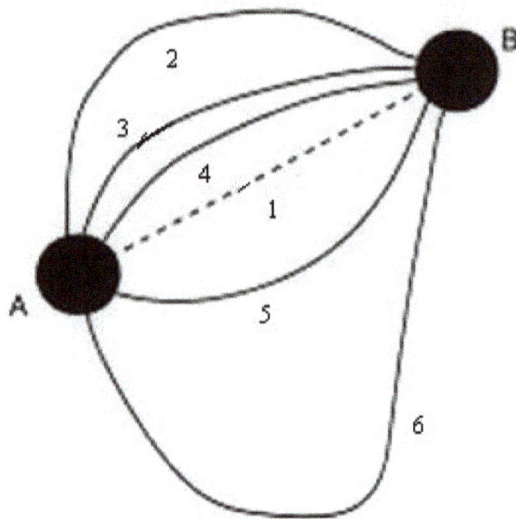

Explanation: Number 1 is the classical, Newtonian path; paths 2, 3, 4, 5 and 6 are other possibilities, each with their own probability

These alternative paths would be impossible in Newtonian mechanics. But as a matter of fact, they are possible, as Allday points out:

[570] Figure adapted from Jonathan Allday, *Quarks, Leptons and the Big Bang* (2002): 65.

"*According to Feynman the ball actually 'sniffs out' all possible paths at the same time. Each path is given an amplitude according to the action along that path. However, and here is the crunch point, it is only along those paths that are similar to the 'actual' classical path that the phase of the amplitudes will be similar. Paths that stray a long way from the classical path will have widely differing phases (positive and negative) which tend to cancel each other out*" and "*only paths that are very close to the classical one will have similar phases*", which would reinforce each other, like, for example path 4.*[571]*

The probability of each path is the square of what Feynman defined as the amplitude, and the amplitude has a magnitude and an angle or phase. In the figures with the graphic representation of the behaviour of the electron, the amplitude is the distance of the peak of a wave from the horizontal axis. The phase or angle is the angle with which the wave starts and the origin of the matrix. A phase shift occurs when the wave enters with a different angle, say angles φ and θ. If we combine two waves (probability functions), with equal amplitudes, these amplitudes can reinforce each other, weaken each other or cancel each other out, depending on both angles. If they cancel each other, the combined amplitude is zero. The combined amplitude of two waves is:

$$(8)\ \lambda = \sqrt{x^2 + y^2 + 2xy Cos(\varphi - \theta)}.$$

In this equation, x and y are the magnitudes of the amplitudes of the two trajectories, and φ and θ the respective angles or phases of these trajectories. Mathematically speaking, two trajectories cancel each other out, when their combined amplitude magnitude sums to zero, which occurs when the cosine of the difference between the two angles is -1, that is when the difference of the angles (the phase shift) is -180^0, so that:

---

[571] Jonathan Allday, *Quarks, Leptons and the Big Bang* (2002): 64–65.

(9) If $[x = y \wedge (\varphi - \theta) = -180^0] \rightarrow Cos(\varphi - \theta) = -1 \rightarrow$

(10) $\lambda = \sqrt{x^2 + y^2 + 2xyCos(\varphi - \theta)} = \sqrt{(x-y)^2} = (x-y) = 0.$

The amplitude can also be represented in terms of the Planck constant and the momentum of the electron:

(11) $\lambda = \dfrac{h}{mv}$.

In the case of an electron, travelling at normal speeds, the amplitude is:

(12) $\lambda_{electron} = \dfrac{(6.626*10^{-34})\, kgm^2s^{-1}}{(3.32*10^{-23}\, kgms^{-1}} = 2*10^{-11}m^2.$

In the case of a tennis ball, which weighs a tenth of a kilo and travels at 10 kilometres per second, the amplitude is much smaller than in the case of the electron. Imagine a graph with myriad waves, many more than the ones shown above, in the case of the tennis ball, but all very small, with very small amplitudes and periods:

(13) $\lambda_{electron} = \dfrac{(6.626*10^{-34})\, kgm^2s^{-1}}{(0.1*10\, kgms^{-1}} = 6.626*10^{-34}\, m^2.$

Now the point here is, that in the case of the tennis ball, there is an almost infinite number of amplitudes, *all very close to each other*, but with quite different phases so that any possible amplitude is almost certainly cancelled out by another one, the difference between the two angles being $\varphi - \theta = -180^0$. For any amplitude with a certain phase, there will be another one with a phase differing from the first one at -180^0.

Since the probability of any path is the square of the amplitude, we most probably end up with a probability close to one for the classical path and almost zero for other paths. This is why Newtonian mechanics functions well for macro-objects with very small amplitudes, like the tennis ball, which does not mean that classical physics is not probabilistic.

Section 7.5 The frontier between science and science fiction

Metaphysical statements and science fiction statements have in common that they cannot be falsified by data from world 1. Although both types of statements share the status of non-falsifiability and therefore, are not scientific, they differ in one very important aspect: metaphysics makes non-falsifiable statements about the interaction of the three worlds, but science fiction makes non-falsifiable statements about phenomena supposed to exist in world 1, though not observed by science.

There is another way to make a theory irrefutable, and therefore place it outside the realm of science, though it is not science fiction. It is what Popper refers to as verificationism.[572] It is typical of verificationism that it does not clarify which are the basic statements that can disprove its universal statements and if other people supply them, verificationism changes the theory, introducing auxiliary hypotheses to ensure that its theory, again, is irrefutable. Verificationism seeks irrefutability, generating a fluid state of the theory, which makes it impossible to put it to the test. Examples of verificationism in the field of social sciences are Freudian psycho-analytic theory[573] and Marxism[574] and, in modern theoretical physics, the multiple and continuously changing theories of eternal inflation, superstrings and the multiverse.[575]

Let us now turn our attention to science fiction and define the difference between science and science-fiction. In essence, science proposes *falsifiable* universal statements about events in world 1, but science fiction produces *non-falsifiable* existential statements about objects in world 1.

---

[572] For a description of verificationism, see Karl Popper, *Conjectures and Refutations* (1962): 34–38.

[573] See John Auping, "Una revisión de la metapsicología freudiana", in: *Una revisión de la teoría psicoanalítica a la luz de la ciencia moderna*, e-book (2000): 19–112.

[574] See Karl Popper, *The open Society and Its Enemies* (1982): 315–380.

[575] See Section 4.4 and 4.5 of Chapter 4.

Science fiction presents itself in the form of existential statements about never observed objects. These statements are not falsifiable. For instance, the statement "extra-terrestrial UFOs exist" is irrefutable, even though countless supposed appearances of extra-terrestrial UFOs have been debunked scientifically as optic illusions or cheating. But UFO watchers can always assume that they nonetheless exist.

Existential statements cannot be falsified for the same reason that universal statements cannot be verified. Any existential statement requires for its refutation the exploration of all possible space-time regions; whose number is always infinite. Since this, by definition, is impossible, the existential statements, by definition, cannot be falsified and, therefore, are non-scientific.

In cosmology and modern physics, there are quite a few statements of this type. For instance, 'monopoles exist', 'superstrings exist', 'additional spatial dimensions beyond the three observables ones exist', 'the multiverse exists'.

Some scholars do not agree with the thesis that existential statements do not belong to science. For example, Penrose, who shares with Popper the philosophy of the three worlds[576] and also, in general, the criterion of falsifiability, thinks, nonetheless, that existential statements have a place in science. As an example of an existential statement, which he thinks is scientific, he proposes the statement about the existence of monopoles, admitting all the way that it is not falsifiable:

"The theory which asserts that such a monopole exists somewhere is distinctly un-Popperian. That theory could be established by the discovery of such a particle, but it appears not to be refutable, as Popper's criterion would require; for, if the theory is wrong, no matter how long

---

[576] Roger Penrose, *The Road to Reality* (2004): 17–23, 1027–1033. See Section 7.2 of this chapter.

experimenters search in vain, their inability to find a monopole would not disprove the theory! Yet the theory is certainly a scientific one, well worthy of serious consideration."[577]

In affirming that the existential statement about the existence of monopoles is part of science, Penrose seems to replace the objective criterion of falsifiability with a subjective criterion, namely, that he feels this theory is 'certainly scientific'. I agree with Penrose science fiction is well 'worthy of consideration', because, together with intuitions, dreams, hunches and strokes of good luck, it belongs to the kitchen where scientific theories are prepared before they are served. Once served, however, to be counted as science they should be falsifiable.

Some scholars propose allowing *some* existential statements to be scientific if we can foresee that one day they might be verified by empirical evidence. Personally, I do not believe that *some* existential statements may be verified one day with empirical evidence. I rather think, that, because of the logical structure of an existential statement, *all* existential statements are verifiable. The problem is not that they are not verifiable, the problem is that they are not falsifiable. The epistemological status of these statements of being verifiable but not falsifiable is precisely the problem.

It might be argued, however, that we may admit *some* existential statements as being part of science, i.e. the *negation of an existential statement* since these are, strictly speaking, falsifiable. A good example, is the negation of the one proposed by Penrose, i.e. the statement 'monopoles do *not* exist'. Once a monopole is found, this statement would be falsified. Likewise, the statement 'UFOs do *not* exist', is a statement that would be falsified, once an UFO is found. The same holds true for the existential statement 'there exists life on other planets of our

---

[577] Roger Penrose, *The Road to Reality* (2004): 1021.

Universe', which can be replaced by its negation, that is certainly falsifiable: 'life on other planets does *not* exist'. Once life on one other planet of our galaxy is found, with no need to look for planets with life in other galaxies, the statement 'life on other planets does *not* exist' is falsified. For this reason, we may admit some existential statements, i.e. those that are from a strictly logical point of view the negation of an existential statement.

Since some existential statements are logically linked to their falsifiable negations, it would be more correct to say that all those existential statements, the negation of which can*not* be falsified, are science fiction. For example, the negation of the statement 'many universes may exist with different physical laws than the ones found in our Universe', is the statement 'no other universes exist'. Since, by definition, if there were other universes, they cannot be observed (even of our own Universe we can observe only a minor part), it follows in this case that neither the existential statement, nor its negation can be falsified, so both belong to the realm of science fiction.

In Sections 4.5 and 4.6 of Chapter 4, I analysed two theories which were shown to be non-testable by observable, real evidence, for which reason they belong to the realm of science fiction, though they are disguised as science. These are the Guth-Linde speculation of eternal inflation and the multiverse; and the speculations on superstrings and the multiverse by Susskind, Kaku and others.

Section 7.6 The fascination with mathematical miracles

Science fiction sometimes disguises its true status by means of what Penrose has called 'mathematical miracles'.[578] If a theory belonging to the realm of science fiction is proven to be mathematically consistent, then some authors feel that this mathematical consistency absolves them from the obligation, fundamental in science, to contrast it with empirical data in order

---

[578] Roger Penrose, *The Road to Reality* (2004):1038–1042.

to be able to decide whether it is false or true. A good example of such a fascination with mathematical miracles is Susskind's philosophy of science:

"Excitingly, all of String Theory's consequences have unfolded in a mathematically consistent way. String Theory is a very complex mathematical theory with very many possibilities for failure. By failure I mean internal inconsistency. It is like a huge high-precision machine, with thousands of parts. Unless they all fit perfectly together in exactly the right way, the whole thing will come to a screeching halt. But they do fit together, sometimes as a consequence of mathematical miracles."[579]

Since he cuts the falsifiability criterion out of his philosophy, he feels compelled to ridicule the *"Popperazzi"* and *"Popperism"*,[580] i.e. Popper's philosophy of science, substituting his own philosophy of science for Popper's: *"Good scientific methodology is.. conditioned by, and determined by the science itself and the scientists who create the science."*[581]

All the way rejecting a philosopher's contribution, Susskind, who has no credentials in philosophy, proposes his own philosophy of science, i.e. scientific methodology is what scientists determine it to be. I think this 'philosophy' is confused, authoritarian and arrogant.

Regarding mathematical miracles, Penrose recommends caution:

"Are such apparent miracles really good guides to the correctness of an approach to a physical theory? This is a deep and difficult question. I can imagine that sometimes they are, but one must be exceedingly cautious about such things ... One thing is certain, however, and that is that such mathematical miracles cannot always be a sure guide."[582]

---

[579] Leonard Susskind, *The Cosmic Landscape* (2005): 124.
[580] *Ibidem*: 192, 195.
[581] *Ibidem*: 194.
[582] *Ibidem*: 1040–41.

For a theory, to be scientific, two conditions must be fulfilled, each of which is necessary, but none of which is in itself, without the other one, sufficient:

1) The theory must be mathematically consistent.
2) The theory must be falsifiable.

If only the first condition is met, but not the second, as is the case in some multiverse theories, then the theory is mathematically consistent science fiction.

CHAPTER 8

THE CAUSE OF THE UNIVERSE

In this chapter, we will explore three problems related to the Universe:
1. An inquiry into the cause of the Universe.
2. The multiverse speculation is in denial of an intelligent cause.
3. What motivates some cosmologists to embrace the multiverse.

Section 8.1 An inquiry into the cause of the Universe

All the models of the origin of the Universe converge on the same question, expressed by Leibnitz and made famous by Martin Heidegger: *"Why are there beings at all instead of nothing?"*[583] Leibniz said:

> *"[It is] true of the different states of the world, [that] the state which follows is, in a sense, copied from the preceding state, though in accordance with certain laws of change. And so, however far back we might go into previous states, we will never find in those states a complete explanation why there is any world at all, and why it is the way it is."*[584]

Today, scientists and philosophers ask the same question that goes beyond the limits of science, for example, Peter Medawar, paraphrasing Popper's question: *"How did everything begin? What are we all here for?"*[585] Before continuing, I will first explain some symbols of logic used in this section.

---

[583] Martin Heidegger repeats this question 15 times, in: *Introduction to Metaphysics* (2000): 1, 4, 5, 7, 8, 10, 13, 22, 24, 26, 30, 34, 35, 41, 215.
[584] Gottfried Wilhelm Leibniz, *Discourse on Metaphysics and Other Essays* (1991): 41.
[585] Peter Medawar, *The Limits of Science* (1985): 66, cited in Alister McGrawth, *The Dawkins Delusion? Atheist Fundamentalism and the Denial of the Divine* (2007): 39

Logical equations box 8.1 Some symbols of logic and set theory	
= is the same as;	∋ contains as an element;
≠ is not;	∃ exists;
⊂ is a subset of;	∄ does not exist;
⊆ is a subset of or is the same as;	| on the condition that;
∧ and (logical conjunction);	⇒ from the previous, it logically follows that;
∨ or (inclusive disjunction);	
⊗ or (exclusive disjunction);	⟶ is the cause of;
∪ and/or	↛ is not the cause of;
⊥ logically contradicts the following;	∀ for all;
	{...} the set of ...
¬ not (negation);	$p(a\|b) = N$ the probability of the event a, given the event b is N (with
∈ belongs to;	
∉ does not belong to;	$0 \leq N \leq 1$)

There is a problem at point $t = 0$ in the history of the Universe. According to science, every event and object in the Universe is part of a chain of cause and effect: "*In the same way, we may have to get used to the idea of an absolute zero of time −a moment in the past beyond which it is in principle impossible to trace any chain of cause and effect.*" [586] In this section, I will carry out an analysis which integrates logic and set theory, to see, firstly, if the Universe is caused by itself, or by an external cause that does not belong to the set of objects and events which constitute the Universe; and secondly, if the cause of the Universe is the cause of itself.

In Logical equations box 8.2, I present five definitions, four axioms belonging to the principle of causality and four theorems derived from these definitions and axioms:

---

[586] Steven Weinberg, *The first three minutes* (1977): 144.

Logical equations box 8.2 Five definitions; four axioms; four theorems

Five definitions

(1) An event is the existence of a real phenomenon e_r or possible phenomenon e_p, or the existence of a *set* of real phenomena $\{e_r\}$ or a set of possible phenomena $\{e_p\}$, in a space-time region of the universe, or in the universe at large.

(2) The events space E is the set of all of the events and sets of events that are real and/or possible: $E = E_r \cup E_p$. The real events space is defined as: $E_r = \{e|e_r\}$ and the possible events space as: $E_p = \{e|e_p\}$. The real events space is a subset of the possible events space: $E_r \subseteq E_p \Leftrightarrow (e \in E_r \Rightarrow e \in E_p)$.

(3) The event A exists if it belongs to the real events space: $A \in E_r \Rightarrow \exists A$ but A does not exist if it belongs to the possible but not the real events space $A \in \{E_p - E_r\} \Rightarrow \nexists A$.

(4) Definitions of different kinds of sets.

A) The set $X = \{A|P(A)\}$ is the collection X of all the elements that satisfy a particular property $P(A)$, such that:

B) if an element A belongs to a set B, the initial time of the event T_A is later than or coincident with the initial time T_B of the set B: if $A \in B \Rightarrow T_A \geq T_B$.

C) A set can be an element of another set, so there are two kinds of sets. We define R as the set of all sets that do not belong to themselves, excluding R: $R = \{A|A \notin A, A \neq R\}$ and S as the set of all sets that belong to themselves, including S: $S = \{A|A \in A\}$.

(5) The universe U is *preliminarily* defined as the set of all events and sets of all events that are real, with the cause of the universe C possibly included in the universe or possibly excluded: $U = \{A|(A \in E_r) \wedge (C \in U \otimes C \notin U)\}$. The 'universe' is not the same as 'universal set' as defined in set theory, represented by the symbol U.

Four axioms of the principle of causality

(6) An event or set of events A, is cause of another event or set of events B, if the event or set of events A is a necessary initial condition (sine qua non), although usually not a sufficient one, for the event or set of events B: $(A \rightarrow B) \Rightarrow (\nexists A \Rightarrow \nexists B)$.

(7) All events and sets of events have a cause, or various causes, i.e. for all B belonging to set E_r, there must exist an event or set of events A that also belongs to E_r such that A is the cause of B: $(\forall B \in E_r) \Rightarrow (\exists A \in E_r | A \rightarrow B)$.

(8) If an event or set of events A is the cause of another event B, the initial time T_A of the cause A is earlier than the initial time T_B of the effect B: $(A \rightarrow B) \wedge (A \in E_r) \Rightarrow T_A < T_B$.

(9) No event or set of events A is the cause of itself, with the possible exceptions of the universe U itself and/or the cause C of the universe: $(A \nrightarrow A) | (A \neq U \wedge A \neq C)$. For the sake of the following logical analysis, the universe is possibly the cause of itself, or possibly it is not; and in case it is not, the cause of the universe is possibly the cause of itself, or possibly it is not. So, there are only three possibilities:

a) the Universe U is the cause C of itself $(C \rightarrow U \wedge C = U)$;

b); C is the cause of the Universe U and C is not the universe and is not the cause of itself $(C \rightarrow U \wedge C \neq U \wedge C \nrightarrow C)$; or

c) the cause C of the universe is not the universe and is the cause of itself $(C \rightarrow U \wedge C \neq U \wedge C \rightarrow C)$.

At the end of this analysis, we shall conclusively determine whether the universe is the cause of itself or not, and if not, whether the cause of the universe is the cause of itself or not. These conclusions will follow from the axioms, and are not axioms themselves.

Four theorems derived from the five definitions and the four axioms of causality

(10) From definition (4) it can be deduced that belonging to the set of sets that belong to themselves excludes belonging to the set of sets that do not belong to themselves and vice versa:

a) $(A \in S \Rightarrow A \in A \Rightarrow A \notin R) \wedge (A \notin R \Rightarrow A \in A \Rightarrow A \in S)$;

b) $(A \notin S \Rightarrow A \notin A \Rightarrow A \in R) \wedge A \in R \Rightarrow A \notin A \Rightarrow A \notin S$,

c) such that $(A \in S \Leftrightarrow A \notin R) \cup (A \notin S \Leftrightarrow A \in R)$.

(11) If it is conjectured that an event is the cause of itself and at the same time we know that this event is not the universe nor the cause of the universe, it can be deduced from axiom 9 that this event does not actually exist: $(A \rightarrow A) \wedge [\neg(A = U \vee A \rightarrow U)] \Rightarrow A \notin E_r$.

(12) If a set of elements with a particular property exists and there is an element that does not have this property, it follows that this element does not belong to that set. A property is a function $P(x)$. If x has the property, then $P(x)$ is true, if x does not have it then $P(x)$ is false. From definition (4), $X = \{A|P(A)\}$, it follows that $\exists A \neg P(A) \Rightarrow A \notin X$.

(13) From (4B) y (8) we obtain $A \in B \underset{4B}{\Rightarrow} T_A \geq T_B$ and $A \rightarrow B \underset{8}{\Rightarrow} T_A < T_B$, it can be deduced that, if an event or set of events A is the cause of the event B, A does not belong to B and, vice versa, if A belongs to B then A is not the cause of B, meaning that a causal relationship between two events precludes belonging to the same set, and belonging to the same set precludes a causal relationship: $(A \rightarrow B \Rightarrow A \notin B)$ and $(A \in B \Rightarrow A \nrightarrow B)$.

With respect to axiom (8), I concur with Jefimenko that the cause precedes the effect in time:

"One of the most important tasks of physics is to establish causal relations between physical phenomena. No physical theory can be complete unless it provides a clear statement and description of causal links involved in the phenomena encompassed by that theory... Causal relations between phenomena are governed by the principle of causality. According to this principle, all present phenomena are exclusively determined by past events. Therefore equations depicting causal relations between physical phenomena must, in general, be equations where a present-time quantity (the effect) relates to one or more quantities (causes) that existed at some previous time... [T]hen, according to the principle of causality, an equation between two or more quantities simultaneous in time but separated in space cannot represent a causal relation between these quantities ... because, according to this principle, the cause must precede its effect."[587]

Axiom (8) might be objected to since Newtonian theory of gravity was originally conceived by Newton as a theory of 'action at a distance' with the coincidence in time of cause and effect. But since Newtonian theory is a limiting case of Einstein's general relativity, the statement on the coincidence in time of cause and effect is consequently abandoned. The time needed by light to travel from one point to another is not 'action at a distance', since in that case the cause (for example, turning on the light) precedes the effect (seeing the light at some distance). And since the speed of a gravitational wave is conceived as being the same as the speed of light, the Newtonian conjecture about the coincidence in time of cause and effect is abandoned.

Niels Bohr's and Werner Heisenberg's interpretation of quantum mechanics also involved a conjecture about 'action at a distance'. This is

---

[587] Oleg Jefimenko, "Presenting Electromagnetic Theory in Accordance with the Principle of Causality", in: *European Journal of Physics,* vol. 25 (2004): 287–288.

not the place to give a detailed refutation of the 'action at a distance' conjecture – I have done so elsewhere, following Einstein and Popper[588] – and shall here limit myself to a brief summary of this criticism. Heisenberg's conjectures about 'action at a distance' start with his confusion related to the experiment of photons directed at a semi-reflective mirror. Observation tells us that 50% will pass through and reach a photographic plate behind the mirror, and 50% will be reflected and not reach the photographic plate. Heisenberg believes he sees action at a distance in this experiment. After a photon has been reflected by the semi-reflective mirror, according to Heisenberg:

> "the probability of finding the photon in the other part of the package immediately becomes zero. The experiment at the position of the reflected part thus exerts a kind of action (reduction of the wave packet) at the distant point occupied by the transmitted packet, and one sees that this action is propagated with a velocity greater than that of light."[589]

Now this is pure confusion, as I shall now explain. Let us call the event of a photon before it reaches the mirror event b; the event of a photon passing through the mirror event a; and the event of a photon being reflected event $-a$. As a matter of fact, the experiment described by Heisenberg is a set of three experiments, i.e. the first experiment is that of a photon before it reaches the semi-reflective mirror (experiment I); the second experiment is the one of a photon after it has been reflected by the semi-reflective mirror (experiment II); and the third experiment is that of a photon after it has passed through the semi-reflective mirror (experiment III). There is no such thing as the collapse of the wave function or 'action at a distance'. There are

---

[588] See Karl Popper, *Quantum Theory and the Schism in Physics* (1982); and "La gran confusión cuántica", in: John Auping, *El Origen y la Evolución del Universo*, e-book (2016): 445–453.

[589] Werner Heisenberg quoted in Karl Popper, *Quantum Theory and the Schism in Physics* (1982): 76–77.

six probability functions for three different experimental setups, two for experiment I, two for experiment II and two for experiment III:

- experiment I: $p(a|b) = 1/2$; $p(-a|b) = 1/2$;
- experiment II: $p(a|-a) = 0$; $p(-a|-a) = 1$;
- experiment III: $p(-a|a) = 0$; $p(a|a) = 1$.

Heisenberg's commentary, revealing that he thinks of action at a distance, is typical for what Popper calls "*the great quantum muddle*".[590] There is no action or communication of the package of the reflected photons to the package of the photons before they reach the semi-reflective mirror, or, for that matter, to the package of photons that have passed through the mirror, neither is there a collapse or change of the wave function through some hypothesised action or communication at a distance. As a matter of fact, the experiments can be realized with *one photon at a time*, one experiment after the other. The probability functions of the experiment with the photon before reaching the semi-reflective mirror, are different from the probability functions of the experiment with the photon after it has been reflected, which are different from the probability functions of the experiment with the photon after it has passed through the semi-reflective mirror. The photons in these three experiments are not acting upon one another at a distance in space or time, nor are they communicating with each other at a distance. On the basis of their thought experiment, Einstein, Podolsky and Rosen reached the same conclusion: there is no such thing as "*spooky action at a distance*".[591]

Another possible objection to Axiom 8 could arise from statements about particles that travel backwards in time, firstly producing the effect – the

---

[590] Karl Popper, *Quantum Theory and the Schism in Physics* (1982): 77.
[591] Albert Einstein, Boris Podolsky & Nathan Rosen, "Can Quantum-Mechanical Description of Physical Reality be Considered Complete?" *Physical Review*, vol. 47 (1935): 777 ss. Einstein, Podolsky and Rosen's thought experiment is commented upon by Einstein in a letter to Popper, which is reproduced in Karl Popper, *The logic of scientific discovery* (2002): 481–488.

emission of a photon – and later the cause – the absorption of a photon – as in Figure 8.1with the Feynman diagram.

Figure 8.1 Feynman diagram: interaction of an electron and a photon[592]

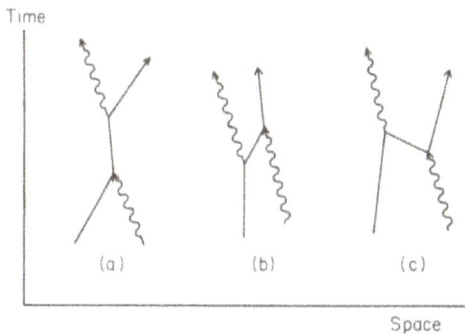

The Feynman diagram corresponding to the apparent case of the cause being later in time then the effect, is the third drawing (c) in Figure 8.1. Feynman comments *"Even more strange is the possibility (c) that the electron emits a photon, then travels backwards in time to absorb a photon, and then proceeds forwards in time again."*[593] What happens in reality, when we observe the sequence of events in the laboratory, is different, as can be appreciated in Figure 8.2 with the alternative Feynman diagram. The electron, apparently moving backwards in time, from T_5 to T_3, in reality is a positron moving forwards in time, from T_3 to T_5. Feynman himself comments upon this figure:

"Looking at example (c) from [the previous] Figure, going only forwards in time (as we are forced to do in the laboratory) from T_0 to T_3, we see the electron [straight line] and photon [wavy line] moving toward each other. All of a sudden, at T_3, the photon "disintegrates" and two particles appear —an electron and a new kind of particle (called a "positron"), which is

---

[592] Adapted from: Richard Feynman, *QED, The Strange Theory of Light and Matter* (2006): 97. 'QED' means 'Quantum Electrodynamics'.
[593] Richard Feynman, *QED. The Strange Theory of Light and Matter* (2006): 97–98.

like an electron going backwards in time, and which appears to move toward the original electron itself! At T_5, the positron annihilates with the original electron to produce a new photon. Meanwhile, the electron created by the earlier photon continues forwards in space-time".[594]

Figure 8.2 Interaction of an electron and a photon[595]

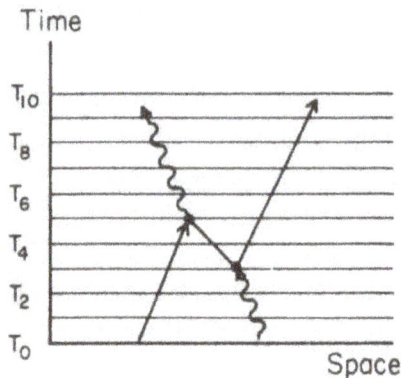

Some may also object to Theorem (13) of Logical equations box 8.2, arguing that, for example, the set of parents is the cause of the set of children and at the same time, belongs to the set of children. This argument is a fallacy, because while it is true that the set of parents is a subset of the set of children, it is false that the set of parents is the cause of the set of children. Rather, the set of fertile sexual relations between a man and a woman is the cause of the set of children.

With the passing of time, B can coincide with A to cause C. For example, an animal species A could cause an ecological change B and later, the coincidence of A and B could be the cause of part of A evolving into C:

---

[594] Richard Feynman, *QED. The Strange Theory of Light and Matter* (2006): 99.
[595] Adapted from *ibidem*: 99.

Figure 8.3 A is the cause of B and A and B are the cause of C

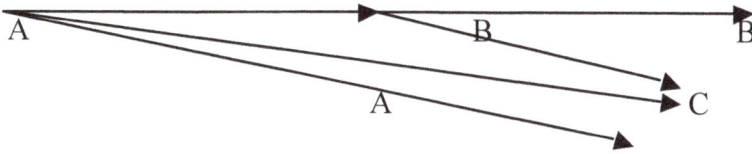

Arrow of time

The reason why S and R, in definition (4C) are defined differently has to do with *Russell's paradox*. It can be proven, using Russell's paradox, that such a thing as 'the set of all the events and sets of events that do not belong to themselves' cannot exist as such: $\nexists R = \{A | A \notin A\}$. Penrose summarises this paradox:

"This paradox proceeds as follows. Consider the set R, consisting of 'all sets that are not members of themselves'. (For the moment, it does not matter whether you are prepared to believe that a set can be a member of itself. If no set belongs to itself, then R is the set of all sets.) We ask the question, what about R itself? Is R a member of itself? Suppose that it is. Then, since it then belongs to the set R of sets which are not members of themselves, it does not belong to itself after all – a contradiction! The alternative supposition is that it does not belong to itself. But then it must be a member of the entire family of sets that are not members of themselves, namely the set R. Thus, R belongs to R, which contradicts the assumption that it does not belong to itself. This is a clear contradiction! What this argument is actually showing is that there is no such thing as the 'set of all sets'." [596]

---

[596] Roger Penrose, *The Road to Reality* (2004): 372.

I agree with Penrose that Russell's paradox proves that 'the set of all sets that are not members of themselves', *as such*, does not exist, being a contradiction. However, 'the set of all sets that are not members of themselves, *excluding the set itself*', does exist. So, we can avoid the contradiction by defining R as 'the set of all sets that do not belong to themselves, excluding R', as per definition (4 C) of Box 8.2.

Axiom (9) means it is not possible that, in the reality surrounding us, some events and sets of events are caused by themselves and others are not. All events and sets of events have a cause external to themselves, except *possibly* the Universe itself, in case the universe is the cause of itself; and in case the Universe is not the cause of itself, the cause of the Universe is *possibly* the cause of itself. We are not *a priori* stating that the Universe or its cause is caused by itself. In the case of the Universe and its cause I leave both possibilities open. The Universe and its cause have a special status, which is not *a priori* defined but rather decided upon *a posteriori* as a result of the logical analysis carried out below.

Given that events and sets of events are not caused by themselves, with the *possible* exception of the Universe and/or its cause, the question arises whether the Universe and its cause are caused by themselves or by a cause external to themselves. Furthermore, another question arises: does the Universe belong to the set of sets that belong to themselves? Or does it belong to the set of sets that do not belong to themselves (excluding the set itself)? At first glance, there appear to be four options in the events space, encapsulated in the following four propositions:

P_A) The Universe belongs to the set of sets that are members of themselves and is the cause of itself.

P_B) The Universe belongs to the set of sets that are members of themselves and is not the cause of itself.

P$_C$) The Universe belongs to the set of sets that are not members of themselves (excluding the set itself) and is the cause of itself.

P$_D$) The Universe belongs to the set of sets that are not members of themselves (excluding the set itself) and is not the cause of itself.

Table 8.1 represents these four possibilities.

Table 8.1 Four propositions about possible properties of the Universe

	The Universe is member of the set of events that are the cause of themselves: $U \to U$	The Universe is member of the set of events that are not cause of themselves: $U \nrightarrow U$
The Universe belongs to the set S of sets that are members of themselves: $U \in U$	$\mathbf{P_A} \equiv [U \to U \wedge U \in U]$.	$\mathbf{P_B} \equiv [U \nrightarrow U \wedge U \in U]$.
The Universe belongs to the set R of sets that are not members of themselves, excluding R: $U \notin U$	$\mathbf{P_C} \equiv [U \to U \wedge U \notin U]$.	$\mathbf{P_D} \equiv [U \nrightarrow U \wedge U \notin U]$.

However, not all four propositions, though conceived as possible, are real in the events space, as we shall see in Logical equations box 8.3.

Logical equations box 8.3 The cause of the Universe: four propositions

P$_A$. According to proposition **P$_A$**, $U \to U \wedge U \in U$. If the Universe were the cause of itself ($U \to U$), then given theorem (13), the Universe would not be a member of itself ($U \notin U$) and therefore, could not belong to the set of sets that are member of themselves ($U \notin S$). Therefore, by theorem (10), it would belong to the set of sets that do not belong to themselves, except for this set itself ($U \in R$). So, $\mathbf{P_A}$ contains a contradiction ($\perp$) and is therefore impossible. In summary, from (4), (10) and (13) respectively, we obtain:

(1) $\left(U \to U \underset{13}{\Rightarrow} U \notin U \underset{4}{\Rightarrow} U \notin S \underset{10}{\Rightarrow} U \in R \right) \perp (U \in S)$.

So, proposition $\mathbf{P_A}$ is false.

$\mathbf{P_B}$. According to proposition $\mathbf{P_B}$, $U \nrightarrow U \wedge U \in U$. The way to prove this proposition is the following. Given that the Universe is the set of all real events and that the Universe is itself a real event, the Universe is a set that is member of itself, and by axiom (13) it cannot be the cause of itself:

(2) $(U \underset{5}{=} E_r \wedge U \in E_r) \Longrightarrow U \in U \underset{13}{\Rightarrow} U \nrightarrow U$.

Given that the proposition $\mathbf{P_B}$ is true, we can provide a complete definition of the universe, which substitutes definition (5): *the Universe is the set of all real events, which include the Universe itself and which have the property of not being the cause of themselves*:

(3) $U = \{A \in E_R | (A \nrightarrow A \wedge U \in U)\}$.

$\mathbf{P_C}$. According to proposition $\mathbf{P_C}$, $[U \to U \wedge U \notin U]$. By definition (5), the Universe is the set of *all* real events:

(4) $U = E_r$.

Given that the Universe exists, the Universe is a real event and so:

(5) $U \in E_r$.

From (4) and (5) and theorem (10), we obtain:

(6) $(U \in U \underset{10}{\Rightarrow} U \notin R) \perp U \in R$.

Given that the statement $U \notin R$ is true, it follows that the statement $U \in R$ is false.

$\mathbf{P_D}$. According to proposition $\mathbf{P_D}$, $[U \nrightarrow U \wedge U \notin U]$. The same reasoning which proved $\mathbf{P_C}$ to be false, proves that $\mathbf{P_D}$ is false. Since it is true that

$U \notin R$ (statement 6), it follows that the statement $U \in R$ is false, so proposition **P**_D is false.

Finally, there is another way to confirm that the proposition we already proved to be true, **P**_B, $U \nrightarrow U \wedge U \in U$, is true. Since it has been proven that propositions **P**_A, **P**_C and **P**_D are false, and since there only exist four possible propositions related to the Universe, i.e., **P**_A, **P**_B, **P**_C and **P**_D, it follows that **P**_B must be true:

(7) $[(\nexists U|P_A) \wedge (\nexists U|P_C) \wedge (\nexists U|P_D)] \Rightarrow (\exists U|P_B)$.

Finally, we must analyse the cause of the cause of the universe. The Logical equations box 8.4 contains this inquiry.

Logical equations box 8.4 The cause of the cause of the Universe

In regard to the cause of the cause of the Universe, there are two options:

P_E. The cause of the Universe is the cause of itself:

(8) $C \rightarrow C$

P_F. The cause of the Universe is not the cause of itself:

(9) $C \nrightarrow C$

I start with the logical analysis of the proposition **P**_F:

P_F. If the cause of the Universe is not the cause of itself, then it follows that the cause of the universe belongs to the universe, because the Universe is the set of all real events and sets of events that are not caused by themselves:

(10) $U \underset{16}{=} \{A \in E_R | (A \nrightarrow A \wedge U \in U)\} \Rightarrow if\ C \nrightarrow C \Rightarrow C \in U$

However, according to theorem (13), if an event or set of events A is the cause of another event or set of events B, A does not belong to B:

$$(11)\,(A \rightarrow B \Rightarrow A \notin B) \underset{13}{\Rightarrow} (C \rightarrow U \Rightarrow C \notin U)$$

It follows from (10) and (11) that proposition (9) is false, because (9) implies a logical contradiction:

$$(12)\,(C \nrightarrow C \Rightarrow C \in U) \perp (C \rightarrow U \Rightarrow C \notin U)$$

PE- Since proposition (9), which says that the cause of the Universe is not the cause of itself, is false, it follows that proposition (8), which states that the cause of the Universe is the cause of itself, is true. If proposition (8) is true, it follows that the cause of the Universe is not part of the universe:

$$(13)\,\left(C \underset{21}{\rightarrow} C\right) \wedge \left(U \underset{P_B}{\nrightarrow} U\right) \Rightarrow (C \neq U) \wedge (C \underset{P_B}{\rightarrow} U) \underset{13}{\Rightarrow} C \notin U$$

So, the cause of the Universe has the following properties:

(14) $\exists U \Rightarrow \exists C | (C \rightarrow U)$ (both the Universe and its cause exist)

(15) $C \notin U$ (the cause of the Universe is not a member of the set of events that constitute the Universe)

(16) $C \rightarrow C$ (the cause of the Universe is the cause of itself)

There are four important conclusions from this inquiry:

I. The Universe is the set of all objects, events and sets of events that actually exist and that are not caused by themselves.

II. The Universe belongs to the set of sets that are members of themselves.

III. The Universe is not caused by itself, but by an external cause that does not belong to the Universe.

IV. The cause of the Universe is the cause of itself.

The logical analysis of the cause the Universe allows us to extend Popper's philosophy of the three worlds to include a fourth world, i.e. the cause of the Universe, which does not belong to the Universe and is the cause of itself, and therefore constitutes a reality *sui generis*, as can be seen in Figure 8.4.

Figure 8.4 The philosophy of four worlds[597]

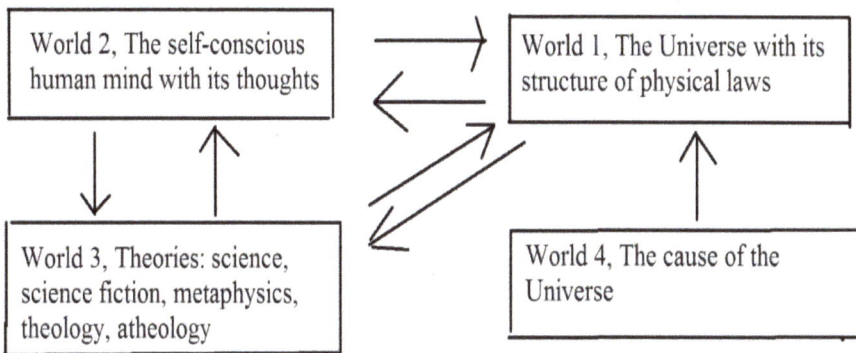

World 2, The self-conscious human mind with its thoughts	→ ←	World 1, The Universe with its structure of physical laws
World 3, Theories: science, science fiction, metaphysics, theology, atheology		World 4, The cause of the Universe

Some cosmologists feel that their theory of an eternal universe has no need for a theory of the cause of the universe. It simply was always there. For example, Stephen Hawking associated the concept of a Universe without a limit in time with the a-theological idea of the absence of a creator-cause of the Universe. At a conference on cosmology organised by the Vatican in 1981, he stated that there exists the possibility that space-time is finite but has no boundary and expressed his conviction that this would mean that there was no beginning, no moment of Creation.[598] And Linde believes that his theory of eternal chaotic inflation, representing an infinite regression of

---

[597] For an explanation of the interactions between world 1, 2 and 3, see Chapter 7.
[598] Stephen Hawking, *The Illustrated A Brief History of Time* (1996).

universes without a beginning in time, frees him from the troubling idea of the creation of the Universe out of nothing, at point $t = 0$.[599]

Others argue that such line of thinking cannot be sustained. Polkinghorne thinks Stephen Hawking is being *"naïve"* when he maintains that the thesis of a finite Universe without a beginning in time has a-theological consequences in the sense of not being in need of creator.[600] Hawley and Holcomb's analysis of Hawking's conjecture makes the same point, developing their argument with great clarity:

"When humans ponder the creation of the universe, generally the question they ask is, 'Why does something exist rather than nothing?' Why is there a universe at all?... How, then, were space and time created?... This issue is sufficiently disturbing to some cosmologists that they attempt to sidestep it by extending the history of the universe into an indefinite, infinite past. If there is no point at which t = 0, the reasoning goes, there is no need for creation. However, the question of existence is not answered by supposing that the universe is infinitely old. Time is physical, and an infinite time would be just another physical attribute of the universe... An infinitely old universe is not nothing, so it must have been created; it was simply created with time that extended infinitely...

Some relativists and cosmologists, most prominently Stephen Hawking, have pointed out that in general relativity, finite space and finite time can form a completely self-contained, finite space-time with no boundary or edge at all. The point we call t = 0 only appears to be a boundary in time because of the way in which we have divided space-time into space and time. Such a universe can be contemplated with the help of an analogy to

---

[599] See Chapter 4 of this book.
[600] John Polkinghorne, *Quarks, Chaos and Christianity* (2006): 51.

the Earth. On the Earth, the North Pole is the limit to how far it is possible to travel in the direction we call north, but it is nevertheless just a point on a continuous, boundary less globe. Similarly, the point t = 0 in a spherical big bang model of the universe represents merely an arbitrary demarcation in time. Without boundaries (spatial or temporal), there is no need [according to Hawking] to imagine the universe to be contained within some meta-universe…

But the presence (or absence) of a t = 0 point in time provides no answer to the mystery of creation, nor does it have implications for the existence of a creator, beyond those provided by the mere fact of existence. There is little, if anything, that can be said about the metaphysical creation of the universe. Since our observations are of physical attributes, and science deals with physical things, the issue of creation, which must necessarily be metaphysical, cannot be addressed [by science]."[601]

Kragh also points out that in the Christian doctrine of creation, an intelligent cause created the world *along with* time rather than *in* time: "*It is generally agreed that God created the world along with time rather than in time. Cosmic creation is primarily about the ontological dependence of the world on God, and not so much about beginning in the conventional temporal sense.*"[602]

Lawrence Krauss, a theoretical physicist, tries another way of avoiding the problem of the cause of the Universe, suggesting there is no cause at all, but that the universe, or rather the multiverse, originated from nothing, equating the vacuum states of quantum fields to nothing. In his view, "*quantum*

---

[601] John Hawley & Catherine Holcomb, *Foundations of Modern Cosmology* (2005): 155.
[602] Helge Kragh, *Entropic Creation* (2010): 281–282.

gravity appears to allow universes to be created from nothing".[603] David Albert, a philosopher and theoretical physicist and an expert in quantum mechanics,[604] criticises Krauss' proposal. In the first place, Krauss does not explain where these quantum fields and the quantum laws that govern them come from:

> "*Where are the laws of quantum mechanics themselves supposed to have come from? Krauss is more or less upfront, as it turns out, about not having a clue about that. He acknowledges (albeit in a parenthesis and just a few pages before the end of the book) that everything he has been talking about simply takes the basic principles of quantum mechanics for granted. "I have no idea if this notion can be usefully dispensed with", he writes, "or at least I don't know of any productive work in this regard"*.[605]... *Forget where the laws came from. Have a look instead at what they say... [T]here is, at the bottom of everything, some basic, elementary, eternally persisting, concrete, physical stuff... And what the fundamental laws of nature are about, and all the fundamental laws of nature are about, and all there is for the fundamental laws of nature to be about, insofar as physics has ever been able to imagine, is how that elementary stuff is arranged. The fundamental laws of nature generally take the form of rules concerning which arrangements of that stuff are physically possible and which aren't, or rules connecting the arrangements of that elementary stuff at later times to its arrangement at earlier times... But the laws have no bearing whatsoever on questions of where the elementary stuff came from, or of why the world should have consisted of the particular elementary stuff it does, as opposed to*

---

[603] Lawrence Krauss, *A Universe from Nothing. Why There Is Something Rather Than Nothing* (2012): 170.

[604] David Albert, *Quantum Mechanics and Experience* (1992).

[605] Albert cites Krauss, *A Universe from Nothing. Why There Is Something Rather Than Nothing* (2012): 176–177.

something else, or to nothing at all. The fundamental physical laws that Krauss is talking about in "A Universe from Nothing" – the laws of relativistic quantum field theories – are no exception to this. The particular, eternally persisting, elementary physical stuff of the world, according to the standard presentations of relativistic quantum field theories, consists... of relativistic quantum fields. And the fundamental laws of this theory take the form of rules concerning which arrangements of those fields are physically possible and which aren't, and rules connecting the arrangements of those fields at later times to their arrangements at earlier times, and so on – and they have nothing whatsoever to say on the subject of where those fields came from, or of why the world should have consisted of the particular kinds of fields it does, or of why it should have consisted of fields at all."[606]

In the second place, Albert argues that Krauss confounds the vacuum states of quantum fields with 'nothing', when, as matter of fact, though a vacuum state does not contain particles, it is not nothing at all, since it is a quantum field:

"There is, as it happens, an interesting difference between relativistic quantum field theories and every previous serious candidate for a fundamental physical theory of the world. Every previous such theory counted material particles among the concrete, fundamental, eternally persisting elementary physical stuff of the world – and relativistic quantum field theories, interestingly and emphatically and unprecedentedly, do not. According to relativistic quantum field theories, particles are to be understood, rather, as specific arrangements of the fields. Certain arrangements of the fields, for instance, correspond to there being 14 particles in the universe, and certain other arrangements correspond to

---

[606] David Albert, "On the Origin of Everything 'A Universe from Nothing,' by Lawrence Krauss", *New York Times*, March 23, 2012.

*there being 276 particles, and certain other arrangements correspond to there being an infinite number of particles, **and certain other arrangements correspond to there being no particles at all**. And those last arrangements are referred to, in the jargon of quantum field theories, for obvious reasons, as 'vacuum" states'. Krauss seems to be thinking that these vacuum states amount to the relativistic-quantum-field-theoretical version of there not being any physical stuff at all...*

But that's just not right. Relativistic-quantum-field-theoretical vacuum states... are particular arrangements of elementary physical stuff. The true relativistic-quantum-field-theoretical equivalent to there not being any physical stuff at all isn't this or that particular arrangement of the fields – what it is... is the simple absence of the fields! The fact that some arrangements of fields happen to correspond to the existence of particles and some don't is not a whit more mysterious than the fact that some of the possible arrangements of my fingers happen to correspond to the existence of a fist and some don't. And the fact that particles can pop in and out of existence, over time, as those fields rearrange themselves, is not a whit more mysterious than the fact that fists can pop in and out of existence, over time, as my fingers rearrange themselves. And none of these poppings... amount to anything even remotely in the neighbourhood of a creation from nothing."[607]

Albert's argues that certain relativistic quantum fields, obeying the laws of quantum mechanics, allow for a certain number of particles to pop up, and certain other fields allow for there being no particles at all, the latter ones receiving the name of vacuum states. Albert criticises Krauss for stating that these vacuum states are equivalent to 'nothing', while they are, as a matter of fact, not nothing at all. Krauss' argument is a fallacy, a play of words. He

---

[607] David Albert, "On the Origin of Everything 'A Universe from Nothing,' by Lawrence Krauss", *New York Times,* March 23, 2012, my emphasis.

does not give an answer to the question 'where do these quantum mechanical fields and the quantum mechanical laws they obey, come from'?

Peter Williams, a Norwegian philosopher, in his criticism of Krauss, essentially argues along the same line as Albert, without displaying, of course, the same expertise in quantum physics.[608]

Up to here, we have discussed the problem of the cause of the Universe. We will next discuss the problem of the cause of the fine-tuning of the physical constants of the Universe.

Section 8.2 The multiverse speculation is in denial of an intelligent cause

Theoretically, there are two mysteries which appear to elude scientific explanation. Firstly, the fact that the Universe exists (*that* the world is), which is the problem I treated in Section 1, and secondly, certain aspects of how the Universe is (*how* the world is), which is the problem we will now analyse:

> "It has been said —to use Wittgenstein's words— 'Not **how** the world is, is the mystical, but **that** it is '[609]. Yet our discussion shows that **how** the world is... seems to be inexplicable in principle and thus 'mystical', if we wish to use this term... The structural homogeneity of the world seems to resist any 'deeper' explanation: it remains a mystery."[610]

The mystery of *how* the world is, is the mystery of the particular values of some physical constants, which appear to have been finely tuned in such a way that long-lived stars with nuclear fusion, and complex, intelligent life on some planet in some solar system, could emerge. In Chapter 3 of this book, I mentioned nine instances of remarkable fine-tuning of physical

---

[608] Peter Williams, *A Universe from Someone. Against Lawrence Krauss*, online, 2013.

[609] Ludwig von Wittgenstein, *Tractatus Logico-Philosophicus* (1981): 6,44.

[610] Karl Popper, *Realism and the Aim of Science* (1994): 150-152, Popper's emphasis.

constants that allowed the emergence of long-lasting stars with nuclear fusion, and complex and intelligent life in some solar system:

1. The initial value of Ω in the Big Bang made it possible for stars and galaxies to form in the Universe.

2. The production and conservation of protons in the Big Bang allowed stable atoms to emerge.

3. Nuclear fusion in stars is possible due to the specific relationship between the gravitational, electromagnetic and nuclear forces.

4. The exact value of the strong nuclear force prevented the diproton disaster and made possible the beginning of nuclear fusion in stars.

5. An 'excited' state of carbon enables the nuclear resonance in the fusion of beryllium and helium and the consequent production of carbon. The lack of nuclear resonance in the fusion of carbon and helium means that not all the carbon is transformed into oxygen.

6. The exact relationship between the gravitational coupling constant α_G, the electromagnetic coupling constant α and the mass ratio β made it possible for solar systems with planets to emerge.

7. The fact that the Universe has three spatial dimensions means that there are stable planetary orbits.

8. The relationship between the neutron, proton and electron masses, and between the strong nuclear, weak nuclear and electromagnetic forces made possible the existence of stable atoms and molecules.

9. The initial value of the immense quantity of matter-energy and the vastness of the Universe made the probability of the event of a solar system with intelligent life come close to unity.

Most cosmologists, both theists and atheists, accept the fact of fine tuning, for example, Barnes,[611] Barrow,[612] Colling,[613] Gonzalez and Richards,[614]

---

[611] Luke Barnes, "The Fine-Tuning of the Universe for Intelligent Life", in: *Publications of the Astronomical Society of Australia*, vol. 29 (2012).

Guth,[615] Hoyle,[616] Kaku,[617] Krauss,[618] Linde,[619] McGrath,[620] Newton,[621] Oberhummer,[622] Penrose,[623] Popper,[624] Polkinghorne,[625] Susskind,[626] Smolin,[627] Tegmark[628] and Wheeler,[629] but not Stenger and Weinberg.[630]

But even when accepting the undeniable fact of the fine tuning of the physical constants of our Universe, some atheist cosmologists who speculate about a multiverse are in denial of an intelligent cause. They appear not to understand, as McGrath has correctly argued, that this fine tuning, *with or without a multiverse*, requires an intelligent cause, which theists believe to be a divine creator.[631] I shall now explain this point.

---

[612] John Barrow & Frank Tipler, *The Anthropic Cosmological Principle* (1986).

[613] Richard Colling, *Random Designer* (2004).

[614] Guillermo Gonzalez & Jay Richards, *The Privileged Plane* (2004).

[615] Alan Guth, *The Inflationary Universe* (1998).

[616] Fred Hoyle, "The mystery of the cosmic helium abundance", in: *Nature*, vol. 203 (1964).

[617] Michio Kaku, *Parallel Worlds* (2005).

[618] Lawrence Krauss, *A Universe from Nothing* (2012).

[619] Andre Linde, *Inflation and Quantum Cosmology* (1990).

[620] Alister McGrath, *The Fine-Tuned Universe* (2009).

[621] Isaac Newton, *Principia Mathematica*, Bernard Cohen, ed. (2008): 490

[622] Heinz Oberhummer, *Kann das alles Zufall sein?* (2008).

[623] Roger Penrose, *The Road to Reality* (2004).

[624] Karl Popper, *Realism and the Aim of Science* (1994).

[625] John Polkinghorne, "Understanding the Universe", in: James Miller, ed., *Cosmic Questions* (2001): 175–182.

[626] Leonard Susskind, *The Cosmic Landscape* (2006).

[627] Lee Smolin, *The Life of the Cosmos* (1997).

[628] Max Tegmark, "Is 'the Theory of Everything' merely the Ultimate Ensemble Theory," in: *Annals of Physics*, vol. 270 (1998):1–51.

[629] John Wheeler, "From relativity to mutability," in: J. Mehra, *The Physicist's Conception of Nature* (1973): 202–247.

[630] Stenger, Victor, *The Fallacy of Fine-Tuning* (2011; Steven Weinberg, "A Universe with No Designer", in: *Cosmic Questions* (2001): 169–174.

[631] Alister McGrath, *The Fine-Tuned Universe* (2009): 124.

According to Rees, there are only three possible explanations of fine-tuning, namely, *"coincidence, providence or multiverse"*.[632] If we dispense with the 'good luck' explanation (what Rees calls 'coincidence'), which is no explanation at all, then there remain only two options: 'providence', which in this context is equivalent to an intelligent cause, and the 'multiverse'.

For example, John Polkinghorne, agreeing with John Leslie,[633] asserts that there are only two possible solutions to explain the fact of fine-tuning, namely, the multiverse or the divine creator:

"Other contributors to this volume have already outlined the many considerations that lead us to conclude that the laws of nature as we observe them in our universe are precisely those that permit the development of carbon-based life, in the sense that even very small changes in intrinsic force strengths would have broken links in the long, delicate and beautiful chain of consequences linking the early universe to the existence of life today here on Earth. I agree with John Leslie's analysis, presented in his book Universes, that suggests, firstly that it would be irrational just to shrug this off as a happy accident, and secondly that there are two broad categories of possible explanation: either many universes with a vast variety of different natural laws instantiated in them, of which ours is the one that by chance has allowed us to appear within its history; or a single universe that is the way it is because it is not "any old world," but a creation that has been endowed by its Creator with just the circumstances that will allow it to have a fruitful history. I simply want to

---

[632] Martin Rees, *Just Six Numbers* (1999): Chapter 11

[633] John Polkinghorne, "Understanding the Universe", in: James Miller, *Cosmic Questions* (2001):178–179 and John Polkinghorne, *Quarks, Chaos & Christianity* (2005): 36–46. See John Leslie, *Universes* (1996): 25–65.

make two comments on this analysis. The first is to emphasize that both proposals are metaphysical in character. The second point is to agree with Leslie that, in relation to the Anthropic Principle, it is a metaphysically even-handed choice between many universes and creation."[634]

Let us start with the multiverse theory. Quite a few authors, who express their views on this matter, attribute the emergence of our particular universe, given an infinite multiplication of universes, to pure chance. For example, authors like Tegmark and Barrow[635], who both strongly support the thesis of fine tuning, attribute this fine tuning to a multiverse.

In what follows, I will argue that the multiverse theorists are in denial of an intelligent cause of the universe's fine tuning, though, from a merely logical point of view, this inference cannot be eluded. I must define the concept 'intelligent'. I define 'intelligent' as 'the property that an object can have that makes it capable of consciously understanding the relationship between cause and effect, to the degree that, if it had the ability or capacity, it could manipulate or create the cause to produce the desired effect'. In the case of the intelligent cause of the fine tuning of the Universe, 'intelligent' specifically means 'the property that an object can have that makes it capable of consciously understanding the relationship between cause (the initial conditions of the Universe and the laws of physics and their

---

[634] John Polkinghorne, "Understanding the Universe", in: James Miller, ed. *Cosmic Questions* (2001): 178–179 and in: *Quarks, Chaos & Christianity* (2005): 36–46.

[635] Max Tegmark, "Is the 'theory of everything' merely the ultimate ensemble theory?" in: *Annals of Physics*, vol. 270 (1998):1–51. Tegmark disregards the option of good luck, so two possible explanations remain, i.e. *"everything that exists mathematically, exists physically"* (which is equivalent to the statement of the multiverse), *ibidem*: 1, or the *"strong anthropic principle"*, that is, "the World was designed by a divine creator so that self-conscious substructures could emerge", *ibidem*: 4. John Barrow & Frank Tipler do not use the term 'multiverse' but rather 'the meta-space of all possible worlds' that counter the 'design arguments', in: *The Anthropic Cosmological Principle* (1986): 248. Also John Barrow, *The Artful Universe Expanded* (2005).

constants, that determine the relationship between cause and effect), and the effect (the fine tuning of the Universe), to the degree that, if it had the ability or capacity, it could consciously manipulate or create the cause (the initial conditions) and the physical laws with their constants to produce the desired effect.

In Chapter 3, we saw that the probability of the fine-tuning of the Universe being the result of good luck is very small: $p \approx 4 * 10^{-267}$, which is close to saying that it is impossible. To avoid this conundrum, the multiverse theorists propose that the fine tuning of physical constants of our Universe is a random event. This chance event is produced by the existence of a mechanism capable of infinite multiplication of Universes, say 10^{500}, each of them with different natural laws and different physical constants. Among this immense multitude of universes, through the law of large numbers, some universes like our own finely tuned for complex and intelligent life should emerge by chance.

This conjecture appears to be rational. There is, however, a problem with the multiverse proposal. The authors who make this proposal do not appear to understand the *necessary relationship between random chance and intelligent cause*. To explain this point, I will next look at the relationship between random chance and intelligent cause in three cases, which are, firstly, the case where the number 14 comes up in a game of *roulette*; secondly, the case where complex and *intelligent life* emerges in our Universe; and thirdly, the case that in a multiverse of trillions of universes, a Universe emerges *with physical constants finely tuned* to produce intelligent life, like our own.

Let us first look at the case of *roulette*. The probability that the number 14 will come up, if we play only once ($N = 1$), is relatively small, $p(14)|(N = 1) = 0.026316$, supposing 36 red and black numbers and two green zeros. But in a set of $N=1000$ games, the number 14 will pop up, on average, some 26 times. So, the probability that in $N=1000$ games, the number 14 will *not* come up at least once is very small indeed, i.e., $p(\neg 14|(N = 1000) = 0.00000000000262$. So, the probability that this mechanism (the roulette) causes the number 14 coming up at least once in a set of $N=1000$ games is close to one. In other words, the law of large numbers means that an event which in itself is quite improbable, acquires a probability of occurring at least once close to one. However, the role of chance and the law of large numbers in this case do not remove the need for an intelligent cause, i.e. the existence of *homo sapiens* who designed and produced this mechanism (the roulette), making it possible for the law of large numbers to function.

Something analogous occurs in the case of the emergence of complex and intelligent life in our Universe. In itself, the emergence of a solar system with intelligent life is highly improbable. No one knows exactly how improbable it is for a solar system with a planet with intelligent life to emerge, though we know it is very improbable.[636] The coincidence of the set of initial conditions, each one of which in itself is necessary but highly improbable, has a probability equal to the product of all of these improbable events, in the same way as the probability that the number 6 comes up twice in a game of one run with two dices is equal to $(1/6)*(1/6) = 1/36$. Among the initial conditions are the following rare Earth factors:[637]

1) Gravitation produces a solar system with only one sun (no binary stars).
2) The first-generation sun explodes in a supernova, producing elements heavier than iron and dispersing all elements.
3) The second-generation sun has planets with carbon, oxygen and all other elements.

---

[636] See Peter Ward & Donald Brownlee, *Rare Earth. Why Complex Life is Uncommon in the Universe*, (2000); and Guillermo Gonzalez & Jay Richards, *The Privileged Planet* (2004). Summary: pp. XVII and XVIII.

[637] Peter Ward & Donald Brownlee, *Rare Earth. Why Complex Life is Uncommon in the Universe* (2000).

4) The right distance from the star allows for liquid water at the surface of the planet.

5) The right mass of the star implies long enough lifetime, with not too much ultraviolet radiation. These factors allow for life to emerge on a planet at the right distance.

6) Giant planets prevent orbital chaos and allow for stable planetary orbits.

7) The right planetary mass retains atmosphere and ocean and produces enough heat for plate tectonics and a solid, molten iron core.

8) The solid iron core produces a magnetic field that functions as a protective shield against solar particles.

9) Plate tectonics allow for CO_2-silicate thermostat, preventing a runaway greenhouse effect. Plate tectonics also allow for the build-up of land mass, mountains, which enhance biotic diversity.

10) Jupiter-like neighbours, not too close and not too far, clear out comets and asteroids.

11) The right volume of oceans, not too much, not too little.

12) A large moon at right distance stabilises planet's tilt.

13) The right tilt allows for seasons not to be too severe.

14) Few giant impacts, if not too big, help evolution of life on the planet, without global sterilising, after an initial period.

15) The right amount of carbon allows for life to emerge, without being too abundant so as to produce a runaway greenhouse effect.

16) The right atmospheric properties maintain adequate temperature, composition and pressure for plants and animals.

17) Amino and nucleic acids, together with right atmospheric conditions create a successful evolutionary pathway to complex plants and animals.

18) The invention of photosynthesis allows for the production of oxygen, not too much, not too little, evolving at the right time.

19) The right kind of galaxy, not too small, not elliptical, neither irregular, allows for the production of enough heavy elements.

20) The right position in the galaxy, not in the centre, where black holes generate collisions, nor at the edge, allows for solar systems to survive.

Now each of these conditions in itself is not that rare. Let us assign for argument's sake a probability of one in ten for the existence of a solar system S with at least one of these conditions c_n: $(p(\exists S|S \ni c_n) = 10^{-1}$. The probability, however, of the existence of a solar system with *all* of these conditions *occurring together* – which is a necessary condition for the emergence of intelligent life – would be very small indeed:

$$p(\exists S|S \ni \{c1 + c2 + c3 + c4 + c5 + c6 + c7 + c8 + c9 + c10 + c11 + c12 + c13 + c14 + c15 + c16 + c17 + c18 + c19 + c20\}) = 10^{-20}.$$

So, the highly improbable coincidence of all these rare Earth factors occurring together makes the emergence of complex, intelligent life equally improbable. But, if we have enough solar systems in a very large universe, finely tuned for the possibility of complex life, the probability that at least one such solar system occurs reaches unity. Since we have an approximate estimate of the number of stars in our Universe, we can make an estimate of that probability. There are some one hundred thousand million galaxies, each with an average of some one hundred thousand million stars, in total some 10^{22} stars. If the probability that a star with a planet with intelligent life emerges were $1/10^{20}$, then by the law of large numbers there would be some $10^{22}*10^{-20}=100$ stars with a planet with intelligent life, so that the probability that at least one solar system with intelligent life emerges would be close to one.

This is what Ludwig Boltzmann referred to in 1895, in *Nature*, and in a later book of his:

"We assume that the whole universe is, and rests for ever, in thermal equilibrium. The probability that one (only one) part of the universe is in a certain state, is the smaller the further this state is from thermal equilibrium; but this probability is greater, the greater is the universe itself. If we assume the universe great enough, we can make the probability of one relatively small part being in any given state (however far from the state of thermal equilibrium), as great as we please. We can also make the probability great that, though the whole universe is in thermal equilibrium, our world is in its present state. It may be said that the world is so far from thermal equilibrium that we cannot imagine the improbability of such a state. But can we imagine, on the other side, how small a part of the whole universe this world is? Assuming the universe great enough, the probability that such a small part of it as our world should be in its present state, is no longer small."[638]

Once the solar system emerges that combines all these rare earth factors, the probability that intelligent life emerge in such a solar system is very high. John Maynard Smith and Eörs Szathmáry have specified the steps needed to get from amino and nucleic acids to self-conscious primates.[639] The first step is the transition from individual molecules that replicate themselves to populations of molecules in compartments that replicate themselves. The second step is from nucleic acid molecules that reproduce independently to chromosomes where molecules are integrally replicated. The third step is the transition from RNA as a gene and enzyme to DNA (the genome or genetic code) and proteins. The fourth step is the transition

---

[638] Ludwig Boltzmann, "On Certain Questions of the Theory of Gases", in: *Nature* (1895): 413-415 & Ludwig Boltzmann, *Theoretical Physics and Philosophical Problems* (1974): 208–209.

[639] John Maynard Smith and Eörs Szathmáry, *Major Transitions in Evolution* (1995).

from ancestors of mitochondria and chloroplast, that live independently as prokaryotes, to eukaryotes, where these organelles live within the host cell. The fifth step is the transition from replication of the eukaryotes by means of asexual cloning to sexual reproduction that enhances genetic diversity. The sixth step is the transition from unicellular protists to multicellular living organisms (animals, plants, fungi). The seventh step is the transition from individual organisms to non-reproductive colonies of animals. The eighth step is the transition from social primates to human societies, with human intelligence and language, which are necessary to succeed in the complexities of human interaction. Karl Popper and John Eccles explain that the eighth transition is not only a socio-cultural, but also a bio-psychological evolution, with the emergence of the self-conscious mind.[640]

In the case of the emergence of solar systems with intelligent life, in a way analogous to the roulette, there must exist an intelligent cause of the universe which makes the probability of a solar system with intelligent life to occur close to unity. Such a universe has two fundamental traits: firstly, its physical constants must be fine-tuned so as to make the emergence of complex, intelligent life *possible*; secondly, this universe must be very large, with a very large number of solar systems, as to make the event of at least one solar system combining all the rare Earth factors necessary to produce intelligent life *probable*, through the law of large numbers. Since our Universe appears to be finely tuned for the occurrence of intelligent life, and also very large, it meets both conditions.

The speculation about a multiverse tries to avoid the conclusion of an intelligent cause: there must be a physical mechanism that multiplies universes, in infinite quantity, with a random variation of physical laws and constants, so that by the law of large numbers the probability of the highly unlikely event of our finely tuned universe, comes close to unity. Where

---

[640] Karl Popper and John Eccles, *The Self and its Brain* (1981).

does that mechanism come from? The multiverse theorists agree on the existence of some kind of mechanism, but they do not address the problem of how this mechanism functions and how it came into being. The following syllogism addresses the problem:

1) An intelligent cause has created a very complicated and sophisticated physical mechanism, with no beginning in time, which incessantly produces and multiplies universes, with physical laws and constants varying from one universe to another.

2) This physical mechanism has already created an almost infinite number of universes, all of them with different physical laws and constants.

3) The probability that at least one universe would emerge with physical constants propitious for intelligent life, like our own, is close to one.

The first statement of this syllogism is logically a necessary *implication* of the multiverse theory, though most cosmologists, with only four exceptions as we shall now see, do not see or do not mention the fact that in the case of a multiverse too, an intelligent cause is needed to create the mechanism that multiplies universes, and make physical laws and constants vary in those universes *ad infinitum*.

Paul Davies was one of the first to mention the fact of the fine tuning of our Universe in modern cosmological discourse.[641] His additional merit is his recognition that the multiverse theory, in whatever form, does not solve the double *mystery* of the existence of the Universe and the fine tuning of its physical constants in favour of the emergence of intelligent life:

"The search for a closed logical scheme that provides a complete and self-consistent explanation for everything is doomed to failure... There will always be mystery at the end of the universe."[642]

---

[641] Paul Davies, *The Accidental Universe* (1984). He later amplified the argument in: *The Goldilocks Enigma* (2006).

[642] Paul Davies, *The Mind of God. The Scientific Basis for a Rational World* (2005): 226.

This is his way of saying that the mechanism that multiplies universes must have a mysterious, intelligent cause. Leslie refers to some of these universe-multiplying mechanisms, suggested by others,[643] and argues that an intelligent cause *"might act through laws which produced an ensemble of universes, relying on chance to generate life-encouraging worlds"*.[644]

Oberhummer makes the same point: *"The large-scale production of the multiverse is more in need of an all-powerful creator than the laborious and painstaking fixing of the different constants of a single Universe."*[645] And so does McGrath, where he argues that *"it seems that substantially the same arguments can be brought to bear for the existence of God in the case of a multiverse as in the case of a universe"*. [646]

Clearly, if in the multiverse there are some 10^{500} universes – this number is equal to the number of possible superstring theories – then, by the law of large numbers, the number of universes with physical laws and constants and initial conditions like our own would be equal to $10^{500}*10^{-267} \approx 10^{233}$. This is what proponents of the multiverse are referring to when they argue that their proposal resolves the problem of the fine tuning of our Universe.

In fact, it does not resolve it. It is true that once we have a mechanism for the multiplication of universes, which randomly varies the physical laws and their constants and initial conditions in them *ad infinitum*, a Universe such as our own has a probability close to unity of emerging by chance. However, *an intelligent cause is needed to create this mechanism for multiplication of universes* that varies the physical laws, constants and initial conditions in them *ad infinitum*, just as an intelligent cause is needed to create the roulette wheel that randomly varies the results from 0 to 36.

---

[643] John Leslie, "Multiple worlds," in: *Universes* (1996): Chapter Four.
[644] John Leslie, *Universes* (1996): 55.
[645] Heinz Oberhummer, *Kann das alles Zufall sein?* (2008): 154.
[646] Alister McGrath, *The Fine-Tuned Universe* (2009): 124.

Table 8.2 summarises the three cases analysed above.

Table 8.2 The interaction between intelligent cause and random chance

Mecha-nism	An intelligent cause C creates a mechanism that makes E possible	Event E is improbable but not impossible	Chance of event E is very small: $p(E\|N = 1) \cong 0$	Law of large numbers N brings probability of event E close to unity: $p(E\|N \approx \infty) = 1$
Roulette	An intelligent cause (*homo sapiens*) produces the roulette wheel which enables number 14 to come up by chance	The ball falls on number 14	$p = 1/38$	$N \geq 10^3$ spins
Universe	An intelligent cause (a creator) produces a Universe which is finely tuned for intelligent life to emerge by chance	Emergence of *homo sapiens* in at least one solar system	$p \approx 1/10^{21}$	$N \geq 10^{22}$ stars
Multiverse	An intelligent cause (a creator) produces a mechanism which multiplies universes, with random variation of physical laws and constants, allowing for the emergence of universes finely tuned for intelligent life	A Universe, finely tuned for intelligent life, emerges	$p \approx 1/10^{267}$	$N \geq 10^{500}$ universes

We see then, that in each of the three cases considered, three things are needed to let the occurrence of a rare random event have a probability close to unity. These are, firstly, an intelligent cause necessary to create a very special mechanism; secondly; a very special mechanism – the roulette, a finely tuned Universe, or a mechanism that multiplies universes – which makes the occurrence of the random event *possible*; and thirdly, a large number of spins, or solar systems, or universes, respectively, such that the law of large numbers operates, by which an intrinsically very improbable event becomes very *probable*, acquiring a probability close to unity.

At which link should we stop going back in this chain of cause and effect? Common sense and the respect for scientific procedure would suggest we stop at the last link that is observable. That would be our finely tuned Universe. The multiverse theorists, however, go back one step further, speculating about a physical mechanism producing an almost infinite number of universes, each one of them with its proper set of physical laws and constants. The problem here is that in doing so they have moved to a theory that cannot be corroborated or refuted by facts. As we saw in Chapter 7, science is defined as the set of theories which contain two properties, each of them necessary but not in itself sufficient: firstly, logical consistency; and secondly, a set of predictions that can be corroborated or refuted by facts. The multiverse theory lacks both, since, firstly, it is logically confused, when confirming, erroneously, that the multiverse is not in need of an intelligent cause; and since, secondly, we never will be able to observe other universes, and the physical laws and constants in these other universes. *The multiverse theory, therefore, is science fiction, not science.*

All this leads us to the logical conclusion that the fine tuning of some physical constants, necessary for the emergence of intelligent life in some solar system of our Universe, can only be explained by the cause of the Universe – or, for that sake, the cause of the multiverse – being *intelligent*. This completes our analysis of the cause of the Universe (Sections 8.1, 8.2):

I.   The Universe is the set of all objects, events and sets of events that actually exist and that are not caused by themselves.

II. The Universe belongs to the set of sets that are members of themselves.

III. The Universe is not caused by itself, but by an external cause that does not belong to the Universe.

IV. The cause of the Universe is also the cause of itself.

V.   The cause of the Universe is intelligent.

Section 8.3 What motivates some cosmologists to embrace a multiverse

Modern science started in astrophysics, with Nicolaus Copernicus (1473–1543), Tycho Brahe (1546–1601), Galileo Galilei (1564–1642), Johann Kepler (1571–1630) and Isaac Newton (1642–1727). [647] From Newton to Einstein most cosmologists explicitly referred to the intelligent design of the Universe and an intelligent designer. According to Newton, the Universe *"could not have arisen without the design and dominion of an intelligent and powerful being"*, whom he called the *"Lord God Pantokrator"*. [648] And Einstein acknowledged that the Universe was mysteriously 'arranged', a circumstance he attributed to God:

> *"I'm not an atheist. The problem involved is too vast for our limited minds. We are in the position of a little child entering a huge library filled with books in many languages… The child dimly suspects a mysterious order in the arrangement of the books but doesn't know what it is. That, it seems to me, is the attitude of even the most intelligent human being toward God. We see the universe marvellously arranged."*[649]

---

[647] See John Auping, *El Origen y la Evolución del Universo*, e-book (2016): 25–74.
[648] Isaac Newton, "General Scholium", in: *Principia Mathematica* (2008): 489–494.
[649] Cited in Walter Isaacson, *Einstein. His Life and Universe* (2007): 386.

It is, therefore, the more surprising that of all sciences, it has been precisely astrophysics, which paved the way for modern science, that has lately – the last 80 years or so – got swamped in the myths of science fiction. How did this come about? In the case of dark matter and dark energy, as I explained in Chapter 2 of this book, the problem is that though many astrophysicists pay lip-service to Einstein, they have not been able or willing to think through all the implications of his general relativity, nor have they been able or willing to use his mathematics in the solution of certain astrophysical problems, applying a Newtonian model, which leads to erroneous conclusions, when, as a matter of fact, general relativity is necessary and sufficient to solve these problems.

In the case of the multiverse, however, something quite different is going on. We saw that the multiverse speculation is no less in need of an intelligent cause than the theory of a unique Universe, very large and finely tuned for intelligent life. But the multiverse theorists, abandoning basic logical reasoning, deny this. One might ask, how can some cosmologists, some of them experts in their field, display such lack of basic logical reasoning, when it comes to the problem of the cause of the Universe? It appears to be that their desire to prove that God does not exist obfuscates their minds. I will now analyse this circumstance in some detail.

Shortly after the discovery of the Big Bang, the desire to eliminate any reference to God, made some cosmologists reject the Big Bang theory as such. Atheist astrophysicists distrusted Lemaître's theory, suspecting that the fact of him being a Catholic priest had influenced his Big Bang cosmology. This atheist faction maintained that the Big Bang theory was no more than a pseudo-scientific justification for a divine creator. For example, the British physicist, William Bonner, suggested that there was a conspiracy behind the Big Bang model: *"The underlying motive [of this model] is, of course, to bring in God as creator. It seems like the opportunity Christian theology has been waiting for ever since science began to depose religion from the minds of rational men in the seventeenth century.*"[650]

Fred Hoyle, like Bonner, condemned the Big Bang model as "*a model built on Judeo-Christian foundations*".[651] After Penzias and Wilson received the Nobel Prize in Physics for discovering the *CMBR*, in 1978, 69% of astronomers supported the Big Bang model, and support for the Steady State model, proposed by Hoyle, had declined to 2%. Even so, Hoyle continued to mock the Big Bang model: "*The passionate frenzy with which the Big Bang cosmology is clutched to the corporate scientific bosom evidently arises from a deep-rooted attachment to the first page of Genesis, religious fundamentalism at its strongest.*"[652] To counter what he perceived as the irrationality of the theist position, Hoyle proposed the Steady State model, a solution that he saw as rational – which it is – but is refuted by the empirical evidence.[653] On the other hand, Hoyle, impressed by the fine tuning of the Universe, the discovery of which owes much to him, spoke about a "*Cosmic Intelligence*"[654] or "*super-intelligence*"[655] which, as an atheist, he was not willing to identify as an attribute of God.

After Paul Davies published his *Accidental Universe,* in 1982, and John Barrow and Frank Tipler their *Anthropic Cosmological Principle,* in 1986, the fine tuning of many physical constants our Universe, calculated and revealed by these authors, made the case for an intelligent cause of the Universe all the more pressing. In response, cosmologists with an a-theological discourse came up with the multiverse. John Barrow pointedly recognised that what motivates the multiverse speculations is the desire to

---

[650] Cited in Simon Singh, *Big Bang. The Origin of the Universe* (2004): 361.

[651] *Ibidem*: 361.

[652] *Ibidem*: 439.

[653] See Chapter 4 of this book.

[654] According to John Polkinghorne, *Quarks, Chaos & Christianity* (2005): 40–41.

[655] Cited in Paul Davies, *The Mind of God. The Scientific Basis for a Rational World* (2005): 229.

eliminate God from the equation: "*the multiverse scenario was suggested by some cosmologists as a way to avoid the conclusion that the Universe was specially designed for life by a Grand Designer*".[656] I will now reveal this motivation in a few cases, without pretending an exhaustive overview.

Alan Guth's argument is paradigmatic for the a-theological position, which presents the multiverse proposal as rational and the intelligent cause proposal as irrational. His multiverse thesis speculates about eternal inflation, which we examined in Chapter 4. Guth remarks:

> "*Today, the dominant view of the origin of the universe, in both Judeo-Christian and scientific contexts, portrays it as a unique event. In the fifth century A.D. Saint Augustine described in his* Confessions *how time itself began with the creation of the universe, and modern scientists frequently refer to the big bang as the beginning of time… However, if the ideas of eternal inflation are correct, then the big bang was not a singular act of creation, but was more like the biological process of cell division… Given the plausibility of eternal inflation, I believe that soon any cosmological theory that does not lead to the eternal reproduction of universes will be considered as unimaginable as [the theory of] a species of bacteria that cannot reproduce.*"[657]

In Guth's argument, eternal inflation and the multiverse liberate us from seeing the big bang, and the fine tuning present in it, as a singular act of creation. Andrei Linde's theory of eternal inflation and multiverse follows the same line of reasoning. At a workshop of 1982, Linde admitted that his idea of eternal inflation, implying a multiverse with no beginning in time dispenses with the worrying idea of creation. It no longer appeared to be necessary to assume that a first mini-universe really appeared out of nothing

---

[656] John Barrow, *The Artful Universe Expanded* (2005): 53.

[657] Alan Guth, *The Inflationary Universe. The Quest for a New Search of Cosmic Origins* (1997): 251–252.

at some moment t=0, so that *"there might be no initial creation to worry about"*.[658] He repeated this in 2002, letting us know that the multiverse resulting from chaotic inflation takes away the 'miracle' of fine-tuning.[659] Linde speculated, as would Max Tegmark after him, that all *possible* universes, by the very fact of being possible, do *really* exist. By elementary logic, though, it is obvious that all that is real must be possible, but not all that is possible is necessarily real (see Section 8.1). Linde admits his theory is just speculation: *"We need to move carefully, constantly keeping in touch with solid... facts, but from time to time allowing ourselves to satisfy our urge to speculate."*[660]

Mario Livio declares himself to be a *"non-religious person"*, but recognizes the fine tuning of our Universe.[661] He resorts to the Guth-Linde theory of eternal inflation and the multiverse to explain it. The fine tuning of the values of the constants in our Universe is either due to a *'fundamental theory'* such as *"string theory or M-theory"*, which would explain why these constants have their exact values,[662] or to a mechanism that varies the physical laws and constants in multiple universes at random: *"Recall that eternal inflation predicts the existence of an infinite ensemble of universes...The laws of physics and/or the values of the universal constants may not be exactly the same in all of these pocket universes."*[663]

Consequently, in some universes life as we know it may emerge by chance. These speculations about a multiverse produced by eternal inflation belong to the realm of science fiction and, even within that realm, do not solve the problem of the intelligent cause. It is just wishful thinking.

---

[658] Cited in Helge Kragh, *Conceptions of Cosmos* (2007): 235.

[659] Andrei Linde, "Inflation, Quantum Cosmology and the Anthropic Principle", arXiv.org/pdf/hep-th/0211048 (November 2002): 16–17.

[660] *Ibidem*: 30.

[661] Mario Livio, *The Accelerating Universe* (2000): 189, 237–241.

[662] *Ibidem*: 190, 246.

[663] *Ibidem*: 189-191, 237–253.

Max Tegmark sustains that all possible universes, by virtue of the fact that they are possible, really exist: "*Everything that exists mathematically exists physically.*" [664] Tegmark does not give any argument for this strange proposition. It is rather like a declaration of a dogma of faith. His belief is in contradiction with my definitions 2 and 3 of box 8.2: "*The events space E is the set of all of the events and sets of events that are real and/or possible: E = $E_r \cup E_p$. The real events space is defined as: $E_r = \{e/e_r\}$ and the possible events space as: $E_p = \{e/e_p\}$. The real events space is a subset of the possible events space: $E_r \subseteq E_p \Leftrightarrow (e \in E_r \Rightarrow e \in E_p)$*". The event A exists if it belongs to the real events space: $A \in E_r \Rightarrow \exists A$ but A does not exist if it belongs to the possible but not the real events space $A \in \{E_p - E_r\} \Rightarrow \nexists A$. This means that not all that is possible really exists. Tegmark's belief is a necessary premise for the multiverse theory, and for him the multiverse is a necessary premise to eliminate 'religion-based' explanations of the fine-tuning of or Universe:

"*Investigation of the effects of varying physical parameters has gradually revealed[665] that virtually no physical parameters can be changed by large amounts without causing radical qualitative changes to the physical world. In other words, the 'island' in parameter space that supports human life appears to be very small. This smallness... has been hailed as support for religion-based theories... Such 'design arguments' stating that the world was designed by a divine creator so as to contain Self-aware Substructures are closely related to... the strong anthropic principle, which states that the Universe must support life.*"[666]

---

[664] Max Tegmark, "Is 'the Theory of Everything' merely the Ultimate Ensemble Theory", *Annals of Physics*, vol. 270 (1998): 1–51.
[665] Tegmark refers here to the previous work of Davies and barrow and Tipler.
[666] Max Tegmark, *Ibidem*.

Leonard Susskind and Steven Weinberg subscribe to the multiverse speculations for the same reasons as Guth, Linde, Livio and Tegmark: the multiverse dispenses with religion-based theories. Susskind asserts that there is no option other than string theory and the multiverse for explaining the fine-tuning of our Universe, because according to him, if one does not accept the multiverse solution, the only remaining solution is design by an intelligent Creator, which to him is an irrational solution. Susskind concludes the epilogue of his book with the phrase: "*Let me then close this book with the words of Pierre-Simon de Laplace that opened it: 'I have no need of this hypothesis'.*"[667] He is referring to the 'hypothesis' of a creator God. This latter concept seems to Susskind to be 'irrational' and an 'illusion', and it is for that reason that his theory of superstrings and the multiverse is so important to him:

"*If String Theory itself is wrong, perhaps because it is mathematically inconsistent, it will fall by the wayside... But if that does happen, then as things stand now, we would be left with no other rational explanation for the illusion of a designed universe.*"[668]

Steven Weinberg presents the theist argument of some physicists in favour of a divine creator and then his a-theological counter-argument:

"*Some physicists have argued that certain constants of nature have values that seem to have been mysteriously fine-tuned to just the values that allow for the possibility of life, in a way that could only be explained by the intervention of a designer with some special concern for life.*"[669]

He further admits, on one hand, that "*physicists won't be able to explain why the laws of nature are what they are*" and on the other hand, denies the

---

[667] Leonard Susskind, *The Cosmic Landscape: String Theory and the Illusion of Intelligent Design* (2006): 380.
[668] *Ibidem*: 355.
[669] Steven Weinberg, "A Universe with No Designer", in: *Cosmic Questions* (2001): 171.

fine tuning in the production of an excited state of the carbon nucleus.[670] The absurdness of this denial, in the face of the evidence, was pointed out in Chapter 3 of this book. Weinberg then proposes the thesis of the multiverse conceived as a Universe with different regions with no mutual contact, with different values of the constants in each region of the Universe:

> *"[T]here would be no difficulty in understanding why these constants take values favourable to intelligent life. There would be a vast number of big bangs in which the constants of nature take values unfavourable for life, and many fewer where life is possible. You don't have to invoke a benevolent designer to explain why we are in one of the parts of the universe where life is possible. In all the other parts of the universe there is no one to raise the question."*[671]

In Chapter 4, I presented the evidence that falsifies the idea, here proposed by Weinberg, of the variation of physical constants in our Universe.

Lawrence Krauss too subscribes to the multiverse theory, connecting it with quantum mechanical fields. I cited David Albert's criticism of this speculations at the end of Section 1 of this Chapter. At the end of his book, Krauss explains that the multiverse is a good substitute for a god-creator, and addresses the question of where the multiverse, or the empty space with its quantum mechanical fields, come from:

> *"When I have thus far described how something almost always can come from 'nothing', I have focused on either the creation of something from pre-existing empty space or the creation of empty space from no space at all... I have not discussed what some may view as the question of First Cause. I have not addressed directly, however, the issues of what might have existed, if anything, before such creation. A simple answer is of*

---

[670] Steven Weinberg, "A Universe with No Designer", in: *Cosmic Questions* (2001): 171.
[671] *Ibidem*: 172.

course that either empty space or the more fundamental nothingness from which empty space may have arisen, pre-existed and is eternal. However, to be fair, this does beg the possible question, which might of course not be answerable, of what, if anything, fixed the rules that governed such creation.... Those who argue that out of nothing, nothing comes seem perfectly content with the notion that somehow God can get around this. But once again, if one requires that the notion of true nothingness requires not even the potential for existence, then surely God cannot work his wonders, because, if he does cause existence from nonexistence, there must have been the potential for existence. To simply argue that God can do what nature cannot do is to argue that supernatural potential for existence is somehow different from regular natural potential for existence ... To posit a god who could resolve this conundrum... often is claimed to require that God exists outside the universe and is either timeless or eternal."[672]

Indeed. There is a curious inconsistency in his argument. Krauss mentions correctly some of the attributes the Creator/First-Cause/God should have in order to create a universe from nothing, he then emphatically denies that something like that, being quite different from nature, can exist – which is prejudice, not proof – then posits that nature can do what a creator-god cannot do, i.e. creating something from nothing, and offers as proof his disingenuous confounding of quantum vacuum states with nothing, where, as a matter of fact, they are not nothing at all, as Albert pointed out.[673]

I would like to finish this chapter quoting an atheist turned theist –among other reasons because of his logical acumen in analysing the fine tuning of the Universe – with whose evaluation of the multiverse speculation I fully agree:

---

[672] Lawrence Krauss, *A Universe from Nothing. Why There Is Something Rather Than Nothing* (2012): 174–175.

[673] See the end of Section 8.1 of this chapter.

"The multiverse hypothesis remains little more than a fascinating yet highly speculative mathematical exercise. It has, perhaps unwisely, been adopted by atheists, anxious to undermine the potential theological significance of fine-tuning in the universe. Thus, part of the attraction of the multiverse hypothesis to atheist physicists such as Steven Weinberg and Leonard Susskind is that it appears to avoid any inference of design or divinity. In fact, however, it seems that substantially the same arguments can be brought to bear for the existence of God in the case of a multiverse as in the case of a universe, with the multiverse hypothesis being consistent with, not the intellectual defeat of, a theistic understanding of God."[674]

Section 8.4 The conclusions of this book

I summarise the conclusions of this book in the following statements:

1) The Universe and its evolution started with the Big Bang, some 14 billion years ago.
2) Dark matter and dark energy may actually end up as myths that invaded modern astrophysics: they can be eliminated taking into account the complete framework of general relativity, both theory and mathematics.
3) In the Big Bang, the physical constants of the laws that determine the evolution of the Universe appear to have been finely tuned to make possible the emergence of long living stars with nuclear fusion and the emergence of intelligent life somewhere in the Universe.
4) From the point of view of logic, the multiverse is as much in need of an intelligent cause as is the theory of our Universe being unique and uniquely fine tuned for the emergence of intelligent life. Cosmologies that suggest otherwise are logically confused and appear not to be in touch with reality.

---

[674] Alister McGrath, *The Fine-Tuned Universe* (2009): 124.

5) The multiverse theory is not falsifiable, since we will never be able to observe other universes, neither the supposed variation of physical constants in those universes. In this sense it is not a scientific theory.

6) The Universe is a geometrically open system in eternal expansion and a thermodynamically closed and adiabatic system involved in an irreversible process, with its entropy continuously increasing in time.

7) Though 'gravitational entropy' started very low, and will be very low again at the end of the evolution of the Universe, conformal cyclical cosmology is a highly questionable because a strictly thermodynamical interpretation of the second law shows the 'thermodynamical entropy' of the Universe increasing with time until reaching a maximum in the end.

8) Popper's philosophy of science is the basis for distinguishing scientific fact from *myth*.

9) The hidden, orderly structure of physical reality reveals a network of causal relations between phenomena that can be represented by a set of physical-mathematical equations that pretend to be compatible with reality in terms of our theories, but are not automatically true, and must be corroborated or falsified by empirical evidence. This causality in the Universe is indeterministic, both scientifically and metaphysically.

10) The physical Universe is the set of objects and events that are not the cause of itself and the Universe itself is not the cause of itself and belongs to the set of objects and events that are not the cause of itself. The cause of the Universe is not part of the Universe, and is also the cause of itself.

11) The cause of the Universe appears to be intelligent.

12) The multiverse speculation, being a mixture of wishful thinking, cumbersome reasoning and non-falsifiable statements, seems to be motivated, in the case of some cosmologists, by a desire to avoid the idea of God as the creator of the Universe.

BIBLIOGRAPHY

Albert, David, "On the Origin of Everything 'A Universe from Nothing,' by Lawrence Krauss", *New York Times,* March 23, 2012

Albert, David, *Quantum Mechanics and Experience*, Harvard University Press, 1992

Allday, Jonathan, *Quarks, Leptons and the Big Bang*, 2nd. ed., Taylor & Francis, New York, London, 2002

Alpher, Ralph, James Follin & Robert Herman, "Physical Conditions in the Initial Stages of the Expanding Universe", in: *Physical Review*, vol. 92, 1953, pp. 1347–1361

Alpher, Ralph, & Robert Herman, "Evolution of the Universe", in: *Nature*, vol. 162 (1948): 774–775

Amsterdamski, Stefan *et al.*, *La querelle du déterminisme*, Editions Gallimard, Paris, 1990

Arnold, M. & Kazataka Takahashi, "The Synthesis of the Nuclides Heavier than Iron: Where do we stand?", in: Nicos Prantzos *et al.*, eds, *Origin and Evolution of the Elements* (1994): 395–411

Auping, John, *El Origen y la Evolución del Universo*, UIA, e–book. 2nd. ed., 2016

Auping, John, "Putting the standard ΛCDM model and the relativistic model in historical context", in: John Auping & Alfredo Sandoval, eds, *Proceedings of the International Conference on Two Cosmological Models,* UIA, P & V, 2012, pp. 35–115

Baade, Walter, "A Revision of the Extra–galactic Distance Scale", in: *Transactions of the International Astronomical Union*, vol. 8, 1952, pp. 397–398

Baade, Walter & Rudolph Minkowski, "Identification of the Radio Sources in Casssiopeia, Cygnus A, and Puppis A", in: *Astrophysical Journal*, vol. 119, 1954, pp. 206–214

Bahcall, Neta, "Clusters and Cosmology", in: *Physics Reports*, vol. 33, 2000, pp. 233–244

Banks, Tom, Willy Fischler, Stephen Shenker & Leonard Susskind, "*M* Theory as a Matrix Model: A conjecture", in: *Physical Review D*, 1997, pp. 5112–5128

Barnes, Luke, *The Fine-Tuning of the Universe for Intelligent Life*, arxiv:1112.4647v2, June 2012

Barnett, Michael *et al.*, *The Charm of Strange Quarks: Mysteries and Revolutions of Particle Physics*, AIP Press, Springer, 2000

Barrow, John, *The Artful Universe Expanded*, Oxford University Press, 2005

Barrow, John, *The Constants of Nature*, Pantheon Books, New York, 2002

Barrow, John & Frank Tipler, *The Anthropic Cosmological Principle*, Oxford University Press, 1986

Barrow, John & John Webb, "Inconstant Constants", in: *Scientific American*, vol. 16, 2006, pp. 72–81

Beck, Alexander, Michal Hanasz, Harald Lesch, Rhea–Silvi Remus, Federico Stasyszyn "On the Magnetic Fields in Voids", in: *Monthly Notices of the Royal Astronomical Society*, vol. 429, 2013, pp. L60–L64

Bennett, Charles *et al.*, "First–Year Wilkinson Microwave Anisotropy Probe (WMAP). Observations: foreground emission", in: *Astrophysical Journal Supplement*. vol. 148, 2003, pp. 97–117

Bennett, Charles *et al.*, "The Microwave Anisotropy Probe (MAP) Mission", in: *Astrophysical Journal*, vol. 583, 2003a, pp. 1–23

Bernstein, Jeremy, *The Kinetic Theory in the Expanding Universe*, Cambridge University Press, 1988

Bertotti, Bruno *et al.*, eds, *General Relativity and Gravitation*, Reidel Publications, Dordrecht, 1984

Bethe, Hans & Robert Bacher, "Nuclear Physics A. Stationary States of Nuclei", in: *Reviews of Modern Physics*, vol. 8 (1936): 82–229

Bethke, Siegfried, "α_s at Zinnowitz 2004", hep-ex/0407021v1

Bohm, David & Basil Hiley, *The Undivided Universe. An Ontological Interpretation of Quantum Theory*, Routledge, 2003

Bohr, Niels, "Neutron Capture and Constitution", in: *Nature*, vol. 137, 1936, pp. 344–348

Bolejko, Krzysztof, Ahsan Nazer and David Wiltshire, "Differential Cosmic Expansion and the Hubble Flow Anisotropy", in: *Journal of Cosmology and Astroparticle Physics*, issue 6, vol. 2016, 2016, pp. 35–64

Boltzmann, Ludwig, "On Certain Questions of the Theory of Gases", in: *Nature*, 1895, pp. 413–415

Boltzmann, Ludwig, *Theoretical Physics and Philosophical Problems*, Reiding Publications, Dordrecht, Boston, 1974

Bondi, Hermann & Thomas Gold, "The Steady State Theory of the Expanding Universe", in: *Monthly Notices of the Royal Astronomical Society*, vol. 108, 1948, pp. 252–270

Borde, Arvind & Alexander Vilenkin, "Eternal Inflation and the Initial Singularity", in: *Physical Review Letters*, vol. 72, 1994, pp. 3305–3309

Bradac, Marusa *et al.*, "Revealing the Properties of Dark Matter in the Merging Cluster *MACS J*0025.4 − 1222 ", in: *The Astrophysical Journal*, vol. 687, 2008, pp. 959–967

Brumfiel, Geoff, "Outrageous Fortune", in: *Nature*, vol. 439, 2006, pp. 10–12

Buchert, Thomas, "Averaging Inhomogeneous Newtonian Cosmologies", in: *Astronomical Astrophysics*, vol. 320, 1997, pp. 1–7

Buchert, Thomas, "On Average Properties of Inhomogeneous Cosmologies", in: *Proceedings of 9th Workshop on General Relativity and Gravitation*, 1999, Hiroshima, Japan, pp. 306–321

Burles, Scott, Kenneth Nollett & Michael Turner, "What is the Big Bang Nucleosynthesis Prediction for the Baryon Density and How Reliable Is It?", in: *Physical Review D* (2001): 63–69

Candelas, Philip, Gary Horowitz, Andrew Strominger & Edward Witten, "Vacuum Configurations for Superstrings", in: *Nuclear Physics B*, vol.258, 1985, pp. 46–74

Carr, Bernard & Martin Rees, "The Anthropic Principle and the Structure of the Physical World", in: *Nature,* vol. 278, 1979, pp. 610 ss.

Carrick, John and Fred Cooperstock, "General Relativistic Dynamics Applied to the Rotation Curves of Galaxies", in: *Astrophysics and Space Science*, vol. 337, 2012, pp. 321–329

Carter, Brandon, "Large Number Coincidences and the Anthropic Principle in Cosmology", in: Malcolm Longair, ed., *Confrontation of Cosmological Theories with Observational Data*, 1974

Chadwick, James, "Bakerian Lecture. The Neutron", in: *Proceedings of the Royal Society A: Mathematical, Physical and Engineering Sciences*. vol. 142, 1933

Clowe, Douglas *et al.*, "A Direct Empirical Proof of the Existence of Dark Matter", in: *Astrophysical Journal Letters*, vol. 648, 2006, pp. L109–L113

Clowe, Douglas *et al.*, "Catching a Bullet: Direct Evidence for the Existence of Dark Matter", in: *Nuclear Physics Supplement*, vol. 173, 2007, pp. 28–31

Clowe, Douglas, "Colliding Clusters Shed Light on Dark Matter," in: *Scientific American,* August 22, 2006

Colling, Richard, *Random Designer*, Browning Press, Bourbonnais, Ill., 2004

Conlon, Joseph, *Why String Theory*, CRC Press, 2016

Cooperstock, Fred, "Clusters of Galaxies", in: *General Relativistic Dynamics*, World Scientific Publications, 2009, Chapter 10

Cooperstock, Fred, *General Relativistic Dynamics. Extending Einstein's Legacy throughout the Universe*, World Scientific Publications, 2009

Cooperstock, Fred, & Steven Tieu, "Galactic Dynamics via General Relativity", in: *International Journal of Modern Physics A*, vol. 22, 2007, pp. 2293–2325

Cooperstock, Fred & Steven Tieu, "General Relativity Resolves Galactic Rotation Without Exotic Dark Matter", arXiv:astro-ph/0507619, 2005

Cooperstock, Fred & Steven Tieu, "General Relativistic Velocity: the Alternative to Dark Matter", in: *Modern Physics Letters A*, vol. 23, 2008, pp. 1745–1755

Cooperstock, Fred & Steven Tieu, "Perspectives on Galactic Dynamics via General Relativity," arXiv:astro-ph/0512048

Davies, Paul, *The Accidental Universe,* Cambridge University Press, London, 1982

Davies, Paul, *The Goldilocks Enigma: Why is the Universe Just Right for Life?*, Allen Lane, 2006

Davies, Paul, *The Mind of God. The Scientific Basis for a Rational World*, Simon & Schuster, New York, 2005

Davies, Paul, "The Variation of the Coupling Constants", in: *Journal of Physics*, vol. 5, 1972, pp. 1296–1304

Davies, Paul & Julian Brown, eds, *Superstrings*, Cambridge University Press, 2000

Davis, Tamara *et al.*, "Scrutinizing Exotic Cosmological Models using ESSENCE Supernova Data", in: *Astrophysical Journal,* vol.662, 2007, pp. 716–725

De Broglie, Louis, "La mécanique ondulatoire et la structure atomique de la matiére et du rayonnement", in: *Le Journal de Physique*, Tome VIII, 1927, pp. 225–241

De Broglie, Louis, *Ondes et mouvements*, Gauthier–Villars, Paris, 1926

De Broglie, Louis, "Recherches sur la théorie de quanta", in: *Annales de Physique*, Tome III, 1925

Dekel, Avishai, David Burnstein & Simon White, "Measuring Omega", in: Neil Turok ed., *Critical Dialogues in Cosmology*, 1997, pp. 175–192

Dicke, Robert, James Peebles *et al.*, "Cosmic Black–body Radiation", in: *Astrophysical Journal*, vol. 142, 1965, pp. 414–419

Dirac, Paul, "The Cosmological Constants", Letter to Editor, in: *Nature*, vol. 139, 1937, p. 323; also in: Kenneth Lang & Owen Gingerich, eds, *A Source Book in Astronomy and Astrophysics, 1900–1975*, Harvard University Press, 1979, pp. 851–852

Dyson, Freeman, "The Fundamental Constants and Their Time Variation", in: Abdus Salam & Eugene Wigner, eds, *Aspects of Quantum Theory*, Cambridge University Press, 1972, pp. 235–236

Ehrenfest, Paul, "Which role does the three–dimensionality of space play in the basic laws of physics?", in: *Paul Ehrenfest. Collected Scientific Papers*, ed. by Martin Klein, North Holland Publications, Amsterdam, New York, 1959

Einstein, Albert, "Concerning a Heuristic Point of View Toward the Emission and Transformation of Light", in: *Annalen der Physik*, vol. 17, 1905, pp. 132–148; English translation in: *American Journal of Physics,* vol. 33, 1965, pp. 367 ss.

Einstein, Albert, *The Collected Works of Albert Einstein*, Pergamon Media, 1989

Einstein, Albert, "The Foundation of the General Theory of Relativity", in: *Annalen der Physik*, vol. 49, 1916, see also: *The Collected Works of Albert Einstein*, Pergamon Media, vol. 6, 1989

Einstein, Albert, Boris Podolsky & Nathan Rosen, "A letter from Albert Einstein, 1935", in: Karl Popper, *The logic of scientific discovery*, 2002, pp. 481–488

Einstein, Albert, Boris Podolsky & Nathan Rosen, "Can Quantum–mechanical Description of Physical Reality be Considered Complete?", in: *Physical Review*, vol. 47, 1935, pp. 777 ss.

Elgaroy, Øystein & Tuomas Multamäki, "Bayesian analysis of Friedmannless cosmologies", arXiv:astroph/ 0603053

Ellis, George, "Physics ain't what it used to be", in: *Nature*, vol. 438, 2005, pp. 739–740

Ellis, George, "Relativistic Cosmology: Its Nature, Aims and Problems", in: Bruno Bertotti *et al.*, eds, *General Relativity and Gravitation*, 1984, pp. 215–288

Fedorov, D. V. & A. Jensen, "The Three–Body Continuum Coulomb Problem and the 3α Structure of ^{12}C", in: *Physics Letters B*, vol. 389, 1996, pp. 631–636

Feigenbaum, Mitchell, "Quantitative Universality for a Class of Nonlinear Transformations", in: *Journal of Statistical Physics*, vol. 19, 1978, pp. 25–52

Feigenbaum, Mitchell, "The Universal Metric Properties of Nonlinear Transformations", in: *Journal of Statistical Physics*, vol. 21, 1979, pp. 669–706

Ferrière, Katia, "The Interstellar Environment of our Galaxy", in: *Review of Modern Physics*, vol. 73, 2001, pp. 1031–1086

Feynman, Richard, *The Character of Physical Law*, MIT Press, Cambridge, Ms., 1967

Feynman, Richard, *QED. The Strange Theory of Light and Matter*, Princeton University, 1988

Fixsen, Dale *et al.*, "The Cosmic Microwave Background Spectrum from the Full COBE FIRAS Data Set", in: *Astrophysical Journal*, vol. 473 (1996): 576–587

Freedman, Wendy, "Determination of the Hubble Constant", in: Neil Turok ed., *Critical Dialogues in Cosmology*, World Scientific, Singapore, 1997, pp. 92–129

Freedman, Wendy *et al.*, "Final Results from the Hubble Space Telescope Key Project to Measure the Hubble Constant", in: *Astrophysical Journal*, vol. 533, 2001, pp. 47–72

Freedman, Wendy, & Michael Turner, "Measuring and Understanding the Universe", in: *Reviews of Modern Physics*, vol. 75, 2003, pp. 1433–1447

Friedan, Daniel, "A Tentative Theory of Large Distance Physics", in: *Journal of High Energy Physics,* vol. 2003, 2003, pp. 1–98

Friedrich, Helmut, "Einstein Equations and Conformal Structure: Existence of Anti–de Sitter–type Space–times", in: *Journal of Geometry and Physics*, vol. 17, 1995, pp. 125–184

Friedrich, Helmut, "Einstein's Equation and Conformal Structure", in: S. Huggett *et al.,* eds, *The Geometric Universe: Science, Geometry, and the Work of Roger Penrose,* 1998, pp. 81–98

Friedrich, Helmut, "Einstein's Equation and Geometric Asymptotics", in: *Proceedings of the 15th International Conference on General Relativity and Gravitation*, Pune, India, December 16 – 21, 1997

Friedrich, Helmut, "Existence and Structure of Past Asymptotically Simple Solutions of Einstein's Field Equations with Positive Cosmological Constant", in: *Journal of Geometry and Physics,* vol. 3, 1986, pp. 101–117

Friedrich, Helmut, "On the Existence of n–Geodesically Complete or Future Complete Solutions of Einstein's Field Equations with Smooth Asymptotic Structure", in: *Communications in Mathematical Physics*, Springer–Verlag, 1986

Friedrich, Helmut & Gabriel Nagy, "The Initial Boundary Value Problem for Einstein's Vacuum Field Equation", in: *Communications in Mathematical Physics*, Springer–Verlag, 1999, pp. 619–655

Gamow, George, *Mr. Tompkins in Wonderland*, Cambridge University Press, 1939

Garfinkle, David, "The Need for Dark Matter in Galaxies", in: *Classical and Quantum Gravity*, vol. 23, 2006, number 4

Glashow, Sheldon, *Interactions: A Journey Through the Mind of a Particle Physicist and the Matter of this World*, Warner Books, 1988

Gleick, James, *Chaos. Making a New Science*, Penguin Books, 1988

Gonzalez, Guillermo & Jay Richards, *The Privileged Planet, How Our Place in the Cosmos is Designed for Discovery*, Regnery Publications, Washington DC, 2004

Green, Michael & John Schwarz, "Anomaly Cancellations in Supersymmetric D=10 Gauge Theory and Superstring Theory", in: *Physics Letters B*, vol. 149, 1984, pp. 117–122

Greenfield, Susan, *Journey to the Centers of the Mind,* Freeman, 1995

Grünbaum, Adolf, *Philosophical Problems of Space and Time*, Reidel Publications, 1998

Grünbaum, Adolf, *Philosophy of Science in Action*, Oxford University Press, 1994

Gurevich, L. & V. Mostepanenko, "On the Existence of Atoms in n–dimensional Space", in: *Physics Letters A*, vol. 35, 1971, 201–202

Gurney, Ronald, & Edward Condon, "Quantum Mechanics and Radioactive Disintegration", in: *Physical Review*, vol 33, 1929, pp. 127–140

Guth, Alan, *The Inflationary Universe*, Helix Books, Reading, Ms., 1998

Hacohen, Malachi, *Karl Popper: The Formative Years, 1902–1945: Politics and Philosophy in Interwar Vienna*, Cambridge University Press, 2000

Hawking, Stephen, *The Universe in a Nutshell*, Bantam Books, 2001

Hawking, Stephen, *The Illustrated A Brief History of Time*, Bantam Books,1996

Hawking, Stephen, Malcolm Perry & Andrew Strominger, "Soft Hair on Black Holes", in: *Physical Review Letters*, vol. 116, 2016, 231301, pp. 1–9

Hawley, John & Katherine Holcomb, *Foundations of Modern Cosmology*, Oxford University Press, 1998

Heidegger, Martin, *Introduction to Metaphysics*, Yale University Press, 2000

Higgs, Peter, "Broken Symmetries, Massless Particles and Gauge Fields", in: *Physics Letters*, vol. 12, 1964, pp. 132–133

Higgs, Peter, "Spontaneous Symmetry Breakdown without Massless Bosons", in: *Physical Review*, vol. 145, 1966, pp. 1156–1164

Hinshaw, Gary *et al.*, "Five–Year Wilkinson Microwave Anisotropy Probe Observations: Data processing, sky maps, and Basic Results", in: *The Astrophysical Journal Supplement*, vol 180, 2009, pp. 225–245

Hinshaw, Gary *et al.*, "Three–Year Wilkinson Microwave Anisotropy Probe Observations: Temperature Analysis", in: *Astrophysical Journal Supplement*, vol. 170, 2007, pp. 288–334

Holt, Stephen, & George Sonneborn, eds, "Cosmic Abundances", in: *Astronomical Society of the Pacific Conference Series*, vol 99, 1996, pp. 48 ss.

Hong, Suk-ho & Suk-yoon Lee, "Alpha Chain Structure in ^{12}C", in: *Journal of Korean Physics*, vol. 35, 1999, pp. 46–48

Horgan, John, *Rational Mysticism. Dispatches from the Border Between Science and Spirituality,* Houghton Mifflin, New York, 2003

Houtermans, Fritz & Robert Atkinson, "Zur Frage der Aufbaumöglichkeit der Elemente in Sternen", in: *Zeitschrift für Physik*, vol. 54, 1929, pp. 656–665

Hoyle, Fred, "A New Model for the Expanding Universe", in: *Monthly Notices of the Royal Astronomical Society*. vol. 108, 1948, pp. 372–382

Hoyle, Fred, "The Synthesis of the Elements from Carbon to Nickel", in: *Astrophysical Journal Supplement*, vol. 121, 1954, pp. 121–146

Hoyle, Fred & Roger Tayler, "The Mystery of the Cosmic Helium Abundance", in: *Nature*, vol. 203, 1964, p. 1108–1110

Hubble, Edwin, "Extragalactic Nebulae", in: *Astrophysical Journal*, vol. 64, 1926, pp. 321–369

Hubble, Edwin, "A Relation between Distance and Radial Velocity among Extra–Galactic Nebulae", in: *Proceedings of the National Academy of Sciences of the United States of America,* vol. 15, 1929, pp. 168–173

Hubble Edwin & Milton Humason, "The Velocity–Distance Relation among Extra–Galactic Nebulae", in: *Astrophysical Journal*, vol. 74, 1931, pp. 43–80

Huheey, James, Ellen Keiter & Richard Keiter, *Inorganic Chemistry: Principles of Structure and Reactivity*, Prentice Hall, 2007

Isaacson, Walter, *Einstein. His Life and Universe*, Simon & Schuster, 2007

Ishak, Mustapha, Roberto Sussman *et al.*, "Dark Energy or Apparent Acceleration Due to a Relativistic Cosmological Model More Complex than FLRW?", in: *Bulletin of the American Astronomical Society*, vol. 39, 2007, pp. 948–952; and in: *Physical Review D,* vol. 78, 2008, 123531

Jefimenko, Oleg, "Presenting Electromagnetic Theory in Accordance with the Principle of Causality", in: *European Journal of Physics,* vol. 25, 2004, pp. 287–288

Jeffreys, Harold, *Theory of Probability*, Oxford University Press, 1998

Kaku, Michio, *Parallel Worlds. The Science of Alternative Universes and our Future in the Cosmos*, Allan Lane, 2005

Kalnajs, A. J., *Internal Kinematics of Galaxies*, IAU Symp. 100, E. Athanassoula ed, Reidel Publications, 1983

Kamiokande, Super-, "Search for Trilepton Nucleon Decay via $p \rightarrow e^{+}\nu\nu$ and $p \rightarrow \mu^{+}\nu\nu$ in the Super-Kamiokande Experiment", in: *Physical Review Letters*, vol. 113 (2014): 101801, pp. 1-6

Karttunen, Hannu, Pekka Kröger, Heikki Oj, Markku Poutanen, Karl Johann Donner, *Fundamental Astronomy*, 6th ed., Springer, Berlin, New York, 2003

Kent, Stephen, "Dark Matter in Spiral Galaxies. I. Galaxies with Optical Rotation Curves", in: *The Astronomical Journal,* vol. 91, 1986, pp. 1301–1327

Kent, Stephen, "Dark Matter in Spiral Galaxies. II. Galaxies with H1 Rotation Curves", in: *The Astronomical Journal,* vol. 93, 1987, pp. 816–832

Kent, Stephen, "Dark Matter in Spiral Galaxies. III. The Sa Galaxies", in: *The Astronomical Journal,* vol. 96, 1988, pp. 514–527

Kirshner, Robert, *The Extravagant Universe*, Princeton University Press, 2002

Knop, Rob, Saul Perlmutter *et al.*, "New Constraints on Ω_M , Ω_Λ and w from an Independent Set of Eleven High–Redshift Supernovae Observed with the HST", in: *Astrophysical Journal*, vol. 598, 2003, pp. 102–137

Kolb, Edward & Michael Turner, *The Early Universe*, Addison–Wesley Publications, Reading Ms., 1990

Kolb, Edward, Sabino Matarrese & Antonio Riotto, "On Cosmic Acceleration without Dark Energy", in: *New Journal of Physics,* 2006, pp. 322–346

Koranda, Scott & Bruce Allen, "CBR Anisotropy from Primordial Gravitational Waves in Two–component Inflationary Cosmology", in: *Physical Review D52,* 1995, pp. 1902–1919

Korzynski, Nikolaj, "Singular Disk of Matter in the Cooperstock–Tieu Galaxy Model", arXiv:astro-ph/0508377

Kragh, Helge, *Conceptions of Cosmos. From Myth to Accelerating Universe: A History of Cosmology*, Oxford University Press, 2007

Kragh, Helge, "Entropic Creation: Religious Contexts of Thermodynamics and Cosmology"*, in: *Journal of Religion and Science*, vol. 45, 2010, pp. 281–282

Krauss, Lawrence, Preface by Richard Dawkins, *A Universe from Nothing. Why There Is Something Rather Than Nothing*, Atria, New York, London, 2012

Kuhn, Thomas, *The Structure of Scientific Revolutions*, University of Chicago Press, 2012

Kwan, Juliana, Matthew Francis & Geraint Lewis, "Fractal Bubble Cosmology: A concordant cosmological model?", in: *Monthly Notices of the Royal Astronomical Society Letters,* vol. 399, 2009, pp. L6–L10

Landau, L. & E. Lifshitz, *The Classical Theory of Fields*, Pergamon, 4th ed., 2002

Lang, Kenneth, & Owen Gingerich, eds, *A Source Book in Astronomy and Astrophysics, 1900–1975*, Harvard University Press, 1979

Laplace, Pierre Simon, *A Philosophical Essay on Probabilities*, Chronicon Books, 2015

Lara, Juan, "Deuterium and ^{7}Li Concordance in Inhomogeneous Big Bang Nucleosynthesis models", in: *Frontier in Astroparticle Physics and Cosmology: Proceedings of the 6th International Symposium held at the University of Tokyo*, K. Sato and S. Nagataki, eds, Universal Academy Press, Tokyo, 2004, pp. 87 ss.

Lara, Juan, "Neutron Diffusion and Nucleosynthesis in an Inhomogeneous Big Bang Model", arxiv.org/abs/astro-ph/0506364v2

Leavitt, Henrietta & Edward Pickering, "Periods of 25 Variable Stars in the Small Magellanic Cloud", in: *Harvard College Observatory Circular*, vol. 173, 1912, pp. 1–3

Leibniz, Gottfried, *Discourse on Metaphysics and Other Essays*, Hackett, 1989

Leith, Ben, Cindy Ng & David Wiltshire, "Gravitational Energy as Dark Energy: Concordance of cosmological tests", in: *Astrophysical Journal Letters*, vol. 672, 2008, pp. L91–94

Lemaître, George, "A Homogeneous Universe of Constant Mass and Growing Radius Accounting for the Radial Velocity of Extragalactic Nebulae", in: *Monthly Notices of the Royal Astronomical Society*, vol. 91 (1931): 483–490

Lemaître, George, *A homogeneous universe of constant mass and growing radius accounting for the radial velocity of extra–galactic nebulae*, Springer online, New York, 2013, pp. 1635–1646

Lemaître, George, *L'Hypothèse de l' atome primitive. Essay de cosmogenie*, Ed. Du Griffon, Neuchâtel, 1946

Lemaître, George "The Evolution of the Universe: Discussion", in: *Nature,* vol. 128, nr. 3234, 1931, p. 699–701

Lemaître, George, *The Primeval Atom. An Essay on Cosmogony*, Van Nostrand Co., 1950

Lemaître, George, "Un univers homogène de masse constante et de rayon croissant rendant compte de la vitesse radiale des nébuleuses extragalactiques", in: *Annales de la Société Scientifique de Bruxelles*, vol. 47A, 1927, pp. 49–59

Leslie, John, *Universes*, Routledge, London & New York, 1996

Lide, David, ed., *CRC Handbook of Chemistry and Physics*, CRC Press, 1994–95

Linde, Andrei, *Inflation and Quantum Cosmology,* Academic Press, Boston, 1990

Linde, Andrei, "Inflation, Quantum Cosmology and the Anthropic Principle", arXiv.org/pdf/hep-th/0211048, November 2002

Livio, Mario, *The Accelerating Universe*, John Wiley, New York, 2000

Livio, Mario, D. Hollowell & J. Truram, "The Anthropic Significance of the Existence of an Excited State of ^{12}C", in: *Nature*, vol. 340, 1989, pp. 281–284

Lonergan, Bernard, *Insight: A Study of Human Understanding*, in: *Collected Works,* vol. 3, University of Toronto Press, 1993

Longair, Malcolm, ed., *Confrontation of Cosmological Theories with Observational Data*, Reidel Publications, Dordrecht, Netherlands, 1974

Longair, Malcolm, *Galaxy Formation,* Springer, 2008

Longair, Malcolm, *The Cosmic Century*, Cambridge University Press, 2006

Lorenz, Edward, *The Essence of Chaos*, University of Washington Press, 1995

Mandelbrot, Benoît, *Fractals and Chaos*, Springer, 2004

Mann, Anthony, T. Kafka & W. Leeson, "The Atmospheric Flux ν_μ/ν_e Anomaly as Manifestation of Proton Decay $p \rightarrow e^+\nu\nu$", in: *Physics Letters B*, vol. 291, 1992, pp. 200–205

Massey, Richard *et al.*, "Dark Matter Maps Reveal Cosmic Scaffolding", in: *Nature* online, January 2008

Massey, Richard *et al.*, "Probing Dark Matter and Dark Energy with Space–Based Weak Lensing", arXiv:astroph/0403229

McGrath, Alister, *A Fine-Tuned Universe.* Westminster John Knox Press, 2009

McGrath, Alister & Joanna Collicutt McGrath, *The Dawkins Delusion. Atheist Fundamentalism and the Denial of the Divine*, Intervarsity Press Books, 2007

Medawar, Peter, *The Limits of Science*, Oxford University Press, 1985

McKee, Trudy & James McKee, *Biochemistry*, Oxford University Press, 2015

Mehra, Jagdish, ed., *The Physicist's Conception of Nature*, Reidel Publications, 1973

Mendoza, Sergio, *Astrofísica relativista*, MS, www.mendozza.org/sergio, 2007

Miller, James, ed., *Cosmic Questions*, New York Academy of Science, 2001

Misner, Charles, Kip Thorne & John Wheeler, *Gravitation*, Freeman & Cy., San Francisco, 1973

Misner, Charles, Kip Thorne & John Wheeler, "Is the Gravitational Constant Constant?", in: *Gravitation*, Freeman & Cy., San Francisco, 1973, pp. 1122–1226

Newton, Isaac, *Principia Mathematica. Mathematical Principles of Natural Philosophy*, Bernard Cohen, translator & editor, The Folio Society, 2008

New York Academy of Sciences, "Cosmic Questions", in: *Annals of the New York Academy of Sciences*, vol. 950, December 2001, pp. 1–319

Oberhummer, Heinz, *Kann das alles Zufall sein?*, Ecowin Verlag, Sallzburg, 2008

Oberhummer, Heinz, Attila Csótó, & Helmut Schlattl, "Stellar Production Rates of Carbon and its Abundance in the Universe", in: *Science*, vol. 289, 2000, pp. 88–90

Oberhummer, Heinz, Rudolf Pichler & Attila Csótó, "The Triple–Alpha Process and Its Anthropic Significance", in: Nikos Prantzos & Sotoris Harissopoulis, eds, *Nuclei in the Cosmos, V. Proceedings of the International Symposium on Nuclear Astrophysics*, 1998, pp. 119–122

Ohanian, Hans, *Einstein's Mistakes. The Human Failings of a Genius*, Norton, New York, London, 2008

Pais, Abraham, *Subtle is the Lord. The Science and the Life of Albert Einstein*, Oxford University Press, 2008

Particle Data Group, 2006, "Review of Particle Physics", in: *Journal of Physics G. Nuclear and Particle Physics*, vol. 33, 2006

Pauli, Wolfgang, "Über den Zusammenhang des Abschlusses der Elektronen–gruppen im Atom mit der Komplexstruktur der Spektren", in: *Zeitschrift für Physik*, vol. 31, 1925, pp. 765–783

Peacock, John, "Measurement of the Cosmological Mass Density from Clustering in the 2nd Galaxy Redshift Survey", in: *Nature*, vol. 410, 2001, pp. 169–173

Peebles, James, *Principles of Physical Cosmology*, Princeton University Press, 1993

Peimbert, Manuel, Valentina Luridiana & Antonio Peimbert, "Revised Primordial Helium Abundance Based on New Atomic Data", in: *Astrophysical Journal*, vol. 666, 2007, pp. 636–646

Penrose, Roger, *Fashion, Faith and Fantasy: on the New Physics of the Universe*, Princeton University Press, 2016

Penrose, Roger, *The Emperor's New Mind. Concerning Computers, Minds, and the Laws of Physics*, Penguin Books, 1991

Penrose, Roger, *The Road to Reality. A Complete Guide to the Laws of the Universe*, Alfred Knopf, New York, 2004

Penzias, Arno, & Robert Wilson, "Measurement of Excess Antenna Temperature at 4080 MHz", in: *Astrophysical Journal*, vol. 142, 1965, pp. 419–421

Perlmutter, Saul, "Measurements of Omega and Lambda from 42 High–Redshift Supernovae", in: *Astrophysical Journal*, vol. 517, 1999, pp. 565–586

Pichler, Rudolf, Heinz Oberhummer, Attila Csótó & S. Moszkowski, "Three Alpha Structures in ^{12}C", in: *Nuclear Physics A*, vol. 618, 1997, pp. 55–64

Pinker, Steven, *How the Mind Works*, Norton, 1997

Planck, Max, "Über irreversible Strahlungsvorgänge", in: *Sitzungsberichte der Königlich Preußischen Akademie der Wissenschaften*, 1899, pp. 440–480

Polkinghorne, John, *Exploring Reality. The Intertwining of Science and Religion*, Yale University Press, 2005

Polkinghorne, John, *Quarks, Chaos & Christianity: Questions to Science And Religion*, Crossroad Publications, New York, 2006

Polkinghorne, John, "Understanding the Universe", in: James Miller, ed., *Cosmic Questions*, New York Academy of Science, 2001, pp. 175–182

Popper, Karl, *Conjectures and Refutations. The Growth of Scientific Knowledge*, Basic Books, London, 1962; available as e–book

Popper, Karl, *Quantum Theory and the Schism in Physics*, Rowman & Littlefeld,, N.J., 1982

Popper, Karl, *Realism and the Aim of Science*, Routledge, New York, London, 2000

Popper, Karl, *The Logic of Scientific Discovery*, Routledge Classics, 2002; Taylor & Francis e–Library, 2005

Popper, Karl, *The Open Society and Its Enemies*, 2 vols., Princeton University Press, 1971

Popper, Karl, *The Open Universe. An Argument for Indeterminism*, Routledge, 2000

Popper, Karl, *The Poverty of Historicism*, Routledge, London, New York, 1999

Popper, Karl & John Eccles, *The Self and Its Brain*, Springer Verlag, New York, London, 1981

Povh, Bogdan *et al.*, *Particles and Nuclei*, Springer Verlag, Berlin, 2002

Prantzos, Nicos, Elisabeth Vangioni-Flam & Michel Cassé, eds, *Origin and Evolution of the Elements,* Cambridge University Press, 1993

Prantzos, Nikos & Sotoris Harissopoulis, eds, *Nuclei in the Cosmos, V. Proceedings of the International Symposium on Nuclear Astrophysics*, Ed. Frontiéres, Paris, 1998

Prigogine, Ilya, *From Being to Becoming. Time and Complexity in the Physical Sciences*, Freeman & Cy., San Francisco, 1980

Prigogine, Ilya, *Order out of Chaos. Man's New Dialogue with Nature.* Bantam Books, New York, 1984

Qualls, Joshua, *Lectures on Conformal Field Theory*, arXiv:1511.04074v2, May 2016

Ramond, Pierre, "Dual Theory for Free Fermions", *Physical Review D*, vol. 3, 1971, pp. 2415–2418

Recs, Martin, *Just Six Numbers*, Basic Books, New York, 2000

Riess, Adam, "Observational Evidence from Supernovae for an Accelerating Universe and a Cosmological Constant", in: *The Astronomical Journal*, vol. 116, 1998, pp. 1009–1038

Riess, Adam *et al.*, "Type Ia Supernova Discoveries at $z < 1$ from the Hubble Space Telescope: Evidence for past deceleration and constraints on dark energy Evolution", in: *The Astrophysical Journal*, vol. 607, 2004, pp. 665–738

Rohlfs, K. & J. Kreitschmann, "A Two Component Mass Model for M81/NGC3031", in: *Astronomy & Astrophysics*, vol. 87, 1980, pp. 175–182

Rowan–Robinson, Michael, *The Nine Numbers of the Cosmos*, Oxford University Press, 1999

Rubin, Vera, Kent Ford *et al.*, "Extended Rotation Curves of High–luminosity Spiral Galaxies. I The angle between the rotation axis of the nucleus and the outer disk of NGC 3672", in: *The Astrophysical Journal*, vol. 217, 1977, pp. L1–L4

Rubin, Vera, Kent Ford *et al.*, "Extended Rotation Curves of High–luminosity Spiral Galaxies. II The anemic Sa galaxy NGC 4378", *ibidem,* vol. 224, 1978, pp. 782–795

Rubin, Vera, Kent Ford *et al.*, "Rotation Curves of High–luminosity Spiral Galaxies. III. The spiral galaxy NGC 7217", *ibidem,* vol. 226, 1978, pp. 770–776

Rubin, Vera, Kent Ford *et al.*, "Rotation Curves of High–luminosity Spiral Galaxies. IV. Systematic dynamical properties", *ibidem,* vol. 225, 1978, pp. L107–L111

Rubin, Vera, Kent Ford *et al.*, "Rotation Curves of High–luminosity Spiral Galaxies. V. NGC 1961, The most massive spiral known", *ibidem,* vol. 225, 1979, pp. 35–39

Rubin, Vera, Kent Ford *et al.*, "Rotational properties of 21 Sc Galaxies with a Large Range of Luminosities and Radii, from NGC 4605 (R=4 lpc) to UGC 2885 (R=122 kpc)", *ibidem,* vol. 238, 1980, pp. 471–487

Rubin, Vera, Kent Ford *et al.*, "Rotational Properties of 23 Sb Galaxies", *ibidem,* vol. 261, 1982, pp. 439–456

Rubin, Vera, Kent Ford *et al.*, "Rotation and Mass of the Inner 5 Kiloparsecs of the SO Galaxy NGC 3115", *ibidem,* vol. 239, 1980, pp. 50–53

Rubin, Vera, Kent Ford *et al.*, "Rotation Velocities of 16 Sa Galaxies and a Comparison of Sa, Sb, and Sc Rotation Properties", *ibidem,* vol. 289, 1985, pp. 81–104

Rubin, Vera, "Dark Matter in Spiral Galaxies", in: *Scientific American* vol. 248, 1983, pp. 98 ss.

Rubin, Vera & George Coyne, eds, *Large Scale Motions in the Universe,* Princeton University Press, 1988

Ruelle, David, *Chance and Chaos,* Princeton University Press, 1991

Ruelle, David & Floris Takens, "On the Nature of Turbulence", in: *Communications of Mathematical Physics,* vol. 20, 1971, pp. 167–192

Rutherford, Ernest, *The Scattering of Alpha and Beta Particles by Matter and the Structure of the Atom*, Taylor & Francis, 1911

Ryle, Martin, "Radio Stars and their Cosmological Significance", in: *The Observatory*, vol. 75, 1955, pp. 137–147

Salam, Abdus & Eugene Wigner, eds, *Aspects of Quantum Theory*, Cambridge University Press, 1972

Sandage, Allan, & Gustav Tammann, "The Evidence for the Long Distance Scale with H_0 <65", in: Neil Turok ed., *Critical Dialogues in Cosmology*, World Scientific, Singapore, 1997, pp. 130–155

Scherk, Joel & John Schwarz, "Dual Models for Non–Hadrons", in: *Nuclear Physics B*, vol. 51, 1974, pp. 118–144

Schmelling, Michael, "Status of the Strong Coupling Constant", arXiv:hep-ex/9701002v1 1996

Shapiro, Charles & Michael Turner, "What do we Really Know about Cosmic Acceleration?", in: *The Astrophysical Journal* vol. 649, 2006, pp. 563–569

Singh, Simon, *The Big Bang, The Origin of the Universe*, Harper Collins, 2004

Sklar, Lawrence, "Probability", in: *Physics and Chance. Philosophical Issues in the Foundations of Mechanical Statistics*, Cambridge University Press, 1993, pp. 90–127

Smale, Peter & David Wiltshire, "Supernova Tests of the Time–scape Cosmology", in: *Monthly Notices of the Royal Astronomical Society*, vol. 413, 2011, pp. 367–385

Smith, John Maynard & Eörs Szathmáry, *Major Transitions in Evolution*, Oxford University Press, 1995

Smolin, Lee, *The Life of the Cosmos*, Oxford University Press, 1997

Smolin, Lee, *The Trouble with Physics. The Rise of String Theory, the Fall of Science, and What Comes Next*, Houghton Mifflin Cy., Boston. New York, 2006

Smoot, George, John Mather, Charles Bennett *et al.*, "Structure in the COBE Differential Microwave Radiometer First–year Maps", in: *Astrophysical Journal*, vol. 396 (September 1992): L1–L5

Smoot, George & Keay Davidson, *Wrinkles in Time. The Imprint of Creation*, Little, Brown & Cy., New York, 1993

Sontag, Richard & Gordon Van Wylen, *Introduction to Thermodynamics. Classical and Statistical* 3rd ed., John Wiley, 1991

Steigman, Gary, "Testing Big Bang Nucleosynthesis", in: *Cosmic Abundances*, Stephen Holt and George Sonneborn, eds, Astronomical Society of the Pacific Conference Series, vol 99, 1996, pp. 48 ss.

Stenger, Victor, *The Fallacy of Fine-Tuning: Why the Universe is Not Designed for Us,* Prometheus Books, 2011

Strömberg, Gustaf, "Analysis of Radial Velocities of Globular Clusters and Non–galactic Nebulae", in: *Astrophysical Journal*, vol. 61, 1925, pp. 353–362

Susskind, Leonard, *The Cosmic Landscape*: *String Theory and the Illusion of Intelligent Design*, Little Brown, New York, 2005

Szekeres, P., "A Class of Inhomogeneous Cosmological Models", in: *Communications in Mathematical Physics*, vol. 41, 1975, pp. 55–64

Tangherlini, Frank, "Atoms in Higher Dimensions", *Nuovo Cimento,* vol. 27, 1963

Tegmark, Max *et al.,* "Cosmological Parameters from SDSS and WMAP", in: *Physical Review D*, vol. 69 (2004): 1–28

Tegmark, Max, "Is 'the Theory of Everything' Merely the Ultimate Ensemble Theory", in: *Annals of Physics*, vol. 270, 1998, pp. 1–51

Thielemann, Friedrich, Ken–ichi Nomoto & Michio Hashimoto, "Explosive Nucleosynthesis in Supernovae", in: Nicos Prantzos *et al.*, eds, *Origin and Evolution of the Elements* (1994): 297–309

Thom, René, "Halte au hazard, silence au bruit", in: Stefan Amsterdamski, *La querelle du déterminisme,* Editions Gallimard, Paris, 1990, pp. 61–78

't Hooft, Gerard, *In search of the ultimate building blocks*, Cambridge University Press, 1997

Tripp, Robert, & David Branch, "Determination of the Hubble Constant Using a Two–Parameter Luminosity Correction for Type Ia Supernovae", in: *The Astrophysical Journal*, vol. 525, 1999, pp. 209–214

Trotta, Roberto, "Bayes in the Sky: Bayesian inference and model selection in cosmology", *Contemporary Physics*, vol. 49, 2008, pp. 71–104

Turok, Neil, ed., *Critical Dialogues in Cosmology,* World Scientific, 1997

Tytler, David, John O'Meara, Nao Suzuki & Dan Lubin, "Deuterium in Quasar Spectra", in: "Review of Big Bang Nucleosynthesis and Primordial Abundances", in: *Physica Scripta* (2000): T 85–103

Uzan, Jean–Philippe, "The Fundamental Constants and their Variation: Observational and theoretical status", in: *Reviews of Modern Physics*, vol. 75, 2003, pp. 403–455

Veneziano, Gabriele, "Construction of a Crossing Symmetric Regge–Behaved Amplitude for Linearly Rising Regge Trajectories", *Nuovo Cimento*, vol. 59, 1968, pp. 190–197

Vogt, Daniel & Patricio Letelier, "Presence of Exotic Matter in the Cooperstock and Tieu Galaxy Model", arXiv:astro-ph/0510750

Von Weizsäcker, Carl Friedrich, *Die Atomkerne, Grundlagen und Anwendungen ihrer Theorie*, Akademische Verlagsgesellschaft, Heidelberg, Germany, 1937

Wagoner, Robert, "Big Bang Nucleosynthesis Revisited", in: *The Astrophysical Journal*, vol. 179, 1973, ps 343–360

Wallace, David, *Gravity, Entropy, and Cosmology: in Search of Clarity*, 2009

Ward, Peter & Donald Brownlee, *Rare Earth. Why Complex Life is Uncommon in the Universe*, Springer Verlag, 2000

Weatherall, James Owen, *Void: The Strange Physics of Nothing*, Yale University Press, 2016

Weinberg, Steven, "A Universe with No Designer", in: *Annals of the New York Academy of Sciences*, vol. 950, *Cosmic Questions*, 2001, pp. 169–174

Weinberg, Steven, *Cosmology*, Oxford University Press, 2009

Weinberg, Steven, "The Cosmological Constant Problem", in: *Review of Modern Physics*, vol. 61, January 1989, pp. 6–23

Weinberg, Steven, *The First Three Minutes of the Universe: A modern view of the origin of the universe*, Flamingo, Fontana Paperbacks, 1977

Weinberg, Steven, *The Quantum Theory of Fields, Volume I Foundations*, Cambridge University Press, 2005

Weinberg, Steven, *The Quantum Theory of Fields, Volume II Modern Applications*, Cambridge University Press, 2005

Weinberg, Steven, *The Quantum Theory of Fields, Volume III Supersymmetry*, Cambridge University Press, 2005

Wheeler, John, "From Relativity to Mutability", in: Jagdish Mehra ed., *The Physicist's Conception of Nature*, Reidel Publications, 1973, pp. 202–247

Wiltshire, David, "Cosmic Clocks, Cosmic Variance and Cosmic Averages", in: *New Journal of Physics*, vol. 9, 2007, pp. 377–442

Wiltshire, David, "Cosmological Equivalence Principle and the Weak Field Limit", in: *Physical Review D*, vol. 78 (2008): 8–9

Wiltshire, David, "Exact Solution to the Averaging Problem in Cosmology", in: *Physical Review Letters*, vol. 99, 2007, 251101

Wiltshire, David, "Gravitational Energy as Dark Energy: Cosmic structure and apparent acceleration", in: John Auping & Alfredo Sandoval, eds, in: *Proceedings of the International Conference on Two Cosmological Models,* UIA, Plaza & Valdés, 2012, pp. 361–384

Wittgenstein, Ludwig, *Tractatus Logico–Philosophicus*, Routledge, 1981

Woit, Peter, *Not Even Wrong. The Failure of String Theory and the Search for Unity in Physical Law*, e–book, Jonathan Cape, London, 2006

Wood–Vasey, Michael *et al.*, "Observational Constraints on the Nature of the Dark Energy: First Cosmological Results from the ESSENCE Supernova Survey", in: *Astrophysical Journal*, vol. 666 (2007): 694–715

Woosley, Stanford, William David Arnett & Donald Clayton, "The Explosive Burning of Oxygen and Silicon", in: *The Astrophysical Journal Supplement*, vol. 26, 1973, pp. 231–312

Zalaletnidov, Roustam, "Averaging out the Einstein's Equations", *General Relativity and Gravitation*, vol. 24, 1992, pp. 1015–1031

Zalaletnidov, Roustam, "Averaging Problem in General Relativity: macroscopic gravity and using Einstein's equations in cosmology", in: *Bulletin of the Astronomical Society of India*, 1997, pp. 401–416

Zaritsky, Dennis, "Colliding Clusters Shed Light on Dark Matter", in: *Scientific American*, August 22, 2006

Zubiri, Xavier, *Inteligencia Sentiente: Inteligencia y Realidad*, Alianza, 1980

Zwicky, Fritz, "Die Rotverschiebung von extragalaktischen Nebeln", in: *Helvetica Physica Acta,* vol. 6, 1933, pp. 110–127

Zwicky, Fritz, "On the Masses of Nebulae and of Clusters of Nebulae", in: *The Astrophysical Journal,* vol. 86, 1937, pp. 217 –246

CREDITS

Permission for reproduction of images and tables, where requested, was obtained, and paid for when required, by the Department of Publications, *Universidad Iberoamericana,* Santa Fe

P[1]	P[2]	Image or table	Source
/	31	Lemaître discovered Hubble's law	The author
184	34	Hubble, recession velocity and distance	*Proceedings of the National Academy of Sciences of the USA,* vol. 15 (1929): 168-173
189	44	Observable universe	The author
195	56	Wagoner, Abundance of light elements produced in big bang	*The Astrophysical Journal,* vol. 179 (1973): 353
197	58	Steigman, Baryon/photon ratio	*International Journal of Modern Physics E* vol. 15 (2006):1–36
198	59	Peimbert, Baryon/photon ratio η and the baryon density Ω_B	*Astrophysical Journal,* vol. 666 (2007): 636–646
322	63	CMBR and its anisotropy	NASA, http:// lambda.gsfc.nasa.gov/product/cobe
/	64	eXtreme Deep Field	NASA, ESA, G. Illingworth, D. Magee, and P. Oesch & R. Bouwens, & HUDF09 Team. Image 09/25/12
298	69	Ratio of the electromagnetic and the gravitational force	Author
298	69	Ratio of the strong nuclear and the electromagnetic force	Author
314	82	Creation of elements in star with mass 25 times mass Sun	Singh, *Big Bang* (2004): 388
307	84	Interstellar gas clouds and the birth of new stars in M16	Hubble telescope, J. Hester and P. Scowen, NASA/ESA/STScl
309	86	Layers of a star with two nuclear fusion processes	Adapted by the author from Habbal Astro 110-01, page 10, 3/18/09
310	87	The M57 ring nebula in Lyra	NASA/ESA/Hubble Heritage, & Habbal Astro 110-01, p. 12, 3/18/09

[1] Page in Spanish edition, John Auping, *El Origen y la Evolución del Universo* (2009)
[2] Page in English edition, John Auping, *The cause and evolution of the Universe* (2017)

311	89	Massive star is a factory of elements	Wikipedia, Stellar evolution, author/user: Rursus
313	91	Stages in the lifecycle of a star in millions of years	Hannu Karttunen, *Fundamental Astronomy*, (2003): 264 (Table 12.1)
212	98	Spiral galaxy M31 with Rubin's flat rotation curve	Longair, *Galaxy Formation* (2008): 67
218	100	Observed and expected Newtonian rotation velocity	Author
213	106	Dark matter in galaxies dissociated from plasma	Douglas Clowe *et al.*, in: *Astrophysical Journal Letters* (2006)
222	111	Relativistic prediction of rotational velocity *Milky Way*	Fred Cooperstock, arXiv:astro-ph/0507619 (2005): 8
223	112	Relativistic prediction of rotational velocity *Milky Way*	The author
227	117	The spiral galaxy NGC 6946	NGC 6946. Photo by John Duncan, *Astronomía* (2007): 223
287	141	The value Ω_Q of as a function of redshift z	Courtesy of David Wiltshire, in a private e-mail, dated May 5th, 2009
289	151	Both ΛCDM and relativistic model fit supernovae 1a data	Mustapha Ishak, Roberto Sussman *et al.*, arXiv:0708.2943 (2008): 5
332	156	A very special universe	Roger Penrose, *Road to Reality* (2004): 730
333	158	Friedmann-Lemaître model, three simulations according to value Ω_M	Adapted from Guillermo Gonzalez & Jay Richards, *The Privileged Planet* (2004): 185
335	162	Balance between expansion of Universe and gravitational collapse	Adapted from: Martin Rees, *Just Six Numbers* (2000): 98
343	176	The limits imposed on α_s and α by the fact that we are here	Max Tegmark, in: *Annals of Physics*, vol. 270 (1998): 15
344	178	Triple alpha nuclear reaction at a temperature of 10^8 K as function of nuclear force p	Heinz Oberhummer *et al.*, in: Nikos Prantzos, ed., *Nuclei in the Cosmos V* (1998): 119-122
345	179	The existence of carbon and oxygen in the Universe	Heinz Oberhummer *et al.*, *Science*, vol. 289 (2000)
348	185	2-body problem in universe of $N \geq 4$ spatial dimensions	Max Tegmark, in: *Annals of Physics*, vol. 270 (1998): 17

352	191	Stability of atomic nuclei strong nuclear force, and fine-structure constant	Paul Davies, in: *Journal of Physics* vol. 5 (1972):1300
376	239	The weak interaction mediated by bosons W^+, W^- and Z^0	Richard Feynman, *QED. The strange theory of light and matter* (2006)
721	241	Table. Standard model: bosons	The author
721	242	Table. standard model: fermions	The author
721	243	Table: two baryons	The author
456	270	Figure. Carnot thermal machine	Erich Starke, Universidad Iberoamericana
/	281	Image. Earth conceived as a thermal machine	The author
402	296	Table. Falsifiability and non-falsifiability	The author
406	304	Figure. Philosophy of 2 worlds idealism and naive realism	The author
407	305	Figure. The philosophy of the 3 worlds	The author
408	310	Image. Picture of water and sugar	Joaquín Cabeza
408	310	Figure. Molecular structure of water	Trudy & James McKee, *Biochemistry,* 3rd ed. (2003): 66
409	311	Figure. Molecular structure glucose	The author
435	321	The light cones of observer $P1$	Stephen Hawking, *The Illustrated A Brief History of Time* (1996): 36
431	324	Phase changes of dynamical systems	James Gleick, *Chaos* (1988): 50
109	338	Probability of an electron's orbits in hydrogen atom	The author
109 442	338	Probability of electron's or-its in excited hydrogen	The author
110	339	Probability of an electron's or-bits in excited hydrogen atom	The author
443	340	Six possible trajectories of a tennis ball	Allday, *Quarks, Leptons and the Big Bang* (2002): 65

471	357	Figure. Interaction of an electron and a photon	Richard Feynman, *QED, The strange theory of light and matter* (2006): 97
471	358	Figure. Interaction of an electron and a photon	Richard Feynman, *QED, The strange theory of light and matter* (2006): 99
472	359	Figure. A is the cause of B and A and B are the cause of C	The author
473	361	Table. Four propositions about properties of the universe	The author
483	365	The philosophy of 4 worlds	The author
482	384	Table. Interaction between intelligent cause and random chance	The author

NAME AND AUTHOR INDEX

MATH BOX INDEX

INDEX OF TABLES

INDEX OF GRAPHS, FIGURES AND IMAGES

www.ingramcontent.com/pod-product-compliance
Lightning Source LLC
Chambersburg PA
CBHW050104220326
41598CB00043B/7378